SYMMETRY PRINCIPLES AND MAGNETIC SYMMETRY IN SOLID STATE PHYSICS

Graduate Student Series in Physics
Other books in the series

Weak Interactions
D Bailin

Collective Effects in Solids and Liquids
N H March and M Parrinello

Hadron Interactions
P D B Collins and A D Martin

Supersymmetry, Superfields and Supergravity: an Introduction
P P Srivastava

Introduction to Gauge Field Theory
D Bailin and A Love

Mechanics of Deformable Media
A B Bhatia (deceased) and R N Singh

Physics of Structurally Disordered Matter: an Introduction
N E Cusack

Gauge Theories in Particle Physics (2nd edition)
I J R Aitchison and A J G Hey

Geometry, Topology and Physics
M Nakahara

Superfluidity and Superconductivity (3rd edition)
D R Tilley and J Tilley

GRADUATE STUDENT SERIES IN PHYSICS

Series Editor: Professor Douglas F Brewer, M.A., D.Phil.
Professor of Experimental Physics, University of Sussex

SYMMETRY PRINCIPLES AND MAGNETIC SYMMETRY IN SOLID STATE PHYSICS

S J JOSHUA
Department of Physics, University of Zimbabwe

ADAM HILGER
BRISTOL, PHILADELPHIA AND NEW YORK

British Library Cataloguing in Publication Data

Joshua, S J
Symmetry principles and magnetic symmetry in solid state physics.
1. Physics. Applications of symmetry
I. Title II. Series
530

ISBN 0-7503-0070-1 (hbk)
0-7503-0071-X (pbk)

US Library of Congress Cataloging-in-Publication Data

Joshua, S J
Symmetry principles and magnetic symmetry in solid state physics/S J Joshua.
p. 24 cm.—(Graduate student series in physics)
Includes bibliographical references (p.).
Includes index
ISBN 0-7503-0070-1 (hard).—ISBN 0-7503-0071-X (pbk.)
1. Solid state physics. 2. Symmetry (Physics) 3. Crystals—Magnetic properties. 4. Crystallography. Mathematical. I. Title. II. Series.
QC176.J65 1991
530.4′1—dc20 90-5338

Published under the Adam Hilger imprint by
IOP Publishing Ltd
Techno House, Redcliffe Way, Bristol BS1 6NX, England
335 East 45th Street, New York, NY 10017-3483, USA
US Editorial Office: 1411 Walnut Street, Suite 200, Philadelphia, PA 19102

Typeset by P & R Typesetters Ltd, Salisbury, Wilts, UK
Printed in Great Britain by Billing and Sons Ltd, Worcester

Let each do as he pleases; but let there be symmetry

Joe Rosen

To Prema and to my girls
Asha, Anita, Sunita and Lalitha

CONTENTS

PART I: INTRODUCTION TO SYMMETRY PRINCIPLES IN SOLID STATE PHYSICS

PREFACE

This work arises from two sources and is therefore written in two parts. The first part has grown out of 20 lectures given to senior undergraduates and fresh graduate students of the University of the West Indies, Trinidad, and the University of Port Harcourt, Nigeria. A generous invitation to the first Caribbean Physics Meeting, sponsored by the Organization of American States and held at the Jamaica Campus of the University of the West Indies, was the motivating factor for adding the second part, which is based on a paper presented at the meeting.

The book is written at the level and style of lecture notes on the applications of group theory to solid state physics. The first part is intended for beginners in the field and as such a more complete treatment of the topics discussed here must be sought in the references given at the end of the text. This part is aimed at showing, via the extensive use of character tables, how symmetry arguments can be used to give a detailed insight into the physical properties of crystals closely linked with structures. An elementary course in solid state physics and quantum mechanics is considered a necessary prerequisite for this course.

Since the mathematical language of symmetry concepts is group theory, which is really a branch of pure mathematics, it is usually developed with reference to abstract entities. However, this is not the approach taken in Part I. All elegant proofs etc have been omitted. Experience has shown that for most beginners a better approach is to present first (as far as possible) the applications of symmetry and only afterwards should the formal aspects of the theory be introduced, if they need to be introduced at all at this stage!

Part I includes the following basic areas:

(i) introduction to group theoretical concepts and techniques;
(ii) basic symmetry operations present in the solid state;
(iii) illustrative examples.

I am sure that those who have the patience to go through this first part would acquire a workmanlike knowledge of group theory and would, with some determination and effort, be in a position to understand the second part of this book which deals with the more recent studies on the symmetry properties of magnetic crystals.

Part I contains a number of exercises at the end of each chapter. Most of these follow some derivation or development from within the text. All these problems have either detailed solutions or hints, given at the end of the book, which readers will find quite useful in understanding the subject matter.

It is hoped that this book will provide the necessary basis for further detailed courses on the subject and as such the selection of topics in Part I is deliberately restricted.

Part II is meant to be self-contained and can be read without reference to Part I by readers who are familiar with elementary group theory. It should also prove useful to graduate students and research workers beginning the study of magnetic crystals. It is hoped that it will provide a useful adjunct to any advanced course on the applications of symmetry principles in solid state physics.

S J Joshua
March 1990

ACKNOWLEDGMENTS

I am particularly grateful to Professor A P Cracknell whose lectures on applied group theory provided the initial stimulus to write this book. I am also grateful to him for the many valuable discussions and suggestions on the subject matter contained in Part II.

I should also like to express my sincere gratitude to Professor A F Gibson FRS, for having given me the opportunity and providing me with financial assistance to work in the Department of Physics at the University of Essex where most of this work was carried out. Furthermore, of the many members of the staff at the University who have aided and encouraged me during the course of this work, I wish to thank especially Professor R Loudon FRS, Professor David Tilley, the late Dr J A Reissland and Dr P Kapadia.

It is also with great pleasure that I acknowledge the invaluable comments and assistance provided by my students, especially S Deonarine and F D Morgan.

Special thanks are also due to Mr Jim Revill of Adam Hilger for his many helpful comments and suggestions on the presentation and for his invaluable assistance in preparing the final figures.

Finally, I should like to thank Mrs Luigina Njuki, Secretary of the Department of Physics, Moi University, for her ceaseless assistance in typing and preparing the manuscript. The responsibility for any errors or shortcomings is of course entirely mine.

S J Joshua
March 1990

PART I

INTRODUCTION TO SYMMETRY PRINCIPLES IN SOLID STATE PHYSICS

1

ELEMENTS OF SYMMETRY THEORY

1.1 Introduction

Crystals play a particularly important role in modern-day solid state electronic devices. In the study of the behaviour and properties of crystalline materials, one requires a good exposure to:

(i) quantum mechanics, which is needed to study the behaviour and properties of individual atoms;
(ii) statistical mechanics, which is needed to describe the average properties of a large ensemble of atoms; and
(iii) symmetry theory (or group theory), which is needed to describe and understand the properties arising from the symmetrical structure of the solid.

In this book we shall be concerned only with (iii). Group theory, however, is really a branch of pure mathematics and as such it is usually developed with reference to abstract entities. Nevertheless, this is not the course we shall adopt, as we shall be mainly concerned with the applications of the theory to problems in crystal physics.

1.2 The role of symmetry

Symmetry arguments not only greatly simplify calculations of the thermal, electrical, optical and magnetic properties of solids etc, but also often give a detailed insight into the physical situation. It is therefore a standard practice amongst research workers to carry out a symmetry analysis before undertaking any calculations or experimental investigations on problems in crystal physics.

Given any crystal system it is prudent to begin by considering what information may be gleaned from a symmetry analysis of the structure. As often happens, even when the empirical work has been done, a symmetry analysis is still very helpful in testing various models put forward.

The symmetry which we aim to exploit via group theory is, in most cases, the symmetry of the Hamiltonian operator. An operator will be a symmetry operator appropriate to a given Hamiltonian, if the Hamiltonian looks the same after applying the operation as it did before; in other words, the Hamiltonian is invariant or unchanged under the operation. Our studies of the symmetry of the Hamiltonian of the system will indicate the range of uses to which symmetry concepts can be put in quantum mechanics. To achieve our goal we must first develop the elements of group theory in a manner suitable for ready applications. Thus, we shall not spend time over fundamental theorems, axioms, proofs, etc which can be found in many excellent textbooks on group theory.

1.3 Symmetry concepts via group theory

1.3.1 Symmetry operations and symmetry elements

The symmetry possessed by a crystalline material is defined in terms of symmetry operations. An operation is said to be a symmetry operation if the body looks exactly the same after the operation as it did before the operation was carried out. The operation could be, for example, rotations through certain angles or reflections in various mirror planes. As an example, consider the equilateral triangle shown in figure 1.1. It is easy to see that if we performed an operation of rotating the triangle through 120° or 240° about the z axis, the triangle would be left in a position which is indistinguishable from the starting position. Such an operation or movement which transforms a body into itself or leaves it invariant is called a symmetry operation and the body is said to possess the appropriate symmetry element. A symmetry element is therefore a real geometrical part of the system and

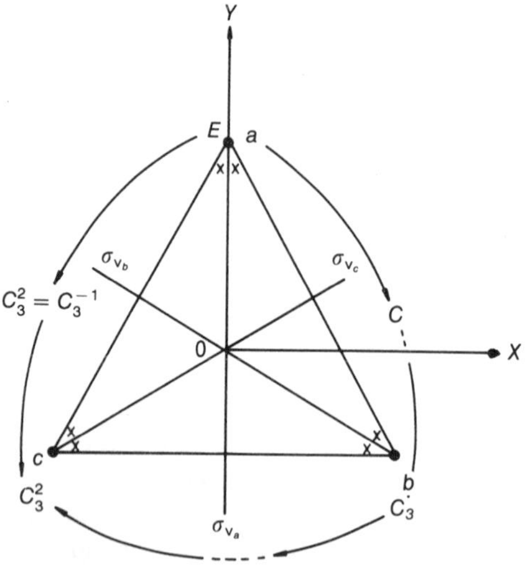

Figure 1.1 Symmetry operations corresponding to the crystal point group C_{3v}. The symbols have the following meanings (explained more fully in Chapter 4): C, cyclic group; σ, reflection plane; subscript 3, three-fold vertical rotation axes located at the central point 0; subscript v, vertical reflection planes.

also an identification symbol of the operation; for example, the symmetry element C_3 of the equilateral triangle implies the following symmetry operations: C_3^1, which corresponds to a clockwise rotation operation through 120° about the z axis; and C_3^2, which corresponds to a clockwise rotation operation through 240° about the z axis. The symbols are defined later on; for the moment it is important to note the difference between the terms symmetry elements and symmetry operations.

1.3.2 Basic types of symmetry elements

In dealing with crystal symmetry (point) groups we need to consider only the symmetry elements described and defined in table 1.1.

It is important to mention here briefly that suitable combinations of the symmetry elements given in table 1.1 define what is technically termed as a crystal point group. A crystal point group is defined as a collection of symmetry operations which, when applied about a point in the crystal, will leave the crystal structure invariant. Note that a crystal point group does not contain any translation operations. These are explained more fully in §4.5.

1.3.3 What is a symmetry group?

A symmetry group is a collection of symmetry elements $E, A, B, C, \ldots, X$ which satisfy the following group properties.

Table 1.1 The basic type of symmetry elements in Schoenflies notation.

No.	Symmetry element and symbol	Symmetry operation(s)
1	Identity element: E	Do nothing operation; like multiplying by 1
2	Mirror plane: σ	Reflection in a mirror
3	Centre of inversion or inversion symmetry: I	Inversion of all coordinates through the centre; I changes (x, y, z) to $(-x, -y, -z)$
4	Proper rotation axis: C_n	n-fold rotations about an appropriate axis. An n-fold axis generates n operations, e.g. the symmetry element C_6 implies the operations: C_6^1, C_6^2, C_6^3, C_6^4, C_6^5 and $C_6^6 = E$
5	Improper rotation axis: S_n	n-fold rotation about an axis followed by a reflection in a plane perpendicular to the axis of rotation, i.e. presence of S_n implies the independent existence of both C_n and σ elements

1.3.4 Group properties

(i) Every symmetry group G must contain a unit element (also called identity element) in the group which commutes with all the other elements in G and leaves them unchanged. We shall always denote this element by the symbol E such that $EX = XE = X$, where X is any element in the group.

(ii) Every element in the group must have its inverse contained in the group such that $XX^{-1} = E$, where X is any element in the group G.

(iii) The product (or combination) of any two elements in the group and the square of each element must be an element in the group. (This is technically known as the closure property of groups.)

(iv) The associative law of multiplication must hold, that is $A(BC) = (AB)C$. The group properties are explained more fully by means of an example given below. It is important to note that, in general, the elements in a group do not commute, that is $AB \neq BA$; but if $AB = BA$ for all the elements A and B in G the group G is said to be an Abelian group, otherwise it is said to be a non-Abelian group.

1.3.5 An example of a symmetry group

As an example we shall consider one of the crystal point groups denoted by the symbol C_{3v} in the Schoenflies notation or 3m in the International notation. These notations are explained more fully in later sections (see table 4.3).

The group C_{3v} contains the following symmetry elements which are sketched in figure 1.1: E, the identity operation which leaves every point unchanged (like multiplying by one); C_3, a clockwise rotation operation about the z axis through 120°; C_3^2, a clockwise rotation operation about the z axis through 240°; σ_{v_a}, a reflection operation in the YZ plane; σ_{v_b}, a vertical reflection operation in the plane passing through b and perpendicular to the line joining a and c; σ_{v_c}, a vertical reflection operation in the plane passing through c and perpendicular to the line joining a and b.

It is useful to emphasize here that, throughout this book, clockwise rotations are taken as positive and anticlockwise rotations are taken as negative for the set of axes chosen as in figure 1.1 (see also [2, 4, 16]). Most students have found this notation easy to remember. However, an alternative convention (usually adopted by crystallographers) is to consider anticlockwise rotations as positive and clockwise rotations as negative for a right-handed set of axes $0XYZ$ as explained in the solved exercise 1.4 (see also [5, 6]). Here the z axis is perpendicular to the plane of the paper and passes through the origin 0. It should be noted that these labels and conventions do not destroy any symmetry operations present in the solid state itself, but a uniform convention must be adopted even though the labels are merely a useful means for easier reference to the symmetry operations. In the actual applications part (Chapter 6), where the symmetry group character tables (described in

Chapter 3) have been used widely, the convention used for labelling the symmetry operations does not pose any problems. However, it is important for readers to stick to one convention and yet be at ease when following either notation from different books.

It is obvious from figure 1.1 that these symmetry operations correspond to the symmetry operations of an equilateral triangle. Note that other symmetry operations are possible but they are all equivalent to one of the operations given above. For example, as indicated in figure 1.1, an anticlockwise rotation through 120° denoted by C_3^{-1} is a symmetry operation, but it is identical with the operation C_3^2; similarly a rotation through 180° about the Y axis is identical with σ_{v_a} etc. The set of operations E, C_3, C_3^2, σ_{v_a}, σ_{v_b}, σ_{v_c} forms a group (called the C_{3v} group) as it obeys all the group properties. This can readily be seen by forming the group multiplication table as shown in table 1.2.

It is worth observing the following characteristic features of the multiplication table.

(i) Each row (or each column) is really a rearranged list of the symmetry elements present in the group. Furthermore, note that each element occurs once and only once. This is sometimes called the group rearrangement theorem.

(ii) Each entry in the table is given by the product of the column element and the row element. For example, the elements in the last row are obtained as follows:

Column element	Row element	Product	Entry in table
σ_{v_c}	E	$\sigma_{v_c} \times E$	σ_{v_c}
σ_{v_c}	C_3	$\sigma_{v_c} \times C_3$	σ_{v_a}
σ_{v_c}	C_3^2	$\sigma_{v_c} \times C_3^2$	σ_{v_b}
σ_{v_c}	σ_{v_a}	$\sigma_{v_c} \times \sigma_{v_a}$	C_3
σ_{v_c}	σ_{v_b}	$\sigma_{v_c} \times \sigma_{v_b}$	C_3^2
σ_{v_c}	σ_{v_c}	$\sigma_{v_c} \times \sigma_{v_c}$	E

(iii) Note that the group is not Abelian, since, for example, $\sigma_{v_a} \times \sigma_{v_b} \neq \sigma_{v_b} \times \sigma_{v_a}$.

Table 1.2 The group multiplication table for the group C_{3v}.

C_{3v}	E	C_3	C_3^2	σ_{v_a}	σ_{v_b}	σ_{v_c}
E	E	C_3	C_3^2	σ_{v_a}	σ_{v_b}	σ_{v_c}
C_3	C_3	C_3^2	E	σ_{v_c}	σ_{v_a}	σ_{v_b}
C_3^2	C_3^2	E	C_3	σ_{v_b}	σ_{v_c}	σ_{v_a}
σ_{v_a}	σ_{v_a}	σ_{v_b}	σ_{v_c}	E	C_3	C_3^2
σ_{v_b}	σ_{v_b}	σ_{v_c}	σ_{v_a}	C_3^2	E	C_3
σ_{v_c}	σ_{v_c}	σ_{v_a}	σ_{v_b}	C_3	C_3^2	E

1.4 Definitions of group theoretical terms

We shall always refer to the C_{3v} group and its multiplication table (table 1.2) for illustrative purposes.

1.4.1 Order of a group

The number of symmetry elements contained in a group is called its order. If we denote the order of a group by the symbol h, then for the C_{3v} group, order $h = 6$.

1.4.2 Subgroup

A subgroup G of a general group H is a subset of the symmetry elements of H which form an independent group, obeying all the group properties.

Example: For the C_{3v} group, the elements E, C_3 and C_3^2 form a subgroup called the cyclic group C_3 with the multiplication table as shown in table 1.3. For this subgroup, the order is 3.

Similarly we can see that (E, σ_{v_a}) forms a subgroup of order 2 and likewise (E, σ_{v_b}) and (E, σ_{v_c}) both form subgroups of order 2 of the group C_{3v}.

Property 1: If a group H of order h has a subgroup G of order g then h/g must be an integer.

Example: For group C_{3v}, $h = 6$; for subgroup C_3, $g = 3$; so that $h/g = 2$.

1.4.3 Cosets

The product of any element of a group with its subgroup is called a coset. If we have a group H with a subgroup G, then $A \times G$ is called the left coset and $G \times A$ is called the right coset, where A is any element in the group H.

Table 1.3 Multiplication table of the cyclic group C_3 which is a subgroup of the C_{3v} group.

C_3	E	C_3	C_3^2
E	E	C_3	C_3^2
C_3	C_3	C_3^2	E
C_3^2	C_3^2	E	C_3

Example: Consider the subgroup $G = (E, \sigma_{v_a})$ of the group C_{3v}. Then the left cosets of all the elements of the group $H = C_{3v}$ are as follows:

(i) coset $E \times G = E(E, \sigma_{v_a}) = E, \sigma_{v_a} = G$
(ii) coset $C_3 \times G = C_3(E, \sigma_{v_a}) = C_3, \sigma_{v_c}$
(iii) coset $C_3^2 \times G = C_3^2(E, \sigma_{v_a}) = C_3^2, \sigma_{v_b}$
(iv) coset $\sigma_{v_a} \times G = \sigma_{v_a}(E, \sigma_{v_a}) = \sigma_{v_a}, E = G$
(v) coset $\sigma_{v_b} \times G = \sigma_{v_b}(E, \sigma_{v_a}) = \sigma_{v_b}, C_3^2$
(vi) coset $\sigma_{v_c} \times G = \sigma_{v_c}(E, \sigma_{v_a}) = \sigma_{v_c}, C_3$.

Notice the following two important characteristic features of the cosets obtained above.

1. If an element A of the group H occurs in the subgroup G then $A \times G = G$. (See (i) and (iv) above.)
2. Two cosets are either identical or contain no elements in common. For example, cosets (ii) and (iii) have no common elements whilst cosets (i) and (iv) are identical.

Property 2: If every coset repeats k times then h/k is an integer. This integer is called the index of the subgroup G in the group H.

Example: For group $H = C_{3v}$, $h = 6$ and each coset for the subgroup $G = (E, \sigma_{v_a})$ repeats twice, and hence $6/2 = 3$ is the index of G in H. This index gives the number of distinct cosets.

1.4.4 Similarity transformations; conjugate elements; class

If A and X are two elements of a group then $X^{-1}AX$ will be equal to some other element of the group, say B. We then have $B = X^{-1}AX$. We express this relation by saying that: (i) B is the similarity transform of A and (ii) A and B are conjugate elements. We also define a class of elements as a complete set of elements which are conjugate with each other.

Property 3: (i) Every element is conjugate with itself. This means that it must always be possible to find an element X such that $A = X^{-1}AX$. *Proof*: If we multiply this by the left with A^{-1} we get $A^{-1}A = A^{-1}X^{-1}AX$ or $E = (XA)^{-1}AX$. This can only be true if X and A commute. Thus the element may always be E (the identity element) for example, or it could be any other element which commutes with the chosen element A.

(ii) If A is conjugate with B then B is conjugate with A. This means that if $A = X^{-1}BX$ then there must be some element, say Y, in the group for which $B = Y^{-1}AY$. *Proof*: $A = (X^{-1}BX)$ or $XAX^{-1} = X(X^{-1}BX)X^{-1} = EBE = B$. Thus, if $Y = X^{-1}$ and $Y^{-1} = X$ we have proved that $Y^{-1}AY = B$.

(iii) If A is conjugate with B and B is conjugate with C then A is conjugate with C; or if A is conjugate with B and C then B and C are conjugate with each other. This is shown by means of the example given below.

Example: We consider the C_{3v} group. Take any element σ_{v_a} say. Let us find all other elements in the group which are conjugate with it. We find:

$$E^{-1}\sigma_{v_a}E = \sigma_{v_a}$$

$$C_3^{-1}\sigma_{v_a}C_3 = \sigma_{v_c}$$

$$(C_3^2)^{-1}\sigma_{v_a}C_3^2 = \sigma_{v_b}$$

$$\sigma_{v_a}^{-1}\sigma_{v_a}\sigma_{v_a} = \sigma_{v_a}$$

$$\sigma_{v_b}^{-1}\sigma_{v_a}\sigma_{v_b} = \sigma_{v_c}$$

$$\sigma_{v_c}^{-1}\sigma_{v_a}\sigma_{v_c} = \sigma_{v_b}.$$

Thus the elements $\sigma_{v_a}, \sigma_{v_b}, \sigma_{v_c}$ are all conjugates and are therefore members of the same class. Similarly, it can be seen that C_3 and C_3^2 are conjugate and therefore belong to the same class. Also, E forms a class by itself (for all the crystal groups). Hence the C_{3v} group has the following three classes: E; $\{C_3, C_3^2\}$; $\{\sigma_{v_a}, \sigma_{v_b}, \sigma_{v_c}\}$.

Note that: (i) the order of the classes is 1, 2 and 3 respectively and that the order of a group divided by the order of a class is an integer; (ii) a class is not a subgroup.

We shall see the importance of determining the various classes in a group in later chapters.

1.4.5 Class multiplication

If $\mathscr{C}_i$ and $\mathscr{C}_j$ are two classes of a group H then $\mathscr{C}_i\mathscr{C}_j = \Sigma_k c_{ij,k}\mathscr{C}_k$, where the class multiplication coefficient $c_{ij,k}$ is an integer telling us how many times the class $\mathscr{C}_k$ appears as a result of multiplying together the two classes $\mathscr{C}_i$ and $\mathscr{C}_j$. Let us illustrate this by taking the classes from the C_{3v} group.

Let the three classes we determined earlier be represented as follows:

$$\mathscr{C}_1 = E \qquad \mathscr{C}_2 = \{C_3, C_3^2\} \qquad \mathscr{C}_3 = \{\sigma_{v_a}, \sigma_{v_b}, \sigma_{v_c}\}.$$

Consider the product of the two classes $\mathscr{C}_3$ and $\mathscr{C}_2$. We have

$$\begin{aligned}\mathscr{C}_3\mathscr{C}_2 &= \{\sigma_{v_a}, \sigma_{v_b}, \sigma_{v_c}\}\{C_3, C_3^2\}\\ &= \sigma_{v_a}\{C_3, C_3^2\} + \sigma_{v_b}\{C_3, C_3^2\} + \sigma_{v_c}\{C_3, C_3^2\}\\ &= \sigma_{v_a}C_3 + \sigma_{v_a}C_3^2 + \sigma_{v_b}C_3 + \sigma_{v_b}C_3^2 + \sigma_{v_c}C_3 + \sigma_{v_c}C_3^2\\ &= \sigma_{v_b} + \sigma_{v_c} + \sigma_{v_c} + \sigma_{v_a} + \sigma_{v_a} + \sigma_{v_b}\\ &= 2\sigma_{v_a} + 2\sigma_{v_b} + 2\sigma_{v_c}\end{aligned}$$

that is, $\mathscr{C}_3\mathscr{C}_2 = 2\mathscr{C}_3$ so that the coefficient $c_{32,3} = 2$. Similarly we can show that $\mathscr{C}_3\mathscr{C}_3 = 3\mathscr{C}_1 + 3\mathscr{C}_2$, so that the coefficients $c_{33,1} = 3$ and $c_{33,2} = 3$.

1.4.6 Isomorphous and homomorphous groups

Two groups are said to be isomorphous if there is a one-to-one correspondence between the elements of the two groups. This correspondence should be preserved during multiplication. Thus any two groups whose multiplication tables can be made identical are said to be isomorphic or, in the abstract sense, they are the same group.

Example: The group C_{3v} is isomorphic with the group S_3, the permutation group of three elements with the following six operations:

$$S_1S_2S_3 = E \text{ (say)}$$

$$S_1S_3S_2 = A \text{ (say)}$$

$$S_3S_1S_2 = B \text{ (say)}$$

$$S_3S_2S_1 = C \text{ (say)}$$

$$S_2S_3S_1 = D \text{ (say)}$$

$$S_2S_1S_3 = F \text{ (say)}$$

This group would have a multiplication table as shown in table 1.4.

This table is of course identical to table 1.1 as can be readily seen if we make the following correspondence between the symmetry elements of the two groups:

S_3	E	A	B	C	D	F
C_{3v}	E	σ_{v_a}	C_3	σ_{v_b}	C_3^2	σ_{v_c}

Two groups are said to be homomorphic if there is many-to-one correspondence between the elements of the two groups which is preserved during multiplication.

Table 1.4 The group multiplication (combination) table of the permutation group S_3.

S_3	E	A	B	C	D	F
E	E	A	B	C	D	F
A	A	E	C	B	F	D
B	B	F	D	A	E	C
C	C	D	F	E	A	B
D	D	C	E	F	B	A
F	F	B	A	D	C	E

Example: Consider a group C_4 of order 4 and consisting of the following elements:

Element	Operation
C_4^+	Rotation operation through 90° in the clockwise direction about the z axis
C_2	Rotation operation through 180° about the z axis
C_4^-	Rotation operation through 90° in the anticlockwise direction about the z axis
E	Identity element

Its group multiplication table is shown in table 1.5.

Consider now another group of order 2 for which the elements are:

Element	Operation
I	Multiplication by 1
A	Multiplication by -1

If we define the group multiplication table as ordinary algebraic multiplication we would get table 1.6.

We can now readily see that the two groups that we so defined are homomorphic if we make the correspondence: E, C_2: I and C_4^+, C_4^-: A; we note that the multiplication equation is preserved on going from one group to another.

Table 1.5 Multiplication table of the group C_4.

C_4	E	C_4^+	C_2	C_4^-
E	E	C_4^+	C_2	C_4^-
C_4^+	C_4^+	C_2	C_4^-	E
C_2	C_2	C_4^-	E	C_4^+
C_4^-	C_4^-	E	C_4^+	C_2

Table 1.6 Multiplication table of the group of order 2 as defined in the text.

	I	A
I	I	A
A	A	I

Exercises

1.1 Write down the symmetry elements present in the following chemical compounds:

(a) 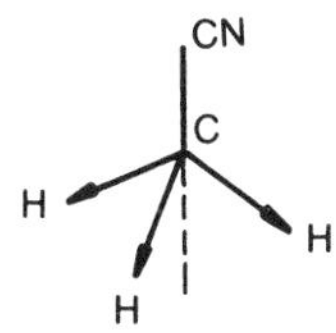

(b) 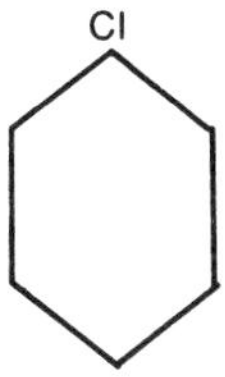

(c) 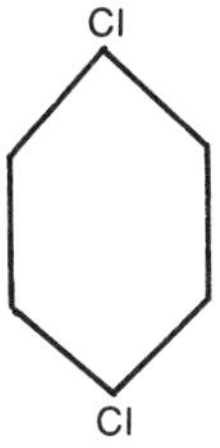

(d) 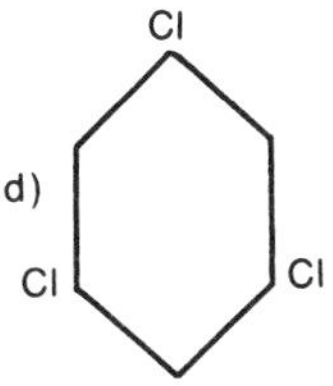

(e) UF_6.

1.2 Show by systematic application of the group properties whether or not the following sets form a group. Are the groups Abelian?
(a) [1, −1, i, −i] under multiplication
(b) [1, 0, −1] under arithmetic addition
(c) E, A, B, C, D, F under matrix multiplication where

$$E=\begin{pmatrix}1&0\\0&1\end{pmatrix}\quad A=\begin{pmatrix}E&0\\0&E^2\end{pmatrix}\quad B=\begin{pmatrix}E^2&0\\0&E\end{pmatrix}$$

$$C=\begin{pmatrix}0&1\\1&0\end{pmatrix}\quad D=\begin{pmatrix}0&E^2\\E&0\end{pmatrix}\quad F=\begin{pmatrix}0&E\\E^2&0\end{pmatrix}$$

and $E^3 = 1$.
(d) $\phi_1(x), \phi_2(x), \phi_3(x), \phi_4(x), \phi_5(x), \phi_6(x)$, under the operation of substitution of a function into another where the six functions are given by $\phi_1(x) = x$, $\phi_2(x) = 1 - x$, $\phi_3(x) = x/(x-1)$, $\phi_4(x) = 1/x$, $\phi_5(x) = 1/(1-x)$, $\phi_6(x) = (x-1)/x$.

1.3 Write down the possible number of permutations of three identical objects labelled a, b, c, and show that the set of permutations forms a group called the permutation group.

1.4 Draw rough sketches and write down in detail all the symmetry operations of the following regular figures in the *alternative convention* as explained in §1.3.5.

(a) The symmetry operations of the equilateral triangle: 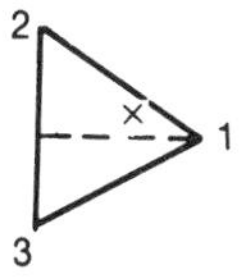

(b) The symmetry operations of the square: 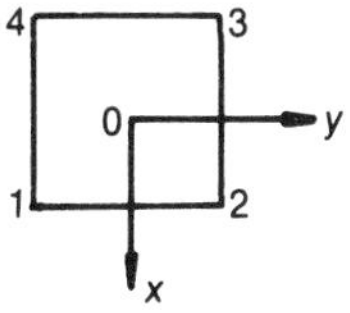

1.5 Write down the multiplication table for **1.4**(b). [We shall denote this group by the symbol C_{4v}.]

1.6 Obtain two subgroups of order 4 and five subgroups of order 2 using the multiplication table of the group C_{4v} (given in **1.5**).

1.7 Find all the classes $\mathscr{C}$ for the C_{4v} group.

1.8 Use the classes given in **1.7** to obtain the coefficients $c_{ij,k}$ of the following class multiplication: $c_{12,2}$, $c_{23,3}$, $c_{33,2}$, $c_{45,2}$, $c_{55,1}$.

2

ELEMENTARY THEORY OF GROUP REPRESENTATIONS

2.1 Group representations

In this section we shall be concerned with matrix algebra and as such a working knowledge of matrices is considered a prerequisite (see references under mathematical books [21–25]).

A set of matrices each corresponding to a single operation in the group and which can be combined amongst themselves in a manner similar to the multiplication table of the group are called a (matrix) representation of the group.

Consider a complete set of orthogonal unit vectors ϕ_r, $r = 1, 2, 3, \ldots, n$. The orthonormality condition requires that the inner product (or the scalar product):

$$\langle \phi_i | \phi_j \rangle = \delta_{ij} \tag{2.1}$$

where the Kronecker delta $\delta_{ij} = 1$ if $i = j$ and 0 if $i \neq j$. If S is any operator in the group then $S\phi_i$ can be expressed as a linear combination of ϕ_r since the ϕ_r form a complete set, that is

$$S\phi_i = \sum_r \phi_r \mathbf{D}(S)_{ri} \tag{2.2}$$

where the coefficients in the linear expansion are denoted by $D(S)_{ri}$. The equation (2.2) applied for $i = 1, 2, \ldots, n$ defines a matrix which we call the representative of the operator S. If we denote the row vector components ϕ_1, ϕ_2, ϕ_n with the symbol $\langle \phi |$ we have

$$S\langle \phi | = \langle \phi | \mathbf{D}(S). \tag{2.3}$$

The matrix $\mathbf{D}(S)$ varies if we define it with respect to a different set of unit vectors. The vector $\langle \phi |$ on which S operates is called the basis of the representation.

If the set of matrices $\mathbf{D}(S)$ correspond to operations in the group then they must multiply according to the group multiplication table. Mathematically, this means that if $RS = T$, where R, S and T are operations in the group then

$$\mathbf{D}(RS) = \mathbf{D}(T) = \mathbf{D}(R) \times \mathbf{D}(S). \tag{2.4}$$

Table 2.1 Matrix representations of the C_{3v} group.

Representation label	E	C_3	C_3^2	σ_{v_a}	σ_{v_b}	σ_{v_c}
Γ_1	(1)	(1)	(1)	(1)	(1)	(1)
Γ_2	(1)	(1)	(1)	(−1)	(−1)	(−1)
Γ_3	$\begin{pmatrix} 1 & 0 \\ 0 & 1 \end{pmatrix}$	$\begin{pmatrix} -\frac{1}{2} & \frac{\sqrt{3}}{2} \\ -\frac{\sqrt{3}}{2} & -\frac{1}{2} \end{pmatrix}$	$\begin{pmatrix} -\frac{1}{2} & -\frac{\sqrt{3}}{2} \\ \frac{\sqrt{3}}{2} & -\frac{1}{2} \end{pmatrix}$	$\begin{pmatrix} -1 & 0 \\ 0 & 1 \end{pmatrix}$	$\begin{pmatrix} \frac{1}{2} & -\frac{\sqrt{3}}{2} \\ -\frac{\sqrt{3}}{2} & -\frac{1}{2} \end{pmatrix}$	$\begin{pmatrix} \frac{1}{2} & \frac{\sqrt{3}}{2} \\ \frac{\sqrt{3}}{2} & -\frac{1}{2} \end{pmatrix}$

We can show this as follows:

$$S\phi_i = \sum_r \phi_r \mathbf{D}(S)_{ri}$$

$$R\phi_r = \sum_s \phi_s \mathbf{D}(R)_{sr}$$

$$\begin{aligned} RS\phi_i = R(S\phi_i) &= R \sum_r \phi_r \mathbf{D}(S)_{ri} \\ &= \sum_r R\phi_r \mathbf{D}(S)_{ri} \\ &= \sum_s \phi_s \mathbf{D}(R)_{sr} \sum_r \mathbf{D}(S)_{ri} \\ &= \sum_s \phi_s \sum_r \mathbf{D}(R)_{sr} \mathbf{D}(S)_{ri} \end{aligned}$$

that is

$$RS\phi_i = \sum_s \phi_s [\mathbf{D}(R)\mathbf{D}(S)]_{si}. \tag{2.5}$$

Equation (2.5) implies that the matrix representatives multiply in the same way as do the operators themselves. As an example consider the matrices given in table 2.1 which form representations of the C_{3v} group. Observe that the group of matrices belonging to each representation multiply in the same way as in table 1.1.

2.2 Symmetry operations in active and passive interpretations

If we define a symmetry operation R to be an operation which rotates a point q through an angle α in the anticlockwise direction, then we can define R^{-1} to be an operation which would rotate the point q through an angle α in the clockwise direction. Hence the operation shown in figure 2.1(a), which involves moving the original point q to q', can be written as

$$R^{-1}q = q'. \tag{2.6}$$

Such symmetry operations where the axes (XY) are kept fixed and the points (or vectors) are made to change positions are said to be symmetry operations in the active interpretation. We shall often use this interpretation. However, if we look at figure 2.1(b) we see that we could have achieved the same result by keeping the point q fixed but rotating the axes through an angle α in the anticlockwise direction from the XY position to the $X'Y'$ position. Such an operation is said to be an operation in the passive interpretation [7].

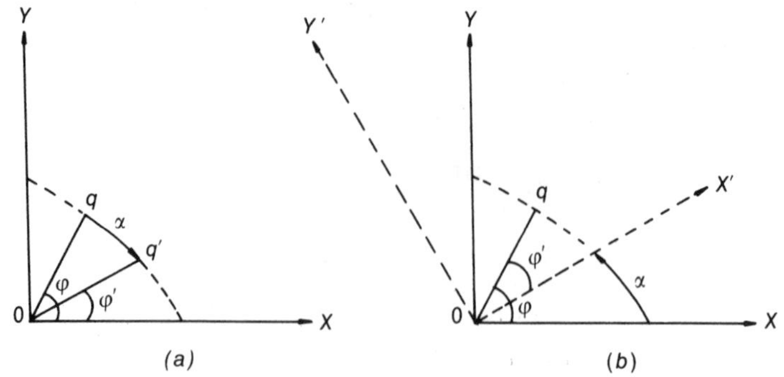

Figure 2.1 Symmetry operations in (a) active and (b) passive interpretation.

Suppose that we have a group G of symmetry operators $E, R, S, T, \dots$ and a function f on which these operators can act. For the sake of convenience, let $f = f(q)$ and let the symmetry operations be rotations. Then the rule to relate the action of the operators on the functions is expressed mathematically as:

$$Rf(q) = f(R^{-1}q) \tag{2.7}$$

and the multiplication rule for two operators acting on $f(q)$ is

$$\begin{aligned} RSf(q) = R[Sf(q)] &= Sf(R^{-1}q) \\ &= f(S^{-1}R^{-1}q) \\ &= f(T^{-1}q) \\ &= Tf(q) \end{aligned}$$

that is

$$RSf(q) = Tf(q) \tag{2.8}$$

where R, S and T are operations in the group.

2.3 Construction of matrix representation of rotation and reflection operations

We shall use the symbol C to denote a rotation operation and the symbol σ to denote a reflection operation.

Let C_α be a rotation operator which corresponds to a rotation of α about the z axis. We consider as our basis functions the well known $2p_x$ and $2p_y$ orbitals defined as

$$p_x = \sin\theta \cos\varphi \qquad p_y = \sin\theta \sin\varphi. \tag{2.9}$$

Here we have omitted the normalization constant which is not important as far as symmetry operations are concerned and θ and φ are angles in spherical polar coordinates. Equation (2.9) is in fact the angular part of the solutions of the Schrödinger equation for a central field system. Using equation (2.7) we have

$$\begin{aligned} C_\alpha p_x &= \sin\theta\cos(C_\alpha^{-1}\varphi) \\ &= \sin\theta\cos(\varphi+\alpha) \\ &= \sin\theta(\cos\varphi\cos\alpha - \sin\varphi\sin\alpha) \\ &= p_x\cos\alpha - p_y\sin\alpha. \end{aligned} \tag{2.10}$$

Similarly

$$C_\alpha p_y = p_x\sin\alpha + p_y\cos\alpha. \tag{2.11}$$

These two equations give us the matrix which represents the transformation of these basis functions by the rotation operator C_α. We can express (2.10) and (2.11) as

$$C_\alpha\langle p_x, p_y| = \langle p_x, p_y|\begin{pmatrix} \cos\alpha & \sin\alpha \\ -\sin\alpha & \cos\alpha \end{pmatrix} \tag{2.12}$$

so that the matrix representing the rotation operator C_α is

$$\mathbf{D}(C_\alpha) = \begin{pmatrix} \cos\alpha & \sin\alpha \\ -\sin\alpha & \cos\alpha \end{pmatrix}. \tag{2.13}$$

Similarly, if we consider a reflection operator, say σ_{v_y} ($\equiv \sigma_{v_a}$ of figure 1.1), then

$$\begin{aligned} \sigma_{v_y} p_x &= \sin\theta\cos(\sigma_{v_y}^{-1}\varphi) \\ &= \sin\theta\cos(\varphi+\pi) \\ &= \sin\theta\cos\varphi \\ &= -p_x \end{aligned} \tag{2.14}$$

and

$$\sigma_{v_y} p_y = p_y \tag{2.15}$$

so that the two equations can be written as

$$\sigma_{v_y}\langle p_x, p_y| = \langle p_x, p_y|\begin{pmatrix} -1 & 0 \\ 0 & 1 \end{pmatrix} \tag{2.16}$$

and therefore

$$\mathbf{D}(\sigma_{v_y}) = \begin{pmatrix} -1 & 0 \\ 0 & 1 \end{pmatrix}. \tag{2.17}$$

In general the matrix representing a reflection operation is given by

$$\mathbf{D}(\sigma) = \begin{pmatrix} \cos 2\Phi & \sin 2\Phi \\ \sin 2\Phi & -\cos 2\Phi \end{pmatrix} \tag{2.18}$$

where Φ is the azimuth of the mirror plane.

2.4 Construction of matrix representations of the C_{3v} group using various choices of basis

2.4.1 *Choice 1: basis functions defined by equation (2.9)*

Using equations (2.13) and (2.18) the matrices representing all the operations of the C_{3v} group are as follows:

$$\mathbf{D}(E) = \begin{pmatrix} 1 & 0 \\ 0 & 1 \end{pmatrix} \quad \mathbf{D}(C_3) = \begin{pmatrix} -\frac{1}{2} & \sqrt{3}/2 \\ -\sqrt{3}/2 & -\frac{1}{2} \end{pmatrix}$$

$$\mathbf{D}(C_3^2) = \begin{pmatrix} -\frac{1}{2} & -\sqrt{3}/2 \\ \sqrt{3}/2 & -\frac{1}{2} \end{pmatrix} \quad \mathbf{D}(\sigma_{v_a}) = \begin{pmatrix} -1 & 0 \\ 0 & 1 \end{pmatrix} \tag{2.19}$$

$$\mathbf{D}(\sigma_{v_b}) = \begin{pmatrix} \frac{1}{2} & -\sqrt{3}/2 \\ -\sqrt{3}/2 & -\frac{1}{2} \end{pmatrix} \quad \mathbf{D}(\sigma_{v_c}) = \begin{pmatrix} \frac{1}{2} & \sqrt{3}/2 \\ \sqrt{3}/2 & -\frac{1}{2} \end{pmatrix}.$$

2.4.2 *Choice 2: rectangular axes*

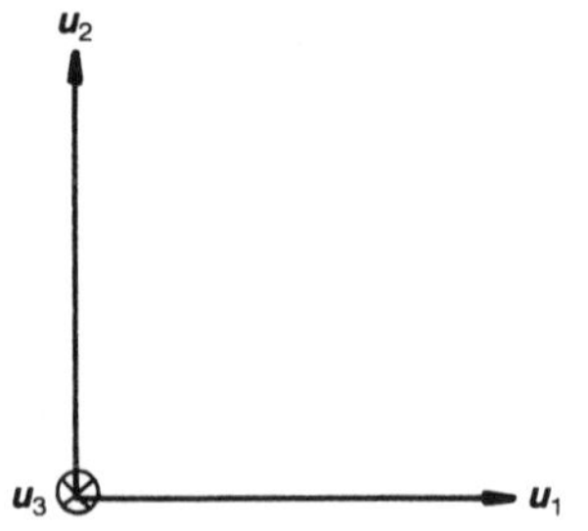

The identity element E transforms as $E(\boldsymbol{u}_1, \boldsymbol{u}_2, \boldsymbol{u}_3) = \boldsymbol{u}_1, \boldsymbol{u}_2, \boldsymbol{u}_3$ and therefore

$$\mathbf{D}(E) = \begin{pmatrix} 1 & 0 & 0 \\ 0 & 1 & 0 \\ 0 & 0 & 1 \end{pmatrix}$$

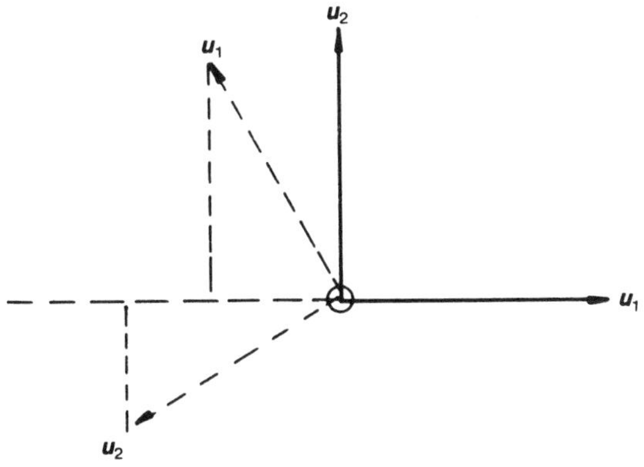

C_3 transforms as

$$C_3(\boldsymbol{u}_1, \boldsymbol{u}_2, \boldsymbol{u}_3) = (C_3\boldsymbol{u}_1,\ C_3\boldsymbol{u}_2,\ C_3\boldsymbol{u}_3)$$

$$= (\boldsymbol{u}_1,\ \boldsymbol{u}_2,\ \boldsymbol{u}_3)\begin{pmatrix} -\frac{1}{2} & \sqrt{3}/2 & 0 \\ -\sqrt{3}/2 & -\frac{1}{2} & 0 \\ 0 & 0 & 1 \end{pmatrix}.$$

Similarly we can find how $\boldsymbol{u}_1$, $\boldsymbol{u}_2$, $\boldsymbol{u}_3$ transform with respect to the other operations in the C_{3v} group. The matrix representations using rectangular axes are therefore as follows:

$$\mathbf{D}(E) = \begin{pmatrix} 1 & 0 & 0 \\ 0 & 1 & 0 \\ 0 & 0 & 1 \end{pmatrix} \qquad \mathbf{D}(C_3) = \begin{pmatrix} -\frac{1}{2} & \sqrt{3}/2 & 0 \\ -\sqrt{3}/2 & -\frac{1}{2} & 0 \\ 0 & 0 & 1 \end{pmatrix}$$

$$\mathbf{D}(C_3^2) = \begin{pmatrix} -\frac{1}{2} & -\sqrt{3}/2 & 0 \\ \sqrt{3}/2 & -\frac{1}{2} & 0 \\ 0 & 0 & 1 \end{pmatrix} \qquad \mathbf{D}(\sigma_{v_a}) = \begin{pmatrix} 1 & 0 & 0 \\ 0 & -0 & 0 \\ 0 & 0 & 1 \end{pmatrix} \qquad (2.20)$$

$$\mathbf{D}(\sigma_{v_b}) = \begin{pmatrix} -\frac{1}{2} & \sqrt{3}/2 & 0 \\ \sqrt{3}/2 & \frac{1}{2} & 0 \\ 0 & 0 & 1 \end{pmatrix} \qquad \mathbf{D}(\sigma_{v_c}) = \begin{pmatrix} -\frac{1}{2} & -\sqrt{3}/2 & 0 \\ -\sqrt{3}/2 & \frac{1}{2} & 0 \\ 0 & 0 & 1 \end{pmatrix}.$$

2.4.3 Choice 3: oblique axes

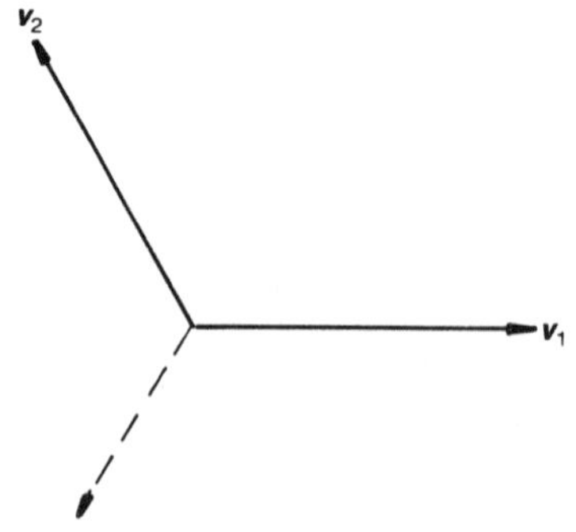

It can be readily seen that the matrix representation of the group C_{3v} using the oblique axes defined by $\boldsymbol{v}_1$ and $\boldsymbol{v}_2$ are as follows:

$$\mathbf{D}(E)=\begin{pmatrix}1 & 0\\ 0 & 1\end{pmatrix} \qquad \mathbf{D}(C_3)=\begin{pmatrix}-1 & 1\\ -1 & 0\end{pmatrix}$$

$$\mathbf{D}(C_3^2)=\begin{pmatrix}0 & -1\\ 1 & -1\end{pmatrix} \qquad \mathbf{D}(\sigma_{v_a})=\begin{pmatrix}1 & -1\\ 0 & -1\end{pmatrix} \qquad (2.21)$$

$$\mathbf{D}(\sigma_{v_b})=\begin{pmatrix}0 & 1\\ 1 & 0\end{pmatrix} \qquad \mathbf{D}(\sigma_{v_c})=\begin{pmatrix}-1 & 0\\ -1 & 1\end{pmatrix}.$$

2.5 The regular representation

We have seen in the above examples that for any given group it is possible to obtain a large number of matrix representations. A particularly interesting representation of any symmetry group and one which can be written down immediately is the so-called regular representation. The method of writing down the regular representation of any symmetry group is as follows.

(i) Write down the group multiplication table.
(ii) Rearrange rows so that they correspond to the inverses of the elements labelling columns (or vice versa). This will give a group table in which the identity element E will always occur along the diagonal position.
(iii) The regular representation for any operation R in the group is obtained by replacing R in the group table by one and putting zero for all other elements in the table.

Example: To find the regular representation of the group C_{3v} we would first rearrange table 1.2 as shown in table 2.2.

Table 2.2 Rearrangement of table 1.2 such that the identity element E appears in the diagonal position.

C_{3v}	E	C_3^2	C_3	σ_{v_a}	σ_{v_b}	σ_{v_c}
$E^{-1} = E$	E	C_3^2	C_3	σ_{v_a}	σ_{v_b}	σ_{v_c}
$(C_3^2)^{-1} = C_3$	C_3	E	C_3^2	σ_{v_c}	σ_{v_a}	σ_{v_b}
$C_3^{-1} = C_3^2$	C_3^2	C_3	E	σ_{v_b}	σ_{v_c}	σ_{v_a}
$\sigma_{v_a}^{-1} = \sigma_{v_a}$	σ_{v_a}	σ_{v_c}	σ_{v_b}	E	C_3	C_3^2
$\sigma_{v_b}^{-1} = \sigma_{v_b}$	σ_{v_b}	σ_{v_a}	σ_{v_c}	C_3^2	E	C_3
$\sigma_{v_c}^{-1} = \sigma_{v_c}$	σ_{v_c}	σ_{v_b}	σ_{v_a}	C_3	C_3^2	E

Table 2.3 The matrices in the regular representation for the symmetry operators E and C_3^2 of the C_{3v} group.

$$\mathbf{D}(E) = \begin{pmatrix} 1 & 0 & 0 & 0 & 0 & 0 \\ 0 & 1 & 0 & 0 & 0 & 0 \\ 0 & 0 & 1 & 0 & 0 & 0 \\ 0 & 0 & 0 & 1 & 0 & 0 \\ 0 & 0 & 0 & 0 & 1 & 0 \\ 0 & 0 & 0 & 0 & 0 & 1 \end{pmatrix}$$

$$\mathbf{D}(C_3^2) = \begin{pmatrix} 0 & 1 & 0 & 0 & 0 & 0 \\ 0 & 0 & 1 & 0 & 0 & 0 \\ 1 & 0 & 0 & 0 & 0 & 0 \\ 0 & 0 & 0 & 0 & 0 & 1 \\ 0 & 0 & 0 & 1 & 0 & 0 \\ 0 & 0 & 0 & 0 & 1 & 0 \end{pmatrix}$$

The matrices for E and C_3^2, for example, in the regular representation would then be given as shown in table 2.3.

2.6 Reducible and irreducible representations

Suppose we have a representation of a group denoted by the symbol Γ with matrices $\mathbf{D}(E)$, $\mathbf{D}(A)$, $\mathbf{D}(B)$, $\mathbf{D}(C), \ldots,$ etc. If now we were able to find a unitary matrix $\mathbf{U}$ such that by performing a single similarity transformation all the matrices of the representation Γ could be put in the block diagonal form as shown below, then the representation Γ is said to be a reducible representation and the new set of matrices $\mathbf{D}(E')$, $\mathbf{D}(A')$, $\mathbf{D}(B')$, ..., etc also form a representation of the group. For example, when $\mathbf{D}(A)$ is transformed to $\mathbf{D}(A')$ using $\mathbf{U}$ we find $\mathbf{D}(A')$ to be a block diagonal matrix of the form:

$$\mathbf{D}_1(A') = \mathbf{U}^{-1}\mathbf{D}(A)\mathbf{U} = \begin{pmatrix} \mathbf{D}_1(A') & & & 0 \\ & \mathbf{D}_2(A') & & \\ & & \mathbf{D}_3(A') & \\ 0 & & & \mathbf{D}_4(A') \end{pmatrix}. \tag{2.22}$$

Similarly

$$\mathbf{D}_1(B') = \mathbf{U}^{-1}\mathbf{D}(B)\mathbf{U} = \begin{pmatrix} \mathbf{D}_1(B') & & & 0 \\ & \mathbf{D}_2(B') & & \\ & & \mathbf{D}_3(B') & \\ 0 & & & \mathbf{D}_4(B') \end{pmatrix} \tag{2.23}$$

and so on for $\mathbf{D}(E)$, $\mathbf{D}(C)$, ..., etc. The information contained in (2.22), (2.23), etc is often written as:

$$\begin{aligned} \mathbf{D}(A') &= \mathbf{D}_1(A') \oplus \mathbf{D}_2(A') \oplus \mathbf{D}_3(A') \oplus \mathbf{D}_4(A') \\ \mathbf{D}(B') &= \mathbf{D}_1(B') \oplus \mathbf{D}_2(B') \oplus \mathbf{D}_3(B') \oplus \mathbf{D}_4(B') \end{aligned} \tag{2.24}$$

where $\oplus$ denotes the direct sum and all the $\mathbf{D}_n$ along the diagonal themselves form representations of the group. Thus:

$$\Gamma_1 = \mathbf{D}_1(E'), \mathbf{D}_1(A'), \mathbf{D}_1(B'), \mathbf{D}_1(C') \ldots$$

$$\Gamma_2 = \mathbf{D}_2(E'), \mathbf{D}_2(A'), \mathbf{D}_2(B'), \mathbf{D}_2(C') \ldots$$

$$\Gamma_3 = \mathbf{D}_3(E'), \mathbf{D}_3(A'), \mathbf{D}_3(B'), \mathbf{D}_3(C') \ldots$$

and so on. The original representation Γ is then called a reducible representation, since it was possible by using a unitary matrix $\mathbf{U}$ to break it into two or more representations Γ_1, Γ_2, Γ_3, etc, and if the new representations (Γ_1, Γ_2, Γ_3, etc) cannot be reduced further by any other unitary matrix, they are said to be irreducible representations of the group. As we shall see later, it is the irreducible representations of a group which are very important in studying the symmetry description of the physical properties of crystals.

2.7 An example of reduction of representations

We shall illustrate briefly how to construct a reducible representation for the C_{3v} group and then how to break it into irreducible representations. We consider figure 2.2 for the C_{3v} group and obtain a transformation table (table 2.4) for all the operations of the group with respect to the basis a_1, a_2, a_3.

We can now readily obtain the matrix representatives of these operators using the fact that $\int a_i a_j \mathrm{d}_\tau = \delta_{ij}$, where τ represents all the variables which occur in a_i and a_j. These are listed in table 2.5 and it is easy to verify that these matrices multiply in the same way as do the operators themselves.

If now we were to find a unitary matrix $\mathbf{U}$ of the form given by

$$\mathbf{U} = \begin{pmatrix} \frac{1}{\sqrt{3}} & \frac{-\sqrt{2}}{\sqrt{3}} & 0 \\ \frac{1}{\sqrt{3}} & \frac{1}{\sqrt{6}} & \frac{-1}{\sqrt{2}} \\ \frac{1}{\sqrt{3}} & \frac{1}{\sqrt{6}} & \frac{1}{\sqrt{2}} \end{pmatrix} \tag{2.25}$$

for which $\mathbf{U}^{-1}$ is

$$\mathbf{U}^{-1} = \begin{pmatrix} \frac{1}{\sqrt{3}} & \frac{1}{\sqrt{3}} & \frac{1}{\sqrt{3}} \\ \frac{-\sqrt{2}}{\sqrt{3}} & \frac{1}{\sqrt{6}} & \frac{1}{\sqrt{6}} \\ 0 & \frac{-1}{\sqrt{2}} & \frac{1}{\sqrt{2}} \end{pmatrix} \tag{2.26}$$

it can be readily seen by finding $\mathbf{U}^{-1}R\mathbf{U}$, where R is a symmetry operation of the group C_{3v}, that table 2.5 will break up in the form shown in table 2.6, that is into 1×1 and 2×2 matrices. We therefore say that we have reduced the representation of table 2.5 into two representations by using a unitary matrix $\mathbf{U}$. It should be observed that the matrices given in table 2.5 are similar to the ones constructed earlier (equation (2.20)) using rectangular axes, and in fact the representations given in table 2.6 correspond to representatives in the basis $\langle z|$ and $\langle xy|$. We shall see later that, in the applications of the symmetry group of crystals, it is the irreducible representations of the group (and its associated properties) which are of fundamental importance.

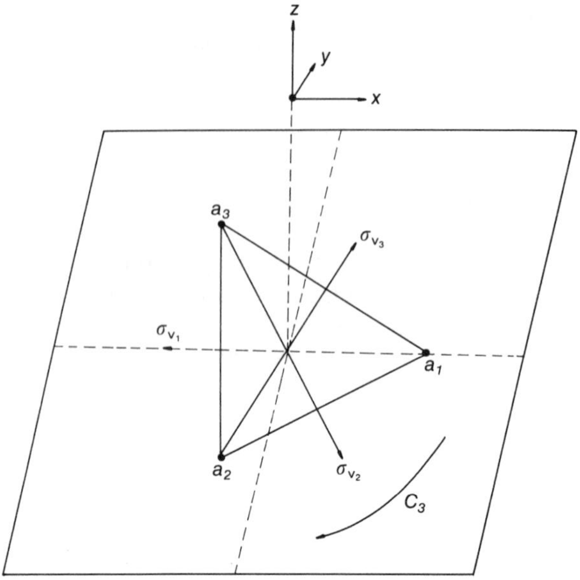

Figure 2.2 The C_{3v} group with two choices of basis, namely a_1, a_2, a_3 and the rectangular axes z, x, y.

Table 2.4 Transformation properties of a_1, a_2, a_3.

Symmetry operation	a_1	a_2	a_3
E	a_1	a_2	a_3
C_3	a_3	a_1	a_2
C_3^2	a_2	a_3	a_1
σ_{v_1}	a_1	a_3	a_2
σ_{v_2}	a_3	a_2	a_1
σ_{v_3}	a_2	a_1	a_3

Table 2.5 Matrix representation of the C_{3v} group using the basis $\langle a_1, a_2, a_3|$.

E	C_3	C_3^2	σ_{v_1}	σ_{v_2}	σ_{v_3}
$\begin{pmatrix}1&0&0\\0&1&0\\0&0&1\end{pmatrix}$	$\begin{pmatrix}0&0&1\\1&0&0\\0&1&0\end{pmatrix}$	$\begin{pmatrix}0&1&0\\0&0&1\\1&0&0\end{pmatrix}$	$\begin{pmatrix}1&0&0\\0&0&1\\0&1&0\end{pmatrix}$	$\begin{pmatrix}0&0&1\\0&1&0\\1&0&0\end{pmatrix}$	$\begin{pmatrix}0&1&0\\1&0&0\\0&0&1\end{pmatrix}$

Table 2.6 Reduction of matrices given in table 2.5 under a unitary transformation using the matrix **U** (see text).

E	C_3	C_3^2	σ_{v_1}	σ_{v_2}	σ_{v_3}
$\begin{pmatrix} 1 & \multicolumn{2}{c}{0} \\ 0 & 1 & 0 \\ & 0 & 1 \end{pmatrix}$	$\begin{pmatrix} 1 & 0 & \\ 0 & -\frac{1}{2} & -\frac{\sqrt{3}}{2} \\ & \frac{\sqrt{3}}{2} & -\frac{1}{2} \end{pmatrix}$	$\begin{pmatrix} 1 & 0 & \\ 0 & -\frac{1}{2} & \frac{\sqrt{3}}{2} \\ & -\frac{\sqrt{3}}{2} & -\frac{1}{2} \end{pmatrix}$	$\begin{pmatrix} 1 & 0 & \\ 0 & 1 & 0 \\ & 0 & -1 \end{pmatrix}$	$\begin{pmatrix} 1 & 0 & \\ 0 & -\frac{1}{2} & -\frac{\sqrt{3}}{2} \\ & -\frac{\sqrt{3}}{2} & \frac{1}{2} \end{pmatrix}$	$\begin{pmatrix} 1 & 0 & \\ 0 & -\frac{1}{2} & \frac{\sqrt{3}}{2} \\ & \frac{\sqrt{3}}{2} & \frac{1}{2} \end{pmatrix}$

Exercises

2.1 Obtain two sets of matrices of dimensions (2×2) and (3×3) which represent all the operations of the C_{4v} group and show that these matrices satisfy the group properties (i)–(iv) of §1.3.4.

2.2 Generate the group C_{4v} using two symmetry elements C_{4z}^+ and σ_{vx}.

2.3 Determine the matrices in the regular representation for C_{2z} and C_{4z}^+ of the group C_{4v}.

2.4 Consider a representation Γ of a group G given by the following matrices:

$$E = \begin{pmatrix} 1 & 0 \\ 0 & 1 \end{pmatrix} \quad A = \begin{pmatrix} \frac{1}{2} - \frac{i\sqrt{3}}{2} & 0 \\ 0 & \frac{1}{2} + \frac{i\sqrt{3}}{2} \end{pmatrix} \quad B = \begin{pmatrix} \frac{1}{2} + \frac{i\sqrt{3}}{2} & 0 \\ 0 & \frac{1}{2} - \frac{i\sqrt{3}}{2} \end{pmatrix}$$

$$C = \begin{pmatrix} 0 & 1 \\ -1 & 0 \end{pmatrix} \quad D = \begin{pmatrix} 0 & \frac{1}{2} - \frac{i\sqrt{3}}{2} \\ -\frac{1}{2} - \frac{i\sqrt{3}}{2} & 0 \end{pmatrix} \quad F = \begin{pmatrix} 0 & \frac{1}{2} + \frac{i\sqrt{3}}{2} \\ -\frac{1}{2} + \frac{i\sqrt{3}}{2} & 0 \end{pmatrix}.$$

Determine whether or not Γ^* and Γ^{-1} are also representations of G.

2.5 Show that the matrix

$$\mathbf{U} = \begin{pmatrix} \frac{1}{2}(1+i) & \frac{i}{\sqrt{3}} & \frac{3+i}{2\sqrt{15}} \\ -\frac{1}{2} & \frac{1}{\sqrt{3}} & \frac{4+3i}{2\sqrt{15}} \\ \frac{1}{2} & \frac{-i}{\sqrt{3}} & \frac{5i}{2\sqrt{15}} \end{pmatrix}$$

is a unitary matrix.

2.6 In the figure shown below, $\hat{x}$ and $\hat{y}$ are unit vectors along the x and y directions and the unit vector $\hat{z}$ at 0 is perpendicular to the plane of the paper. We define the following set of operations (note: the alternative notation is being used here):

C_3^1, anticlockwise rotation through 120° about the z axis
C_3^2, anticlockwise rotation through 240° about the z axis
C_3^3, identity operation $= E$

C'_2, C''_2, C'''_2, rotations through angles of 180° about the three axes in the x–y plane.

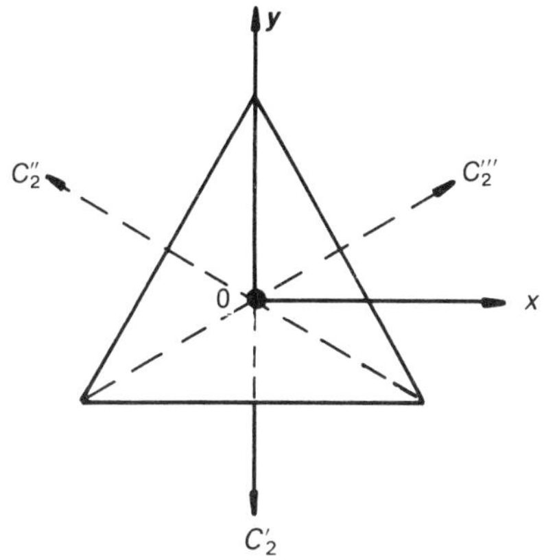

Determine a 3 × 3 matrix representation for this group and then reduce it to obtain a 2 × 2 and a 1 × 1 irreducible representation.

3

IMPORTANT PROPERTIES OF THE IRREDUCIBLE REPRESENTATIONS OF CRYSTAL SYMMETRY GROUPS

In this chapter we shall merely state (without proofs) all the properties of group representations which are important for our purposes and use the C_{3v} group for illustration.

3.1 Character of the irreducible representation (IR)

The sum of the diagonal elements (i.e. trace) of the matrices of IRS are called the characters of the IR and we shall denote it by the symbol χ. Thus, for the representation of the C_{3v} group given by equation (2.19) we have

$$\begin{array}{lll} \chi(E) = 2 & \chi(C_3) = -1 & \chi(C_3^2) = -1 \\ \chi(\sigma_{v_a}) = 0 & \chi(\sigma_{v_b}) = 0 & \chi(\sigma_{v_c}) = 0. \end{array} \qquad (3.1)$$

The complete character table for the C_{3v} group is given in table 3.1.

It is perhaps worth mentioning here that one does not have to find the IR of a group to find the character χ of the representation. Several other methods are available (see for example [5, 20]) but as the character tables have now been tabulated for all the crystal symmetry groups (see for example [1, 4, 5, 10]) we shall simply state their properties and emphasize their applications.

Table 3.1 Character table of the C_{3v} group.

C_{3v}	E	$2C_3 = \{C_3, C_3^2\}$	$3\sigma_v = \{\sigma_{v_a}, \sigma_{v_b}, \sigma_{v_c}\}$
$\mathbf{D}_1$	1	1	1
$\mathbf{D}_2$	1	1	-1
$\mathbf{D}_3$	2	-1	0

3.2 The dimension of an IR

The dimension of an IR is given by the dimension of its matrix representative. For example, if a set of 2 × 2 matrices form an IR of a group then this representation would be said to have a dimension equal to two. In practice the dimension of an IR labelled i, say, is obtained simply by looking at the value of $\chi^i(E)$, since $\chi^i(E)$ is the trace of the unit matrix representing E. Thus, for example, equation (3.1) corresponds to an IR of dimension 2 since $\chi(E) = 2$. We shall denote the dimension of a representation by the symbol l.

3.3 The great orthogonality theorem

We shall express this very important theorem mathematically and study its properties by means of examples. The proof (which is not trivial) may be found in most books on group theory (see for example [2, 14]).

If $\mathbf{D}_i(R)_{mn}$ is the (m, n) matrix element of the ith IR of the element R of a group G, then this theorem states (mathematically):

$$\sum_{\text{all } R \text{ in } G} \mathbf{D}_i(R)^*_{mn} \mathbf{D}_j(R)_{m'n'} = \frac{h}{l_i} \delta_{ij} \delta_{mm'} \delta_{nn'}. \tag{3.2}$$

Here h is the order of G, that is the number of elements contained in G, and l_i is the dimension of the ith IR; δ is the Kronecker delta symbol, that is $\delta_{ij} = 1$ if $i = j$, and $\delta_{ij} = 0$ if $i \neq j$, etc.

The interpretation of equation (3.2) becomes more clear if we split it into three simpler equations as follows [4]:

(a) $$\sum_R \mathbf{D}_i(R)_{mn} \mathbf{D}_j(R)_{mn} = 0 \quad \text{since } i \neq j. \tag{3.3}$$

Example: For the C_{3v} group (table 3.1) we have

$$\sum_R \mathbf{D}_i(R) \mathbf{D}_2(R) = 1 \times 1 + 2(1 \times 1) + 3(1 \times -1)$$

$$= 1 + 2 - 3$$

$$= 0.$$

(b) $$\sum_R \mathbf{D}_i(R)_{mn} \mathbf{D}_i(R)_{m'n'} = 0 \quad \text{since } m \neq m' \text{ and } n \neq n'. \tag{3.4}$$

Example: For the C_{3v} group, we consider the $\mathbf{D}_3$ IR of dimension $l_i = 2$ (table 3.1). To prove equation (3.4) we need to know the matrix elements explicitly of $\mathbf{D}_3$. This is given by the 2 × 2 matrices of table 2.6. Let us take $m = 1$, $n = 2$, that is the 1, 2 elements, and $m' = 2$, $n' = 1$, that is the 2, 1 elements.

We can then express $\mathbf{D}_3$ for all R contained in C_{3v} as follows:

$$\begin{aligned}
&\mathbf{D}_3(E)_{12} = 0 && \mathbf{D}_3(E)_{21} = 0 && \therefore\ \mathbf{D}_3(E)_{12} \times \mathbf{D}_3(E)_{21} = 0\\
&\mathbf{D}_3(C_3)_{12} = -\frac{\sqrt{3}}{2} && \mathbf{D}_3(C_3)_{21} = -\frac{\sqrt{3}}{2} && \therefore\ \mathbf{D}_3(C_3)_{12} \times \mathbf{D}_3(C_3)_{21} = -\tfrac{3}{4}\\
&\mathbf{D}_3(C_3^2)_{12} = -\frac{\sqrt{3}}{2} && \mathbf{D}_3(C_3^2)_{21} = \frac{\sqrt{3}}{2} && \therefore\ \mathbf{D}_3(C_3^2)_{12} \times \mathbf{D}_3(C_3^2)_{21} = -\tfrac{3}{4}\\
&\mathbf{D}_3(\sigma_{v_a})_{12} = 0 && \mathbf{D}_3(\sigma_{v_a})_{21} = 0 && \therefore\ \mathbf{D}_3(\sigma_{v_a})_{12} \times \mathbf{D}_3(\sigma_{v_a})_{21} = 0\\
&\mathbf{D}_3(\sigma_{v_b})_{12} = -\frac{\sqrt{3}}{2} && \mathbf{D}_3(\sigma_{v_b})_{21} = -\frac{\sqrt{3}}{2} && \therefore\ \mathbf{D}_3(\sigma_{v_b})_{12} \times \mathbf{D}_3(\sigma_{v_b})_{21} = \tfrac{3}{4}\\
&\mathbf{D}_3(\sigma_{v_c})_{12} = \frac{\sqrt{3}}{2} && \mathbf{D}_3(\sigma_{v_c})_{21} = \frac{\sqrt{3}}{2} && \therefore\ \mathbf{D}_3(\sigma_{v_c})_{12} \times \mathbf{D}_3(\sigma_{v_c})_{21} = \tfrac{3}{4}.
\end{aligned} \tag{3.5}$$

Hence from the RHS we see that

$$\sum_R \mathbf{D}_3(R)_{12}\mathbf{D}_3(R)_{21} = 0 - \tfrac{3}{4} - \tfrac{3}{4} + 0 + \tfrac{3}{4} + \tfrac{3}{4} = 0.$$

That is, if we take the same IR but different matrix elements then the vectors are orthogonal.

(c)
$$\sum_R \mathbf{D}_i(R)_{mn}\mathbf{D}_i(R)_{mn} = \frac{h}{l_i}. \tag{3.6}$$

Example: Taking $m = 1$, $n = 2$ and $\mathbf{D}_i = \mathbf{D}_3$ of the C_{3v} group we may write, using the values of $\mathbf{D}_3(R)_{12}$ given in equation (3.5),

$$\sum_R \mathbf{D}_3(R)_{12}\mathbf{D}_3(R)_{12} = 0 + \tfrac{3}{4} + \tfrac{3}{4} + 0 + \tfrac{3}{4} + \tfrac{3}{4} = 3$$

which is equal to h/l_i since $h = 6$ for the C_{3v} group and $l_i = 2$ for the $\mathbf{D}_3$ representation.

From the great orthogonality theorem we may deduce the following important properties [4]:

Property 1: The sum of the squares of the dimensions of the IR of a group G is equal to the order of the group. That is

$$\sum_{\text{all IR of G}} l_i^2 = l_1^2 + l_2^2 + \ldots = h. \tag{3.7}$$

Example: Taking the C_{3v} group in table 3.1, we have $1^2 + 1^2 + 2^2 = 6 = h$ (order of group).

Property 2: The sum of the squares of the characters in any IR equals the order of the group. That is

$$\sum_i [\chi_i(R)]^2 = h. \tag{3.8}$$

Example: Using table 3.1 for the C_{3v} group and taking the $\mathbf{D}_3$ representation we see that

$$(1)^2 + 2(1)^2 + 3(-1)^2 = 1 + 2 + 3 = 6 = h.$$

Property 3: The vectors whose components are the characters of two different IRs are orthogonal. That is

$$\sum_R \chi_i(R)\chi_j(R) = 0 \quad \text{since } i \neq j. \tag{3.9}$$

Example: Let us take the characters of the $\mathbf{D}_1$ and $\mathbf{D}_2$ representations from table 3.1. We get

$$\mathbf{D}_1(E) \times \mathbf{D}_2(E) + 2[\mathbf{D}_1(C_3) \times \mathbf{D}_2(C_3)] + 3[\mathbf{D}_1(\sigma_v) \times \mathbf{D}_2(\sigma_v)]$$
$$= 1 \times 1 + 2(1 \times 1) + 3(1 \times -1) = 1 + 2 - 3 = 0.$$

Observe that this property is similar to (3.3).

Property 4: In a given representation (reducible or irreducible) the characters of all matrices representing group elements in the same class are equal.

Example: In the C_{3v} group the two rotation operations C_3 and C_3^2 form a class and the three reflection operations σ_{v_a}, σ_{v_b}, σ_{v_c}, form a class. Hence from table 3.1, we see that the characters of the operations in the same class are identical.

Property 5: The number of IRs of a group is equal to the classes in the group.

Example: The C_{3v} group has three classes (see §1.4.4); it therefore has three IRs as shown in table 3.1.

3.4 Connecting relation between the IR of a group and any given reducible representation of the group

In general any given reducible representation $\mathbf{D}$ may be completely reduced into the block diagonal form by means of some similarity transformations (§2.6). Now each of these diagonal matrices which constitute an IR of the group may occur more than once along the diagonal as shown schematically below:

$$\left(\begin{array}{c|c|cc|cc|ccc}
X & & & & & & & & \\ \hline
 & X & & & & & & & \\ \hline
 & & a_{11} & a_{12} & & & & & \\
 & & a_{21} & a_{22} & & & & & \\ \hline
 & & & & a_{11} & a_{12} & & & \\
 & & & & a_{21} & a_{22} & & & \\ \hline
 & & & & & & b_{11} & b_{12} & b_{13} \\
 & & & & & & b_{21} & b_{22} & b_{23} \\
 & & & & & & b_{31} & b_{32} & b_{33}
\end{array}\right).$$

Letting $\mathbf{D}_1$ be a 1×1 matrix represented by the symbol X, $\mathbf{D}_2$ a 2×2 matrix denoted by

$$\begin{pmatrix} a_{11} & a_{12} \\ a_{21} & a_{22} \end{pmatrix}$$

and $\mathbf{D}_3$ a 3×3 matrix denoted by

$$\begin{pmatrix} b_{11} & b_{12} & b_{13} \\ b_{21} & b_{22} & b_{23} \\ b_{31} & b_{32} & b_{33} \end{pmatrix}$$

then $\mathbf{D} = 2\mathbf{D}_1 + 2\mathbf{D}_2 + \mathbf{D}_3$. Hence in such cases we write

$$\mathbf{D}(R) = \sum_i a_i \mathbf{D}_i(R) \tag{3.10}$$

where $\mathbf{D}(R)$ is the reducible matrix corresponding to the operation R in the group and a_i represents the number of times the IR $\mathbf{D}_i$ appears along the diagonal when $\mathbf{D}$ is completely reduced by the necessary similarity transformations.

For our purposes, as we shall be working mainly with character tables, equation (3.10) can be written as:

$$\chi(R) = \sum_i a_i \chi_i(R). \tag{3.11}$$

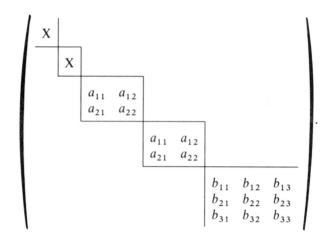

Multiplying each side of (3.11) by $\chi_j(R)$ and summing over all R in the group, we get

$$\sum_R \chi(R)\chi_j(R) = \sum_R \sum_i a_i\chi_i(R)\chi_j(R)$$

$$= \sum_i \sum_R a_i\chi_i(R)\chi_j(R)$$

$$= ha_j$$

for $i = j$, using (3.8). Therefore

$$a_j = \frac{1}{h}\sum_R \chi(R)\chi_j(R)$$

or

$$a_j = \frac{1}{h}\sum_R \chi^*(R)\chi_j(R) \tag{3.12}$$

which takes into consideration the possibility of having a character of an operator which is a complex quantity. Equation (3.12) can also be expressed as follows when summing over all classes ρ (note: not over all elements R) in the group:

$$a_j = \frac{1}{h}\sum_\rho \chi^*(\rho)\chi_j(\rho)g_\rho. \tag{3.13}$$

Here $\chi(\rho)$ is a character of the reducible representation, $\chi_j(\rho)$ is a character of the class and g_ρ is the order of the class. We show below by means of an example how to express a reducible representation $\mathbf{D}$ of the C_{3v} group in terms of the IRS $\mathbf{D}_1$, $\mathbf{D}_2$, $\mathbf{D}_3$ (table 3.2).

Table 3.2 Decomposition of a reducible representation $\mathbf{D}$ in terms of the IR of the group C_{3v}. The representation Γ is explained in the text.

C_{3v}	E	$2C_3$	$3\sigma_v$
$\mathbf{D}_1$	1	1	1
$\mathbf{D}_2$	1	1	−1
$\mathbf{D}_3$	2	−1	0
$\mathbf{D}$	7	1	−3
Γ	3	0	1

Using (3.13) we get

$$a_1 = \tfrac{1}{6}[7 \times 1 \times 1 + 1 \times 1 \times 2 + (-3) \times 1 \times 3] = 0$$

$$a_2 = \tfrac{1}{6}[7 \times 1 \times 1 + 1 \times 1 \times 2 + (-3) \times (-1) \times 3] = 3$$

$$a_3 = \tfrac{1}{6}[7 \times 2 \times 1 + 1 \times (-1) \times 2 + (-3) \times 0 \times 3] = 2.$$

Thus $\mathbf{D} = 3\mathbf{D}_2 + 2\mathbf{D}_3$.

Equation (3.13) is also used to find whether a given representation is reducible or not by finding the scalar product of the representation with itself. If the coefficient a_j is greater than one the representation is reducible.

Example: Consider the representation Γ given in table 3.2. Here

$$a_\Gamma = \tfrac{1}{6}[3 \times 3 \times 1 + 0 \times 0 \times 2 + 1 \times 1 \times 3] = 12/6 = 2.$$

Hence the representation Γ is reducible and it can be seen that $\Gamma = \mathbf{D}_1 + \mathbf{D}_3$.

3.5 Character tables

We shall use the character tables as listed by Koster *et al* [10]. These are now widely used by most solid state physicists. Their notation and symbols are discussed and explained thoroughly in their book as well as the book by Lax [1]. (Note: Chemists and spectroscopists often use a notation called the Mulliken notation and this is described by Cotton [4] and Lax [1].)

Exercises

3.1 Construct the character table for the groups C_{3v} and C_{4v}.

3.2 Use the character table for the C_{4v} group to show that any two representations Γ_i and Γ_j will have no IR in common if their characters are mutually orthogonal.

3.3 Use the character table for the cubic group labelled O (see table 6.1) to verify the following properties of the irreducible representations:
(a) The sum of the squares of the dimensions of the irreducible representations of O is equal to the order of the group.
(b) The sum of the squares of the characters in any irreducible representation equals the order of the group.
(c) The vectors whose components are the characters of two different irreducible representations are orthogonal.

3.4 Decompose the representation denoted by Γ in the table given below into the irreducible representations of the group labelled T_d.

T_d	E	$8C_3$	$3C_2$	$6S_4$	$6\sigma_d$
Γ_1	1	1	1	1	1
Γ_2	1	1	1	−1	−1
Γ_3	2	−1	2	0	0
Γ_4	3	0	−1	1	−1
Γ_5	3	0	−1	−1	1
Γ	9	0	1	−1	3

3.5 Verify that the two-dimensional and one-dimensional reduced representations of exercise 2.6 satisfy the orthogonality relation (3.8).

3.6 By considering the group C_{3v} and the basis function (xyz) show that every operator commutes with each class.

3.7 Consider the following symmetry operations of a square which forms a group labelled D_4:

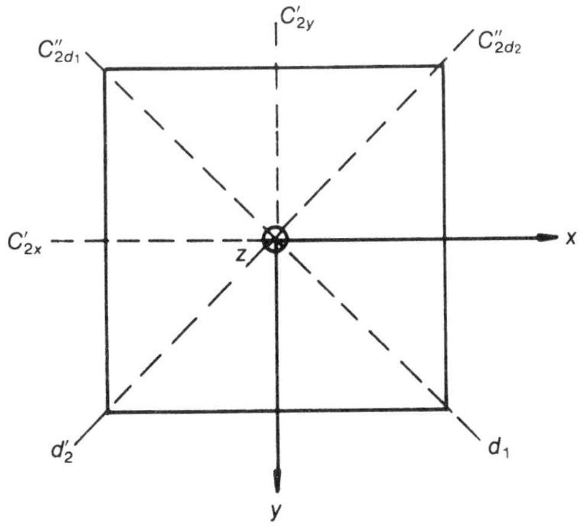

E, identity element
C^1_{4z}, C^2_{4z}, C^3_{4z}, clockwise rotations through 90°, 180°, 270° respectively about the z axis
C'_{2x}, C'_{2y}, 180° rotations about the corresponding x, y axis
C''_{2d_1}, C''_{2d_2}, 180° rotations about the diagonal axes d_1, d_2 respectively.

If the character table for the group is as given below, determine the character table for a group labelled $D_{4h} = D_4 \times I$, where I is a group of order 2 containing only the inversion operation I and the identity element E:

D_4	E	C^1_{4z}, C^3_{4z}	C^2_{4z}	C'_{2x}, C'_{2y}	C''_{2d_1}, C''_{2d_2}
Γ_1	1	1	1	1	1
Γ_2	1	1	1	−1	−1
Γ_3	1	−1	1	1	−1
Γ_4	1	−1	1	−1	1
Γ_5	2	0	−2	0	0

4

SYMMETRY IN THE CRYSTALLINE STATE

4.1 Prerequisites

In order to be able to use the (symmetry) theory that we have developed so far to describe a wide variety of physical properties of crystals, readers should be familiar with the 14 Bravais lattices, the seven crystal systems, the 32 crystal classes, etc; that is, an elementary knowledge of solid state physics or crystallography is considered a necessary prerequisite (see references listed under solid state physics and crystallography). However, we have given below a brief description of some concepts and definitions which are important for the understanding of the later chapters dealing with applications.

4.2 Crystal structures

The study of solids is largely a matter of geometry. All solids can be broadly divided into two general categories:

(i) *Crystalline*: in which the basis units or building blocks (i.e. atoms, ions, molecules or radicals, etc) are arranged in a regular, periodic or symmetric fashion in a three-dimensional (3D) space called the space lattice (defined below); and
(ii) *Amorphous*: in which the building units are arranged randomly or in a haphazard fashion in 3D space.

A good example of a crystalline material is diamond, which exhibits very high symmetry, and good examples of amorphous materials are plastics and glasses. Solids which consist of masses of small crystals randomly oriented with respect to each other are termed polycrystalline. They should not be confused with amorphous materials.

The arrangement of the building units in 3D space, whether in an ordered fashion or disordered fashion, is due to the nature of the binding forces (bond type) between the constituent units. It is these binding forces and the consequent arrangement of the building units (i.e. the structure of the solid) that by and large play a dominant role in determining the behaviour and properties of all solids.

Although the study of bond types is very important in the description of the physical and chemical properties of crystals, it is not our intention to

Table 4.1 The important types of chemical bonding and their associated characteristic properties.

Type of bond	Nature of bond	Properties of bonds	Cation coordination	Gives these characteristic properties to compounds
Ionic (heteropolar electrovalent)	Electrostatic attraction between ions formed by loss or gain of electrons by atoms	Non-directional, never between similar atoms; binding energy from ~6.5 to ~15 eV	Usually ⩾ 6; efficient packing	Mechanically strong, hard, high melting point (e.g. NaCl, LiF); colourless, strong infrared absorption, high thermal expansion; *ionic conductors*
Covalent (homopolar)	Sharing of electron pairs by two atoms	Directional, often between similar atoms; binding energy from ~3.5 to ~12.5 eV	<4; poor packing due to directional forces	Fairly strong and hard (diamond is an exception); lower melting point, short liquid range (e.g. Cl_2, SiC, ZnS); colourless, low expansion; *insulators*
Metallic	Positive nuclei in electron gas; electrons move in discrete levels, bands, zones	Non-directional; binding energy from ~1 to ~9 eV	>6; often 12; perfect packing	Variable strength and hardness; long liquid range (e.g. Na, Cu, Fe); opaque, moderate expansion; *high electrical conductors*
Molecular (Van der Waals bonding)	Weak forces caused by induced dipoles resulting from juxtaposition of molecules	Non-directional; binding energy from ~0.002 to ~0.4 eV/atom	12	Weak, very soft, very low melting point and boiling point (e.g. A, CH_4); very high thermal expansion; *insulators*

Hydrogen bonding: bond by hydrogen between two atoms, e.g. in $[HF_2]^- \rightrightarrows [{}^{\times}_{\times}\overset{\times\times}{\underset{\times\times}{F}}{}^{\times}_{\bullet}\,H{}^{\times}_{\bullet}\overset{\times\times}{\underset{\times\times}{F}}{}^{\times}_{\times}]^-$ or water (ice). Binding energy from ~0.1 to ~0.5 eV

Note: In crystalline materials, *structure* in general is determined by strongest bonds present; *properties* in general are determined by weakest bonds present

consider them here in any detail. We have therefore summarized in table 4.1 the five important bond types and the characteristic properties associated with each type of bonding.

We need to emphasize here that the classification of solids by bond types is not exact and in fact most solids have bonds that overlap two or more of these ideal groups. Another important fact worth mentioning at this point is that the properties possessed by all solids are really the cooperative properties of the ensemble rather than of the individual building units. Of course, it is the properties of the building units (namely, size, charge, polarizability, atomic structure, etc) that are of prime importance, but these are influenced strongly by the environment (structure) and other factors such as defects, impurities present, thermal history of the solids, etc.

In this section we shall be interested mainly in crystalline solids and their structures. When we talk of crystal structure, we refer to the actual arrangement of the individual building units in a three-dimensional space called the space lattice. We define a crystal space lattice as the arrangement of an infinite number of mathematical points in 3D space, which satisfy the important condition that the environment of any one point is exactly similar to the environment of any one of the other points in the 3D space. In two-dimensional (2D) space there are only five such lattices possible and these are illustrated in figure 4.1. In 3D space, it was shown by Bravais in 1849

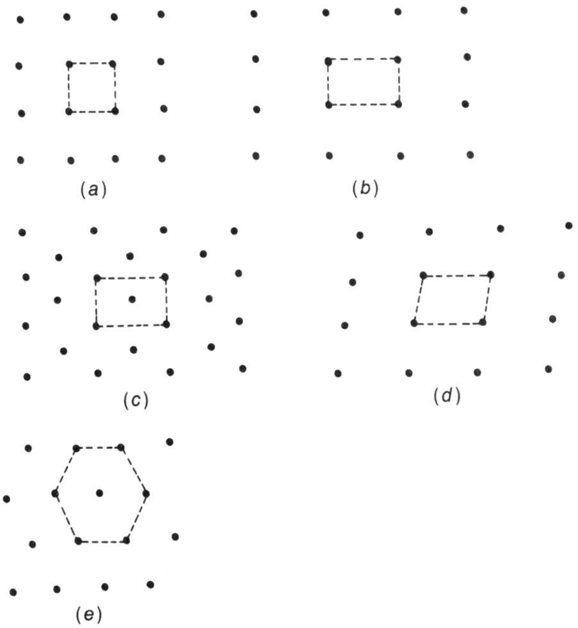

Figure 4.1 The five possible 2D lattices: (*a*) square lattice; (*b*) rectangular lattice; (*c*) centred rectangular lattice; (*d*) oblique lattice; (*e*) hexagonal lattice.

that there are only 14 essentially different ways of arranging a set of points in space which satisfied the important condition mentioned earlier. These 14 distinct arrangements have been consequently called Bravais lattices, and they are illustrated in figure 4.2. The important symmetry properties associated with the Bravais lattices are summarized in §4.8.

It is important to note the difference between a crystal lattice and a crystal structure. A crystal structure is made up of basis atoms (building units) arranged about lattice points in an identical fashion so that

$$\text{lattice} + \text{basis} = \text{structure}.$$

For example, as shown in figure 4.3(a) using the same square lattice, a number of different crystal structures can be obtained depending upon the arrangements of the basis atom(s) associated with every lattice point.

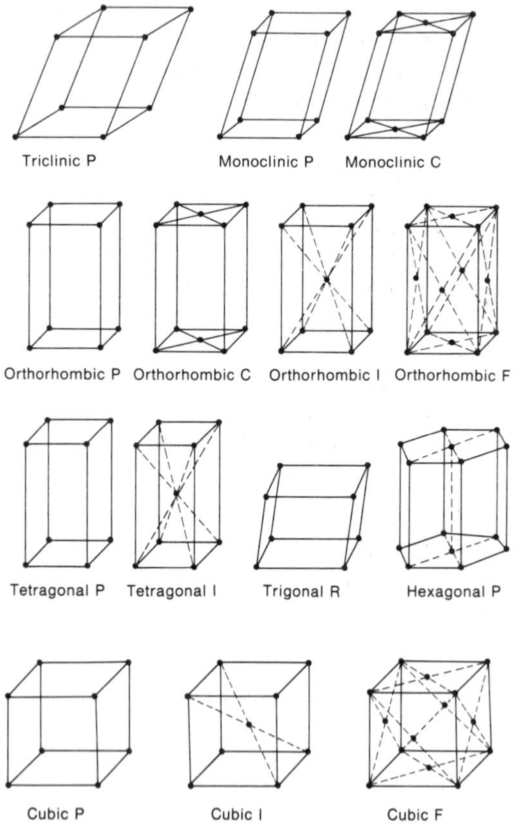

Figure 4.2 The 14 Bravais lattices. The arrangement of points are repeated infinitely throughout 3D space to make up the complete crystal space lattice. P, primitive; I, body centred; F, face centred; C, base centred; R, rhombohedral.

It is perhaps worth re-emphasizing the fact that the properties of all solids are greatly influenced by their structures. Thus, for example, the two forms of carbon exhibiting (a) diamond structure and (b) graphite structure have very different mechanical, electrical, thermal and optical properties. Similarly, silica (SiO_2) in the form of quartz (crystalline structure) and glass (amorphous structure) exhibits remarkedly different properties. Ideally, therefore, in order

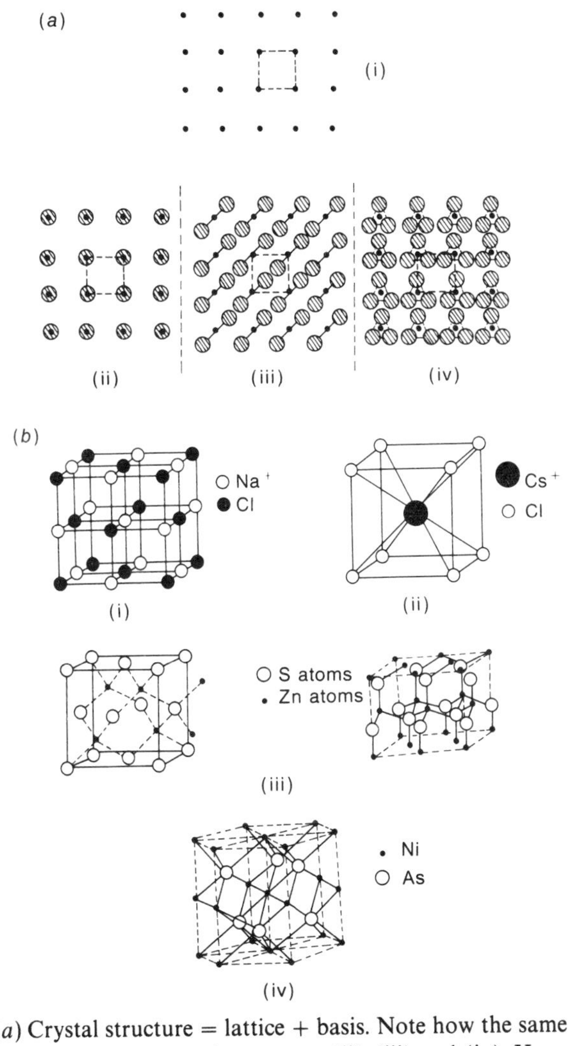

Figure 4.3 (*a*) Crystal structure = lattice + basis. Note how the same square lattice (i) can give rise to different crystal structures (ii), (iii) and (iv). Here points denote lattice sites and shaded circles denote basis atoms associated with lattice points. (*b*) Some important crystal structures: (i) NaCl (rocksalt structure); (ii) CsCl structure; (iii) zincblende and wurtzite structures; (iv) NiAs structure; *continued over.*

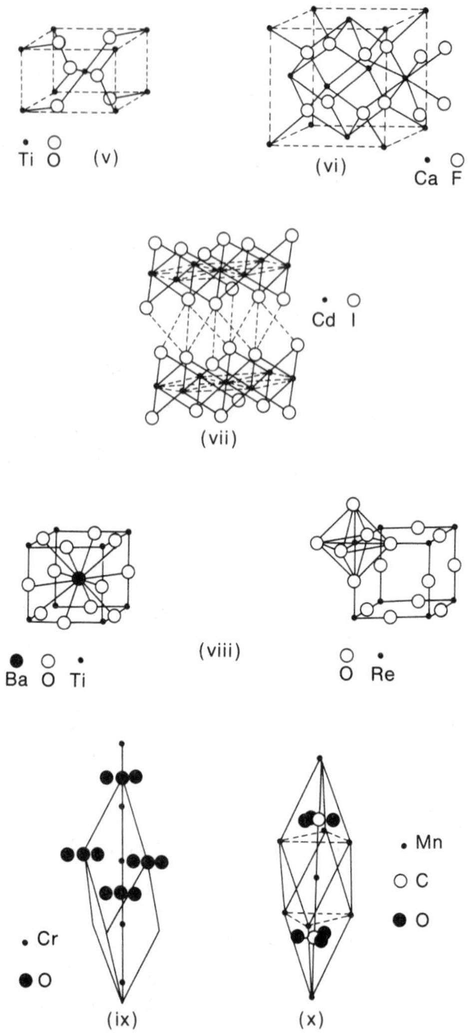

Figure 4.3 *continued*: (v) TiO_2 (rutile) structure; (vi) CaF_2 (fluorite) structure; (vii) CdI_2 (layer) structure; (viii) $BaTiO_3$ (perovskite) structure; (ix) Cr_2O_3 (corundum) structure; (x) $MnCO_3$ structure.

to gain an appreciation of actual crystals and their symmetry properties, it is necessary for us to be thoroughly familiar with at least the important types of crystal structures and their associated characteristic properties. However, a comprehensive account of such structures cannot be undertaken here. We therefore mention in the following section some of the well known crystal structures and suggest guidelines for independent study.

4.3 Some important crystal structures

For the present we are not very much concerned with the structure of the elements of the periodic table. However, it is important to observe that most elements in the periodic table are in fact metals. There are six inert gases (He, Ne, Ar, Kr, Xe, Rn) and 12 or 14 definitely electronegative elements (the halogens, O, S, etc.) leaving approximately 70 elements to be classed as:

(a) *ordinary metals* (Na, Fe, Co, Ni, Cu, Ag, Au, etc)
(b) *peculiar metals* (In, Sn, Sb, Te, etc).

The ordinary metals have 'pure' metallic bonding (see table 4.1), whilst the peculiar metals have a mixture of metallic and covalent bondings, though predominantly they exhibit metallic characteristics.

The ordinary metals crystallize in one of the following three structures:
(i) face-centred cubic (FCC) or cubic close-packed (CCP) structure;
(ii) hexagonal close-packed (HCP) structure;
(iii) body-centred cubic (BCC).

The first two structures (i.e. (i) and (ii)) are alternate ways of packing equal spheres as tightly as possible (see Wert and Thompson [39] or Kittel [34]).

We mention below some important types of crystal structures and suggest the following guidelines for independent study:

(i) Draw a picture of the structure in 3D if possible.
(ii) Identify the lattice type (figure 4.2) and find out the number of atoms per unit cell and their fractional or equivalent coordinates (see Steadman [37]).
(iii) Study the arrangement and the symmetry of the basis about a lattice point (explained later).
(iv) Find the coordination number (CN) of each ion in the structure.
(v) Study the type or types of bonds present.
(vi) Note the characteristic properties associated with the structure. Observe whether the structure could be classified as an open or loose structure, or dense structure or layer structure, etc.
(vii) See if the effect of temperature and pressure could change or modify the structure.
(viii) Note whether similar ions but with different size, charge and polarizability properties could alter the structure. For example, the effect of the radius ratio of the ions causes NaCl to have a different structure from that of CsCl.
(ix) Note how substitution of ions can produce different structures. Also, see what types of foreign (impurity) ions can be (theoretically) introduced into the structure to modify its properties.
(x) See if the compound exhibits more than one type of structure and observe how structures orient properties.

4.3.1 Important crystal structures

Figure 4.3(*b*) shows some of the more common crystal structures which can be categorized as follows.

1. AX structures
Four important structures are listed here:

(a) NaCl structure; also called 'rock salt' structure; 6 CN, A ion in centre of octahedron of X ions.
(b) CsCl structure; 8 CN, A ion in centre of cube of X ions.
(c) ZnS structure; two types:
 (i) sphalerite or CCP structure; 4 CN; diamond-type structure;
 (ii) wurtzite or HCP structure; 4 CN.
(d) NiAs structure; 6 CN.

2. AX_2 structures
We list four important structures here:

(a) TiO_2 (rutile) structure; 6 CN.
(b) CdI_2 (layer) structure; 6 CN.
(c) CaF_2 (fluorite) structure; 8 CN.
(d) SiO_2 structures—quartz and cristobalite.

3. A_2X_3, ABX_3 and A_2BX_4 structures

(a) Cr_2O_3 (corundum) structure; 6 CN, hexagonal.
(b) $BaTiO_3$ (perovskite) structure; A is in 12 CN, B is in 6 CN.
(c) Al_2MgO_4 (spinel) structure; A is in 6 CN, B is in 4 CN, CCP structure.

4.4 Unit cells and translation groups of Bravais lattices

Consider an example of a primitive cubic Bravais lattice (figure 4.4(*a*)). Let *a* be the length of cube edge. We can define a set of vectors

$$\boldsymbol{t}_1 = a\hat{\boldsymbol{i}} \qquad \boldsymbol{t}_2 = a\hat{\boldsymbol{j}} \qquad \boldsymbol{t}_3 = a\hat{\boldsymbol{k}} \tag{4.1}$$

$\hat{\boldsymbol{i}}, \hat{\boldsymbol{j}}, \hat{\boldsymbol{k}}$ being unit vectors along the x, y, z axes respectively, such that any point on the lattice can be specified by a vector $\boldsymbol{T}_n$ given by:

$$\boldsymbol{T}_n = n_1\boldsymbol{t}_1 + n_2\boldsymbol{t}_2 + n_3\boldsymbol{t}_3 \tag{4.2}$$

where n_1, n_2, n_3 are integers.

The three vectors (equation (4.1)) define a unit cell of the crystal space lattice and are called basis vectors or fundamental vectors. If the cell defined by the basis vectors contains only one lattice point/cell then such a cell is called a primitive cell and the basis vectors are also called primitive translation vectors. On the other hand, if the chosen basis vectors define a cell which

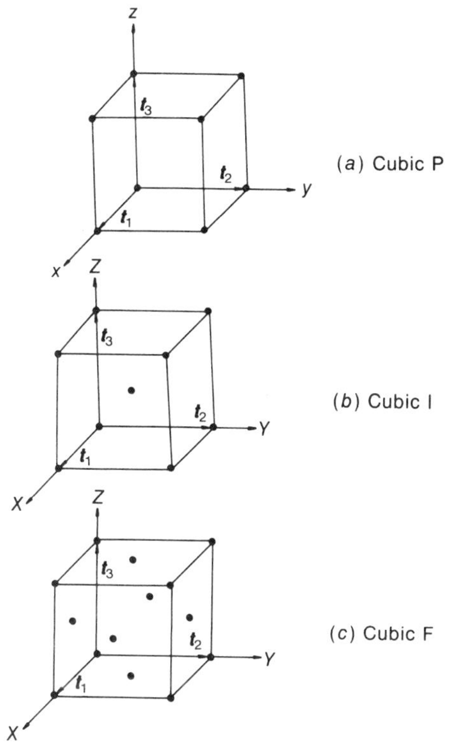

Figure 4.4 The primitive (a) and non-primitive (b, c) translation vectors of the three cubic Bravais lattices.

contains more than one lattice point/cell, then such a cell is called a non-primitive cell and the basis vectors are non-primitive translation vectors. For example, figures 4.4(b, c) define non-primitive cells which contain two lattice points/cell and four lattice points/cell respectively. To count the number of points enclosed in a cell, the following method is used:

corner points contribute one-eighth of a point per cell;
edge points contribute one-quarter of a point per cell;
face points contribute half a point per cell;
interior points contribute one point per cell.

It should be noted that a primitive cell for the BCC lattice (figure 4.4(b)) could be obtained by considering the body-centred lattice point as origin and defining three vectors as follows (figure 4.5(a)):

$$\boldsymbol{t}_1 = \frac{a}{2}(\hat{\boldsymbol{i}} + \hat{\boldsymbol{j}} - \hat{\boldsymbol{k}}) \qquad \boldsymbol{t}_2 = \frac{a}{2}(-\hat{\boldsymbol{i}} + \hat{\boldsymbol{j}} + \hat{\boldsymbol{k}}) \qquad \boldsymbol{t}_3 = \frac{a}{2}(\hat{\boldsymbol{i}} - \hat{\boldsymbol{j}} + \hat{\boldsymbol{k}}). \quad (4.3)$$

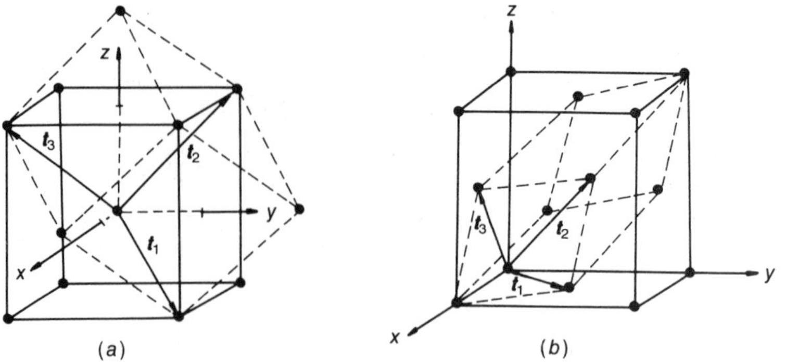

Figure 4.5 The primitive translation vectors of (*a*) a body-centred cubic (BCC) lattice and (*b*) a face-centred cubic (FCC) lattice. The primitive cell for each lattice is shown by broken lines.

Similarly, the primitive cell of an FCC lattice could be defined by (figure 4.5(*b*)):

$$t_1 = \frac{a}{2}(\hat{i} + \hat{j}) \qquad t_2 = \frac{a}{2}(\hat{j} + \hat{k}) \qquad t_3 = \frac{a}{2}(\hat{i} + \hat{k}). \tag{4.4}$$

Another useful primitive cell called the Wigner–Seitz cell or the symmetrical unit cell may be obtained quite generally from the lattice vectors using the following procedure (figure 4.6):

(i) Choose an arbitrary lattice point as origin.
(ii) Draw lattice vectors joining the origin to all the nearby equivalent points.
(iii) Construct planes passing through the midpoints and perpendicular to these vectors.

The smallest volume enclosed by the intersecting planes about the origin is the Wigner–Seitz (primitive) cell. Such a cell is often chosen because it displays the entire symmetry of the lattice.

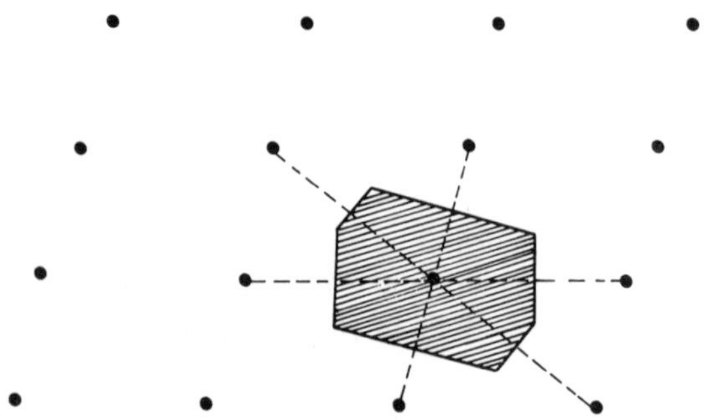

Figure 4.6 The Wigner–Seitz cell for a 2D oblique lattice.

The vector $\boldsymbol{T}_n$ defined by equation (4.2) is known as a translational operator since it defines the operation of displacing a crystal lattice parallel to itself. The totality of all such operations for all values of the integers n_1, n_2, n_3 is known as the translation group of the crystal lattice.

Obviously the translation group operating on a unit cell (defined by $\boldsymbol{t}_1$, $\boldsymbol{t}_2$, $\boldsymbol{t}_3$) will map out the entire crystal lattice. We shall see the importance of the translation group later on when dealing with the applications of symmetry principles in solid state physics.

4.5 The seven crystal systems

The crystal system defines the basic primitive cell which, when repeated infinitely throughout 3D space, maps out the entire crystal space lattice. If we define a primitive cell in terms of the interaxial angles α, β, γ and the lattice constants (i.e. length of primitive vectors) a, b, c, (see figure 4.7) it is found that all the 14 Bravais lattices can be classified amongst the seven crystal systems which are described and defined in table 4.2.

In defining the seven crystal systems, observe that we do not take into consideration the possible symmetrical arrangements of the building units at the various lattice points. However, if we do that, it can be shown that the 14 Bravais lattices classified under the seven crystal systems permit only 32 distinct symmetry arrangements of the building units (basis atoms). These 32 distinct symmetry arrangements are technically known as the 32 crystallographic symmetry point groups. Note that a crystal point group consists of a collection of symmetry operations which, when applied about a fixed point (called the origin), would leave the body (e.g. a molecule, radical, a lattice or a structure) invariant.

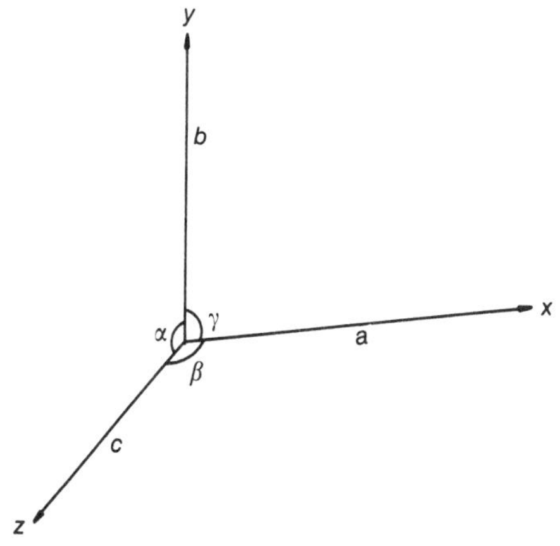

Figure 4.7 Definition of crystal axes and interaxial angles.

Table 4.2 Classification of the 32 crystallographic point groups and the 14 Bravais lattices amongst the seven crystal systems.

Crystal system	Definition of unit cell (see figure 4.7)	Bravais lattices symbol	Point-group symmetry symbol (see tables 1.1 and 4.3)
Triclinic	$\alpha \neq \beta \neq \gamma$ $a \neq b \neq c$	P	$S_2(\equiv C_i)$; C_1
Monoclinic	$\alpha = \beta = 90^\circ \neq \gamma$ $a \neq b \neq c$	P, C	C_{2h}; C_2; C_{1h}
Orthorhombic	$\alpha = \beta = \gamma = 90^\circ$ $a \neq b \neq c$	P, C, I, F	D_{2h}; D_2; C_{2v}
Tetragonal	$\alpha = \beta = \gamma = 90^\circ$ $a = b \neq c$	P, I	D_{4h}; C_{4h}; D_{2d}; C_{4v}; D_4; S_4; C_4
Trigonal	$\alpha = \beta = \gamma \neq 90^\circ < 120^\circ$ $a = b = c$	R	D_{3d}; D_3; C_{3v}; $S_5(\equiv C_{3i})$; C_3
Hexagonal	$\alpha = \beta = 90^\circ; \gamma = 120^\circ$ $a = b \neq c$	P	D_{6h}; C_{6h}; D_{3h}; C_{6v}; D_6; C_{3h}; C_6
Cubic	$\alpha = \beta = \gamma = 90^\circ$ $a = b = c$	P, I, F	O_h; T_h; T_d; O; T

The operations in a point group can include only rotations, reflections, inversion and their combinations. Observe that in a point group all the symmetry elements (i.e. symmetry axes, symmetry planes, centre of inversion) pass through a fixed point and hence the name point group. Furthermore, note that point groups do not include the translation operations defined in §4.4.

All crystals found in nature belong to one of these 32 point groups or crystal classes. Thus the class of a particular crystal is defined as the crystallographic point group to which the particular crystal belongs. This information is summarized in table 4.2†. The crystal point group can be determined by x-ray and neutron diffraction experiments.

† Note that the first point group within the same crystal system is technically known as the holohedry of the system. It represents a point group with the highest number of symmetry elements within the crystal system. Observe from table 4.2 that the groups S_2, C_{2h}, D_{2h}, D_{4h}, D_{3d}, D_{6h} and O_h represent the holohedral groups of the crystal systems. The connection between the holohedral groups and their lower-order subgroups is given by Lax [1]).

4.6 The reciprocal lattice and Brillouin zones

For several applications in solid state physics, crystallography, metallurgy and materials science, etc, it is found that the (direct) lattice vectors that we have been discussing so far are not very convenient to use. For example, when considering the diffraction properties of crystal planes, the important conclusion from Laue's diffraction equations (see Kittel [34]) is that the diffraction geometry can be explained very simply in terms of reciprocal lattice vectors. Again, in the study of the energy band theory of solids or elementary excitations (phonons, magnons, excitons, etc) in crystals, the introduction of reciprocal lattice vectors allows the equivalence of working with a simple system of orthogonal unit vectors. For these reasons we now look at the construction of reciprocal lattice vectors and summarize some of their important properties.

If $\boldsymbol{a}_i$ $(i = 1, 2, 3)$ are the primitive lattice vectors of a crystal in the direct lattice space, then its reciprocal lattice vectors $\boldsymbol{a}_j^*$ $(j = 1, 2, 3)$ are given by the relation

$$\boldsymbol{a}_i \cdot \boldsymbol{a}_j^* = \delta_{ij} \tag{4.5}$$

where $\delta_{ij} = 1$ if $i = j$, and $\delta_{ij} = 0$ if $i \neq j$.

Equation (4.5) expresses the orthogonality of the two vectors, one in the direct lattice space (the $\boldsymbol{a}_i$) and the other in the reciprocal lattice space (the $\boldsymbol{a}_j^*$). Thus the vector $\boldsymbol{a}_1^*$, for example, is perpendicular to $\boldsymbol{a}_2$ and $\boldsymbol{a}_3$ since only $\boldsymbol{a}_1 \cdot \boldsymbol{a}_1^* = 1$. However, $\boldsymbol{a}_2 \cdot \boldsymbol{a}_1^* = 0$. It is easy to see from (4.5) that the explicit expressions for the reciprocal lattice vectors are

$$\boldsymbol{a}_1^* = (\boldsymbol{a}_2 \times \boldsymbol{a}_3)/V \qquad \boldsymbol{a}_2^* = (\boldsymbol{a}_3 \times \boldsymbol{a}_1)/V \qquad \boldsymbol{a}_3^* = (\boldsymbol{a}_1 \times \boldsymbol{a}_2)/V \tag{4.6}$$

where

$$V = \boldsymbol{a}_1 \cdot (\boldsymbol{a}_2 \times \boldsymbol{a}_3) \tag{4.7}$$

is the volume of the unit cell in direct lattice space.

It is important to note that the reciprocal lattice vectors (from which the reciprocal lattice space is built) is an invariant mathematical construction fixed by the direct lattice vectors of the crystal. All the properties possessed by the direct lattice will be inherent in its reciprocal lattice. Thus, for example, the symmetry properties of the Bravais lattices which we have summarized in §4.8 will be equally valid for the reciprocal lattice of a Bravais lattice. Similarly, the reciprocal lattice of a given direct lattice must belong to the same crystal system as the original (direct) lattice. Some of the geometrical properties of reciprocal lattice vectors which will be important to us in future discussions are listed below:

(1) Each vector of the reciprocal lattice is normal to a set of parallel lattice planes of the direct lattice.

(2) The length of $\boldsymbol{a}^*$ is inversely proportional to the spacing of the lattice planes (in the direct lattice) normal to $\boldsymbol{a}^*$. For example, if d_{hkl} is the distance between parallel planes (hkl) in the direct lattice space, then in reciprocal lattice space, $d^*_{hkl} = 1/d_{hkl}$.

(3) The volume of a unit cell of the reciprocal lattice is inversely proportional to the volume of the unit cell of the direct lattice.

(4) The direct lattice is the reciprocal of its own reciprocal lattice.

(5) Whenever a lattice is invariant under the operations of a certain point group in a direct lattice space, it is also invariant under the same point group in the reciprocal space.

To get the proper reciprocal lattice vectors for a given direct lattice, one must use the primitive lattice vectors of the crystal in the direct lattice space. For example, in the case of BCC and FCC lattices, one must use the vectors $\boldsymbol{a}_i$ ($i = 1, 2, 3$) as defined in figure 4.5 and not the non-primitive vectors of figure 4.4.

The method of obtaining these reciprocal lattice points is as follows. For simplicity, we shall illustrate the method using a rectangular lattice (see figure 4.8).

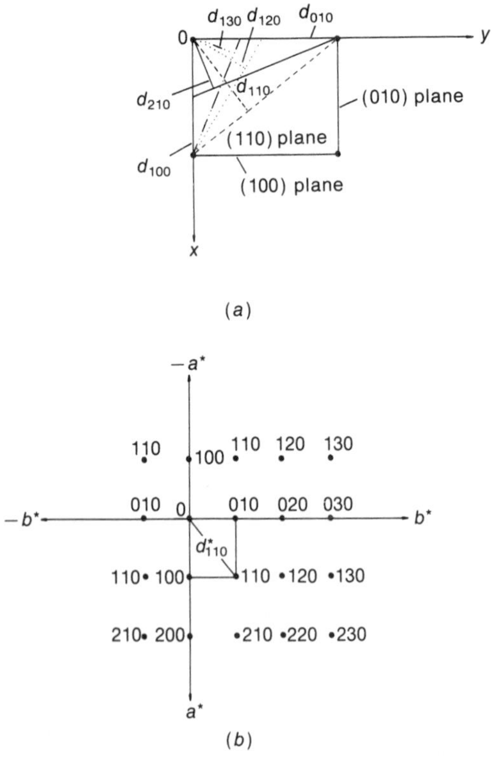

Figure 4.8 Construction of a reciprocal lattice (b) for a rectangular lattice (a).

1. Pick some lattice point as the origin 0 in the direct lattice space.
2. From 0, draw perpendiculars to every set of planes (hkl) in the direct lattice.
3. Measure the lengths d_{hkl} of perpendiculars. For example, for the rectangular lattice shown in figure 4.8(a) we have: $d_{100} = 4$ cm, $d_{200} = 2$ cm, $d_{010} = 5$ cm, $d_{110} = 3.1$ cm, $d_{220} = 1.55$ cm, $d_{120} = 2.1$ cm, $d_{130} = 1.5$ cm, etc.
4. Find $d^*_{hkl} = k/d_{hkl}$, where k in the numerator is a suitable scale factor ($k = 8$ cm in our example) and construct a point at the end of each value of d^*_{hkl}. For example $d^*_{100} = 2$ cm etc.
5. The assembly of points so constructed (figure 4.8(b)) will correspond to the reciprocal lattice points of the direct lattice.

It should be obvious from the method of construction that with each point in the reciprocal lattice space is associated the Miller indices (hkl) of a set of planes in the direct lattice. Thus the set of points (figure 4.8(b)) represents a tabulation of (a) normals to all the direct lattice planes and (b) their interplanar spacings.

The symmetrical unit cell in reciprocal lattice space is called a Brillouin zone. Its construction is exactly similar to the construction of the Wigner–Seitz cell (figure 4.6) except that the points are now reciprocal lattice points. The Brillouin zones for simple cubic, BCC and FCC lattices are shown in figure 4.9. The Brillouin zones for all the Bravais lattices are given in Lax [1].

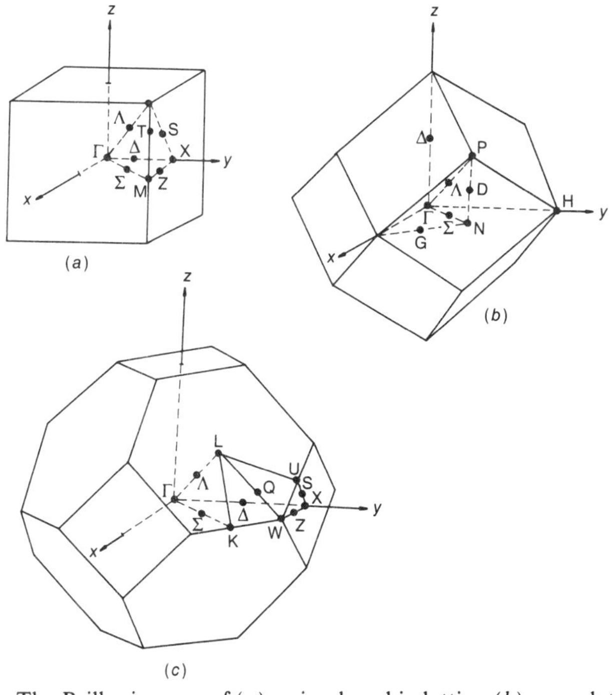

Figure 4.9 The Brillouin zone of (a) a simple cubic lattice, (b) a BCC lattice and (c) an FCC lattice, showing the important points and lines of symmetry (see [1]).

Table 4.3 Summary of the 32 crystallographic point groups in the Schoenflies and International notations.

No.	Symbol of groups	Definition	Point-group symbols: Schoenflies	International: Short	International: Full	No. of point groups
1	C_n	These are the pure rotation groups which contain an *n*-fold principal (proper) axis of rotation (e.g. the *z* axis or *c* axis); note: proper rotation axis involves pure rotations only	C_1 C_2 C_3 C_4 C_6	1 2 3 4 6	1 2 3 4 6	5
2	C_{nv}	These groups contain a σ_v reflection plane in addition to the C_n axis	C_{2v} C_{3v} C_{4v} C_{6v}	2mm 3m 4mm 6mm	2mm 3m 4mm 6mm	4
3	C_{nh}	These groups contain a σ_h reflection plane as well as the C_n axis	C_{1h} C_{2h} C_{3h} C_{4h} C_{6h}	m 2/m $\bar{6}$ 4/m 6/m	m $\frac{2}{m}$ $\bar{6}$ $\frac{4}{m}$ $\frac{6}{m}$	5
4	S_n	These groups contain a $2\pi/n$ improper rotation axis and a reflection plane perpendicular to the improper axis; note: improper rotation axis involves combination of rotation–reflection or rotation–inversion operations	S_2 S_4 S_6	$\bar{1}$ $\bar{4}$ $\bar{3}$	$\bar{1}$ $\bar{4}$ $\bar{3}$	3

5	D_n	These groups, which are called dihedral groups, have $2n$, two-fold rotation axes perpendicular to the principal n-fold axis	D_2 D_3 D_4 D_6	222 32 422 622	222 32 422 622	4
6	D_{nd}	These groups contain the elements of D_n ($n = 2$ or 3) together with diagonal reflection planes σ_d bisecting the angles between the two-fold axes perpendicular to the principal axis	D_{2d} D_{3d}	$\bar{4}2m$ $\bar{3}m$	$\bar{4}2m$ $\bar{3}\frac{2}{m}$	2
7	D_{nh}	These groups contain the elements of D_n ($n = 2, 3, 4$ or 6) together with the horizontal plane of symmetry σ_h	D_{2h} D_{3h} D_{4h} D_{6h}	mmm $\bar{6}m2$ 4/mmm 6/mmm	$\frac{2}{m}\frac{2}{m}\frac{2}{m}$ $\bar{6}m2$ $\frac{4}{m}\frac{2}{m}\frac{2}{m}$ $\frac{6}{m}\frac{2}{m}\frac{2}{m}$	4
8	Cubic	These groups have no unique axis of highest symmetry but rather more than one axis which is at least of three-fold symmetry. Brief individual descriptions are given below. For more details see Tinkham [2] and Koster *et al* [10]	T T_d T_h O O_h	23 $\bar{4}3m$ m3 432 m3m	23 $\bar{4}3m$ $\frac{2}{m}\bar{3}$ 432 $\frac{4}{m}\bar{3}\frac{2}{m}$	5
		The T group: consists of 12 proper rotational operations which send a regular tetrahedron into itself The T_d group: the full tetrahedral group which contains all the covering operations of a regular tetrahedron including reflections; it has 24 symmetry operations The T_h group: formed by taking the direct product of T with the inversion operator I, i.e. $T_h = T \otimes I$; it has 24 operations The O group: contains the set of operations which takes a cube or a regular octahedron into itself; it has 24 symmetry operations The O_h group: represents the full octahedral group and is the direct product of O with the inversion operator I, i.e. $O_h = O \otimes I$; it has 48 symmetry operations				

4.7 Enumeration of the 32 crystallographic point groups

The 32 crystallographic point groups are enumerated and described briefly in table 4.3 using the Schoenflies notation (see table 1.1). The equivalent International symbols (short and full) commonly used by crystallographers are also listed alongside. There would seem little to be gained for our present purpose to give a full description of all these groups including an account of all the symmetry elements contained in each group. Readers who are interested should refer to the recommended references [2, 10].

As an illustrative exercise we have sketched in figure 4.10 a schematic representation of the 24 symmetry elements present in the cubic groups. In sketching these figures we have used the standard graphical symbols given in table 4.4 for the symmetry operations.

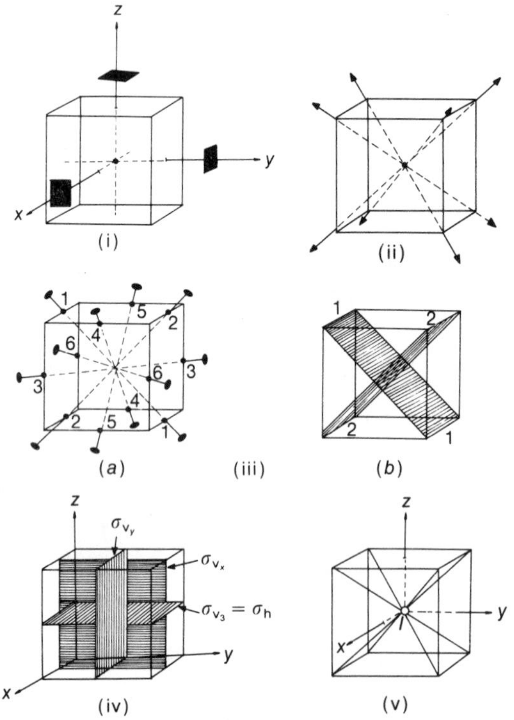

Figure 4.10 Graphical representation of the 24 symmetry elements present in a cube. The cube has: (i) three tetrad axes ($3C_4$); (ii) four triad axes ($4C_3$); (iii) (*a*) six diad axes ($6C_2$) and six diagonal reflection planes ($6\sigma_d$). The diagonal planes are formed by corresponding pairs of opposite parallel edges. The pairs of corresponding edges are shown by the same number; (*b*) an example of diagonal reflection planes; (iv) three vertical reflection planes ($3\sigma_v$); (v) one inversion centre (I). The identity operation (E) is not shown.

Table 4.4 Graphical symbols of the symmetry operations.

Symmetry operation	Graphical symbol	
Rotation: diad (C_2)	▮ or ▬	(two-sided solid figure)
Rotation: triad (C_3)	▲	(three-sided solid figure)
Rotation: tetrad (C_4)	■	(four-sided solid figure)
Rotation: hexad (C_6)	⬢	(six-sided solid figure)
Inversion: (I)	O	(Hollow circle)
Reflection plane: (σ)	—	(Full line; the planes could be horizontal, vertical or diagonal)

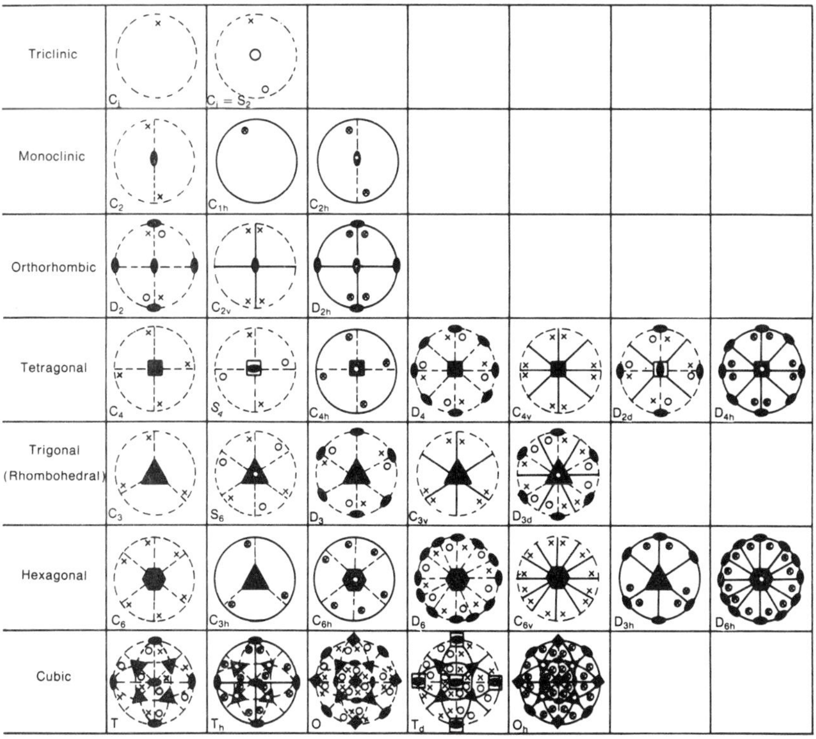

Figure 4.11 Stereograms of the 32 point groups enumerated in table 4.3. The z axis is chosen normal to the plane of the paper. The symmetry symbols are explained in tables 4.4 and 4.5.

Table 4.5 Symbols used in drawing the stereograms (see figure 4.11).

Symbol	Description
Solid outer circle	Indicates the presence of σ_h, i.e. a reflection plane normal to the principal axis (usually the z or c axis)
Broken outer circle	Indicates the absence of σ_h but σ_v, σ_d, etc may be present
n-sided solid figure at centre of circle	Represents the presence of n-fold principal axis of rotation ($n = 2$, 3, 4 or 6 only, for crystals)
n-sided solid figure at centre enclosing an open circle	Represents the presence of inversion symmetry I along with the n-fold principal axis of rotation
	Represents the presence of four-fold rotation–inversion axis, e.g. see group T_d
×	Represents a general point above plane of paper
○	Represents a general point below plane of paper

It is a common practice when enumerating the point groups to present also the stereographic projections (or stereograms for short) of all the 32 point groups. These are illustrated in figure 4.11. The method of constructing these stereograms is given in several books (see for example [36, 38]). The symbols used in drawing these stereograms are explained in table 4.5.

4.8 Symmetry properties of Bravais lattices

As mentioned earlier (§4.2) a Bravais space lattice has a number of characteristic properties associated with it due to the manner of arrangements of the lattice points in space. The useful properties which are important to us are summarized below:

1. Every Bravais lattice point is a centre of inversion symmetry. This means that if there is a point at x, y, z there must be an identical point at $-x$, $-y$, $-z$. Thus if we define any arbitrary lattice point by (see equation (4.2))

$$\boldsymbol{T}(n) = \sum_{i=1}^{3} n_i \boldsymbol{t}_i$$

then

$$T(-n) = \sum_{i=1}^{3} (-n_i)\boldsymbol{t}_i$$

$$= -\sum_{i=1}^{3} n_i \boldsymbol{t}_i$$

$$= -T(n).$$

The existence of a centre of symmetry leads to the occurrence of parallel pairs of faces on opposite sides of the crystal.

2. An axis of two-fold rotation in a Bravais lattice must always be normal to a plane of reflection symmetry, and conversely two-fold rotation + inversion = reflection.
3. The intersection of two planes of reflection symmetry is an $n = \pi/\alpha$-fold axis of rotation symmetry. Here α is the angle between the two planes of intersection. For example, if $\alpha = 45°$, then $n = 180/45$ = four-fold rotation axis.
4. If a Bravais lattice has an axis of n-fold symmetry, it also has n-fold symmetry about any lattice point.
5. There can only be one-, two-, three-, four- or six-fold axes of rotation. A Bravais lattice cannot be built having five-, seven-, eight-fold rotation axes etc.

Further discussions and more information on the associated properties of Bravais lattices can be found in most of the elementary crystallography books suggested for further reading (see for example Buerger [33]).

4.9 Planes, directions and lattice sites

In this section we give a brief account of the method of identifying planes (and faces), directions and coordinates of lattice points in crystals.

4.9.1 Labelling of planes

Since a crystal is bound externally by plane faces arranged in definite geometrical fashion in accordance with the crystal system, it is necessary to have a universal system of notation for labelling the faces of a crystal and for specifying the position and orientation of planes within the crystal. The notation used is called Miller indices. These are three small indices h, k, l which are determined as follows:

(i) Find the intercept of the face or the plane on the three basic axes, in terms of the lattice constants a, b and c (see figure 4.7).

(ii) Take the reciprocals of these numbers and reduce them to the smallest three integers having the same ratio. Enclose the result in parentheses, thus: (hkl).

For example for the plane whose intercepts on the three axes are 4, 1, 2, the reciprocals are $\frac{1}{4}$, 1, $\frac{1}{2}$ and therefore the Miller indices for the plane are (142).

Note that for a plane which is parallel to a particular axis, the intercept is at infinity and therefore the corresponding index is zero. The indices (hkl) denote a single plane or a set of parallel planes. If a plane cuts an axis on the negative side the corresponding index is negative and is indicated by placing a negative sign above the index, for example $(h\bar{k}l)$ which is read as (h, bar k, l). The Miller indices of some important planes in a cubic crystal are illustrated in figure 4.12.

Note that the faces of a cubic crystal are labelled as: (100); (010); (001); $(\bar{1}00)$; $(0\bar{1}0)$; and $(00\bar{1})$. Observe that all these faces are equivalent by symmetry, that is they can be carried or mapped one on to another by means of the symmetry operations present in a cube. Planes equivalent by symmetry are represented thus: {100}, meaning the six planes mentioned above.

4.9.2 Indices of direction

The directions in a crystal are specified in terms of three small numbers which are enclosed in square brackets, thus: $[uvw]$ or $[xyz]$. For example, the x axis is written as [100], the $-y$ axis as $[0\bar{1}0]$ and the z axis as [001],

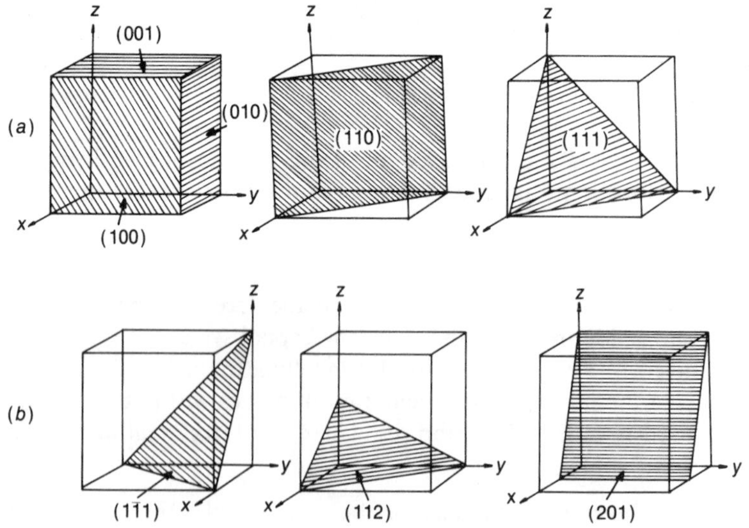

Figure 4.12 The Miller indices of (a) some important planes and (b) some arbitrary planes in a cubic crystal.

etc. A full set of directions which are equivalent by symmetry is denoted by $\langle uvw \rangle$. For example, in a cubic crystal $\langle 111 \rangle$ denotes the following eight directions which are all equivalent by symmetry: $[111]$; $[\bar{1}11]$; $[1\bar{1}1]$; $[11\bar{1}]$; $[\bar{1}\bar{1}1]$; $[\bar{1}1\bar{1}]$; $[1\bar{1}\bar{1}]$; $[\bar{1}\bar{1}\bar{1}]$. Similarly the close-packed directions (i.e. the directions along which the atoms are most densely packed) for cubic structures could be written as follows:

simple cubic (SC)	$\langle 100 \rangle$
BCC:	$\langle 111 \rangle$
FCC:	$\langle 110 \rangle$.

Some important directions are illustrated in figure 4.13.

4.9.3 Lattice sites

The position of lattice points in a unit cell are specified in terms of the lattice constants a, b, c. Thus the coordinates of the central point of a cubic cell is $a(\frac{1}{2}, \frac{1}{2}, \frac{1}{2})$ and face-centred positions are $(\frac{1}{2}, \frac{1}{2}, 0)$; $a(0, \frac{1}{2}, \frac{1}{2})$; $a(\frac{1}{2}, 0, \frac{1}{2})$. Note, however, that whilst the locations of all atoms in a unit cell of NaCl are as follows (see figure 4.3(*b*)(i)): Na $\rightarrow (\frac{1}{2}, 0, 0)$, $(0, \frac{1}{2}, 0)$, $(0, 0, \frac{1}{2})$, $(\frac{1}{2}, \frac{1}{2}, \frac{1}{2})$; Cl $\rightarrow (0, 0, 0)$, $(0, \frac{1}{2}, \frac{1}{2})$, $(\frac{1}{2}, 0, \frac{1}{2})$, $(\frac{1}{2}, \frac{1}{2}, 0)$, the fractional (equivalent) coordinates of the FCC lattice of NaCl are given as $(0, 0, 0)$, $(\frac{1}{2}, \frac{1}{2}, 0)$, $(\frac{1}{2}, 0, \frac{1}{2})$ and $(0, \frac{1}{2}, \frac{1}{2})$. Similarly, both the diamond and the ZnS structure which have an FCC lattice (shown in figure 4.3(*b*)(iii)) have atoms located in a unit cell given by the coordinates $(0, 0, 0)$, $(0, \frac{1}{2}, \frac{1}{2})$, $(\frac{1}{2}, 0, \frac{1}{2})$, $(\frac{1}{2}, \frac{1}{2}, 0)$, $(\frac{1}{4}, \frac{1}{4}, \frac{1}{4})$, $(\frac{1}{4}, \frac{3}{4}, \frac{3}{4})$, $(\frac{3}{4}, \frac{1}{4}, \frac{3}{4})$, $(\frac{3}{4}, \frac{3}{4}, \frac{1}{4})$. Observe that by locating the coordinates of only two lattice sites, that is $(0, 0, 0)$ and $(\frac{1}{4}, \frac{1}{4}, \frac{1}{4})$, and the lattice type, the entire zincblende structure is repeated in three dimensions.

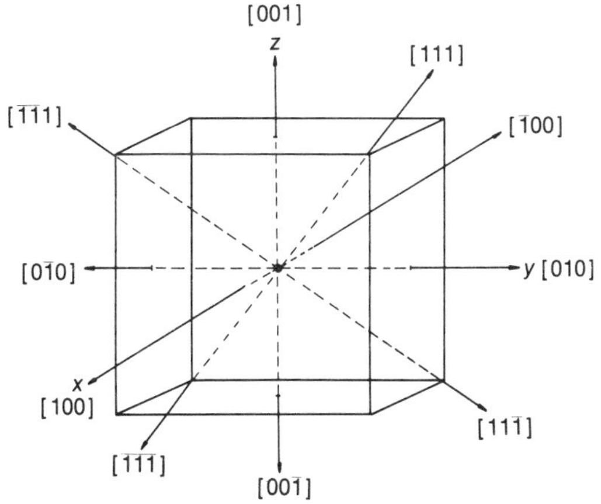

Figure 4.13 Some important indices of direction in a cubic crystal.

4.10 Space groups

In §4.6 we enumerated the 32 crystal point groups which are consistent with lattice translational symmetry. We have also seen that, due to the restrictions imposed on the lattice by the point group symmetry, only 14 distinct lattices are possible which are technically classified under seven crystal systems as described in table 4.2. All this information does not, however, complete a description of the full symmetry of a crystal. The full symmetry group of a crystalline solid is the space group which contains the lattice translational symmetry operations (§4.4) in addition to the operations of a point group. It turns out that when all the point group operations are combined with the various possible translational operations of the Bravais lattices, a total of 230 distinct groups can be formed. These groups are called space groups or Federov groups and are derived in all standard books on crystallography.

As for point groups, we now introduce a systematic notation for all operations in a space group. The notation which is widely used is called the Seitz space-group symbols. In this notation a general space-group operation is denoted by $\{R|\boldsymbol{t}\}$ where R represents a point-group operation such as rotation or reflection and $\boldsymbol{t}$ represents a translation operation of the lattice. The effect of $\{R|\boldsymbol{t}\}$ on a vector $\boldsymbol{x}$ is defined by

$$\{R|\boldsymbol{t}\}\boldsymbol{x} = R\boldsymbol{x} + \boldsymbol{t} = \boldsymbol{x}'. \tag{4.8}$$

From this definition it is easy to see that the multiplication rule for two space-group operations is

$$\{R|\boldsymbol{t}\}\{S|\boldsymbol{t}'\} = \{RS|R\boldsymbol{t}' + \boldsymbol{t}\}. \tag{4.9}$$

The product RS must of course be in the group (closure property; see §1.3.4).

Similarly, the inverse of $\{R|\boldsymbol{t}\}$ is given by

$$\{R|\boldsymbol{t}\}^{-1} = \{R^{-1}|-R^{-1}\boldsymbol{t}\}. \tag{4.10}$$

A space group is said to belong to the same crystal class as the crystal point group to which the operation R belongs. If the translational part of the space-group operations contains non-primitive translations as well, then the space-group operation is represented by

$$\{R|\boldsymbol{t}_n + \boldsymbol{\tau}\} \tag{4.11}$$

where $\boldsymbol{\tau}$ is now a non-primitive translation which is present whenever the group contains screw axes or glide planes. A screw axis is a compound symmetry operation involving a rotation and a simultaneous translation parallel to the axis of rotation whilst a glide plane is a compound operation, combining reflection and then parallel translation. A simple example of a two-fold screw axis operation and a glide reflection operation is shown in figure 4.14. The accompanying parallel translations in such operations are the non-primitive translations $\boldsymbol{\tau}$ as indicated in the figure (see Buerger [33]

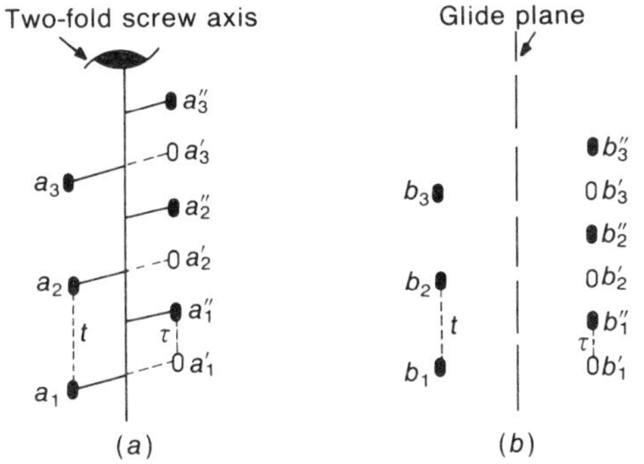

Figure 4.14 The effect of operations of (*a*) a two-fold screw axis and (*b*) a glide plane. In a proper two-fold rotation operation, the atoms labelled a_1, a_2, a_3 would be found at a_1', a_2', a_3' respectively, whilst with the two-fold screw rotation the atoms would be found at a_1'', a_2'', a_3'', each atom having moved upwards by one-half ($\frac{1}{2}$) the repeat distance t, i.e. $\tau = \frac{1}{2}t$. Similarly, for a glide reflection the atoms b_1, b_2, b_3 would be found at b_1'', b_2'', b_3'' and not at b_1', b_2', b_3' respectively, as would have been the case for an ordinary reflection in a mirror plane. Observe that the distance moved upwards by each atom after a glide reflection is $\tau = \frac{1}{2}t$.

for more details). Space groups not containing τ, that is when $\tau = 0$ in equation (4.11), are termed *symmorphic space groups*, of which there are 73.

In the remaining $(230 - 73 = 157)$ space groups for which $\tau \neq 0$ (i.e. non-primitive translations are present) the groups are called non-symmorphic or asymmorphic space groups.

This completes our brief discussion on symmetry in the crystalline state. As mentioned earlier, readers are referred to other articles and texts on the subject for more detailed descriptions of point groups and space groups.

Exercises

4.1 Write down the lattice type and fractional coordinates of the atoms in a unit cell of:
(a) NaCl structure
(b) CsCl structure
(c) ZnS (diamond) structure
(d) metallic Cu which has a cubic close-packed structure.

4.2 Write down the close packed directions for:
(a) SC structure
(b) BCC structure
(c) FCC structure.

4.3 Write down all the symmetry equivalent directions in a cubic system for $\langle 111 \rangle$ and $\langle 110 \rangle$.

4.4 Draw simple figures depicting an SC, a BCC and an FCC lattice and determine how many nearest-neighbour (NN) lattice points a lattice point has in the figures drawn.

4.5 Rigid spheres of radius R are packed together to form the following type of cubic structures:
(a) simple cubic structure
(b) body-centred structure
(c) face-centred structure
If a is the side of the cube edge, show that the fractions f of the volume occupied by the spheres are 52.4%, 68% and 74% respectively for the three types of close-packed structures.

4.6 Construct the primitive cell for the following: (a) an FCC space lattice; (b) a BCC space lattice.

4.7 Explain the differences if any between the following:
(a) a Bravais lattice and a mathematical space lattice
(b) a CCP structure and an HCP structure
(c) a crystal class and a crystal system
(d) a Wigner–Seitz cell and a Brillouin zone.

4.8 Write down the primitive translation vectors for the following lattice structures:
(a) a BCC lattice and a is the length of primitive translation vector
(b) an FCC lattice and $2a$ is the length of primitive translation vector.

4.9 Use the defining relations for reciprocal lattice vectors to show that if $\boldsymbol{a}_i$ $(i = 1, 2, 3)$ are the direct lattice vectors of a crystal lattice then its reciprocal lattice vectors $\boldsymbol{a}_j^*$ $(j = 1, 2, 3)$ are given explicitly as follows:

$$\boldsymbol{a}_1^* = (\boldsymbol{a}_2 \times \boldsymbol{a}_3)/V \qquad \boldsymbol{a}_2^* = (\boldsymbol{a}_3 \times \boldsymbol{a}_1)/V \qquad \boldsymbol{a}_3^* = (\boldsymbol{a}_1 \times \boldsymbol{a}_2)/V$$

where V is the volume of the unit cell.

4.10 Show that a BCC lattice has as its reciprocal lattice an FCC lattice and vice versa.

4.11 Construct the first Brillouin zone for a BCC and an FCC lattice.

4.12 Construct planes with the following Miller indices in an orthorhombic primitive lattice: (a) (123); (b) (103); (c) (030); (d) (110).

4.13 Show that, for a cubic system, the distance d between parallel planes with Miller indices (hkl) is given by

$$\frac{1}{d^2} = \frac{(h^2 + k^2 + l^2)}{a^2}$$

where a is the length of the cube edge.

4.14 The primitive translation vectors of a cubic lattice with lattice parameter a are given by

$$t_1 = a(\hat{i} + \hat{j} - \hat{k}) \qquad t_2 = a(\hat{i} - \hat{j} + \hat{k}) \qquad t_3 = a(-\hat{i} + \hat{j} + \hat{k}).$$

Here $\hat{i}, \hat{j}, \hat{k}$ are unit vectors along the x, y, z, axes respectively. Find:
(a) the length of the primitive translation (lattice) vector
(b) the angle between the primitive lattice vectors.

4.15 If C_{2x}, C_{2y}, C_{2z} denote rotations through 180° about the x, y, z axes respectively, find the product of the following space-group operations:
(a) $\{C_{2x}|(\frac{1}{2}\ \frac{1}{2}\ 0)\}\{C_{2x}|(\frac{1}{2}\ \frac{1}{2}\ 0)\}$ (b) $\{C_{2x}|(x\ y\ z)\}\{C_{2z}|0\}$.

4.16 Determine whether the following space-group operations satisfy the closure property of groups:

$$\{E|(0\ 0\ 0)\} \quad \{C_{2x}|(\tfrac{1}{2}\ \tfrac{1}{2}\ 0)\} \quad \{C_{2y}|(\tfrac{1}{2}\ \tfrac{1}{2}\ 0)\} \quad \{C_{2z}|(0\ 0\ 0)\}.$$

4.17 Show that the product of the space-group operations

$$\{C_{2x}|(x\ y\ z)\}^{-1}\{I|0\}\{C_{2x}|(x\ y\ z)\} = \{I|T\}$$

where $T = t_1 + t_2 + t_3$ and $t_1 = -2x\hat{i}$, $t_2 = 2y\hat{j}$, $t_3 = 2z\hat{k}$; $\hat{i}, \hat{j}$ and $\hat{k}$ are unit vectors along the x, y, z axes respectively and the point-group operations C_{2x} and I have their usual meaning.

4.18 Find the inverse of the space-group operation $\{C_{2x}|\frac{1}{2}(x\ y\ z)\}$ and verify that its product with the inverse is $\{E|0\}$.

5

THE THREE-DIMENSIONAL ROTATION GROUP

5.1 Elementary representation theory

Having briefly considered the various point-group symmetry operations on which crystal structures are based, we also need to know the representation theory of the three-dimensional rotation group, represented generally by the symbol O_3. This is because each of the pure rotation groups listed in table 4.3 is really a subgroup of the full rotation group O_3 and as such each of them contain symmetry operations of some regular three-dimensional solid.

The rotation group O_3 is of infinite order covering all operations of a sphere. It is not our intention to describe the rotational operations of symmetry of a sphere or discuss the properties of a continuous group because the algebra gets quite complicated and unnecessarily obscure for our purposes (see [3, 6, 15] and Appendix A). For the present we shall be interested in considering the basis functions of the representations of the rotation group.

The basis functions of O_3 are the spherical harmonic functions $Y_l^{m_l}$. One usually comes across them in the theory of atomic spectra (e.g. solution of the H atom problem):

$$\left[\frac{1}{\sin\theta}\frac{\partial}{\partial\theta}\left(\sin\theta\,\frac{\partial}{\partial\theta}\right)+\frac{1}{\sin^2\theta}\frac{\partial^2}{\partial\phi^2}\right]Y_l^{m_l}(\theta,\phi)=l(l+1)\,Y_l^{m_l}(\theta,\phi). \quad (5.1)$$

Here θ and ϕ are the polar angles, l takes values zero or positive integers and m_l takes values from $-l$ to $+l$ in steps of one. The operator in equation (5.1) is just the angular part of the Laplacian

$$\nabla^2=\frac{\partial^2}{\partial x^2}+\frac{\partial^2}{\partial y^2}+\frac{\partial^2}{\partial z^2}$$

and is invariant under rotations; hence its eigenfunctions, the Y, form bases of the representation of the rotation group.

The spherical harmonics are defined as follows:

$$Y_l^{m_l}(\theta,\phi)=N_{lm_l}P_l^{m_l}(\cos\theta)\exp(\mathrm{i}m_l\phi). \quad (5.2)$$

Here N_{lm_l} is a normalization constant and $P_l^{m_l}(\cos\theta)$ is a polynomial in $\cos\theta$ called the associated Legendre function. Since m_l takes values from $-l$ to

$+l$, it means the representation of the rotation group denoted by D^l is of $(2l+1)$ dimensions.

5.2 Characters

If we define R_α to be a rotation operation through angle α, then its effect on the $Y_l^{m_l}$ (in the active or passive interpretation; see §2.2) is given by

$$\begin{aligned} R_\alpha Y_l^{m_l}(\theta, \phi) &= Y_l^{m_l}(\theta, \phi - \alpha) \\ &= \exp(-\mathrm{i}m_l\alpha) Y_l^{m_l}(\theta, \phi). \end{aligned} \tag{5.3}$$

Thus the representation for all possible values of m_l from $-l$ to $+l$ is a diagonal matrix:

$$\Gamma^l(\alpha) = \begin{pmatrix} \mathrm{e}^{-\mathrm{i}l\alpha} & 0 & \cdots & 0 \\ 0 & \mathrm{e}^{-\mathrm{i}(l-1)\alpha} & \cdots & 0 \\ \vdots & & & \\ 0 & 0 & \cdots & \mathrm{e}^{\mathrm{i}l\alpha} \end{pmatrix}. \tag{5.4}$$

Hence the character of an operation which represents a rotation through an angle α in the rotation group is

$$\begin{aligned} \chi^l(\alpha) = \operatorname{tr}\Gamma^l(\alpha) &= \exp(-\mathrm{i}l\alpha) + \exp[-\mathrm{i}(l-1)\alpha] + \cdots + \exp(\mathrm{i}l\alpha) \\ &= \frac{\sin(l+\frac{1}{2})\alpha}{\sin\frac{1}{2}\alpha}. \end{aligned} \tag{5.5}$$

The representations thus obtained are of odd dimensionality $(2l+1)$ since l takes only zero or positive integral values.

5.3 The double-group representations

For even-dimensional representations we use the quantum number $j = l + s$, where j can now be half an odd integer if the spin quantum number s has half-integral value. The characters in the j representation are then obtained by replacing l in equation (5.5) by j. Thus

$$\chi^j(\alpha) = \frac{\sin(j+\frac{1}{2})\alpha}{\sin\frac{1}{2}\alpha}. \tag{5.6}$$

If, therefore, j is half an odd integral, then the numerator of equation (5.6) becomes, for a (further) rotation of 2π,

$$\sin[(j+\tfrac{1}{2})(\alpha + 2\pi)] = \sin(j+\tfrac{1}{2})\alpha. \tag{5.7}$$

On the other hand, for integral values of l the numerator of (5.5) becomes

$$\sin[(l+\tfrac{1}{2})(\alpha+2\pi)] = -\sin(l+\tfrac{1}{2})\alpha. \tag{5.8}$$

The denominator in either case, for a rotation of 2π, becomes

$$\sin\tfrac{1}{2}(\alpha+2\pi) = -\sin\tfrac{1}{2}\alpha. \tag{5.9}$$

We therefore have the very important feature of half-integral representations, namely:

$$\chi^j(\alpha+2\pi) = -\chi^j(\alpha) \tag{5.10}$$

whereas

$$\chi^l(\alpha+2\pi) = \chi^l(\alpha). \tag{5.11}$$

Equation (5.10) therefore suggests that for half-integral representations a rotation of 2π is no longer an identity operation; however, a rotation of 4π is now equivalent to the identity operator E. To cope with such a situation Bethe [46] introduced a new symmetry element $\bar{E}$ (read as E bar, or bar E) which represents a rotation by 2π. Introducing $\bar{E}$ in all the point groups enumerated earlier will produce groups which have twice as many elements as the original group. These groups are called double groups. The double group contains more classes and has more irreducible representations than the original point group. Furthermore $\bar{E}$ will commute with all elements in the group, and hence it will form a class by itself. Double-group representations are often known as groups of even-dimensional representations. As an example the character table for C_{3v} given in table 3.1 becomes as shown in table 5.1 for the double group $^dC_{3v}$. The character tables for all the double groups can be found in Koster *et al* [10].

Simplifying rules for obtaining the complete multiplication table of double groups can be found in Slater [18]. As an example we have given in table 5.2 the multiplication table for the double group $^dC_{3v}$. Compare this table with the multiplication table 1.2 for the single group C_{3v} and observe the

Table 5.1 Character table for the double group $^dC_{3v}$ (cf. table 3.1).

$^dC_{3v}$	E	$\bar{E}$	$2C_3$	$2\bar{C}_3$	$3\sigma_v$	$3\bar{\sigma}_v$
Γ_1	1	1	1	1	1	1
Γ_2	1	1	1	1	−1	−1
Γ_3	2	2	−1	−1	0	0
Γ_4	2	−2	1	−1	0	0
Γ_5	1	−1	−1	1	i	−i
Γ_6	1	−1	−1	1	−i	i

Table 5.2 Multiplication table for the double group $^{d}C_{3v}$ (cf. table 1.2).

$^{d}C_{3v}$	E	C_3	C_3^2	σ_{v_a}	σ_{v_b}	σ_{v_c}	$\bar{E}$	$\bar{C}_3$	$\bar{C}_3^2$	$\bar{\sigma}_{v_a}$	$\bar{\sigma}_{v_b}$	$\bar{\sigma}_{v_c}$
E	E	C_3	C_3^2	σ_{v_a}	σ_{v_b}	σ_{v_c}	$\bar{E}$	$\bar{C}_3$	$\bar{C}_3^2$	$\bar{\sigma}_{v_a}$	$\bar{\sigma}_{v_b}$	$\bar{\sigma}_{v_c}$
C_3	C_3	$\bar{C}_3^2$	E	σ_{v_c}	σ_{v_a}	$\bar{\sigma}_{v_b}$	$\bar{C}_3$	C_3^2	$\bar{E}$	$\bar{\sigma}_{v_c}$	$\bar{\sigma}_{v_a}$	σ_{v_b}
C_3^2	C_3^2	E	$\bar{C}_3$	σ_{v_b}	$\bar{\sigma}_{v_c}$	σ_{v_a}	$\bar{C}_3^2$	$\bar{E}$	C_3	$\bar{\sigma}_{v_b}$	σ_{v_c}	$\bar{\sigma}_{v_a}$
σ_{v_a}	σ_{v_a}	σ_{v_b}	σ_{v_c}	$\bar{E}$	$\bar{C}_3$	$\bar{C}_3^2$	$\bar{\sigma}_{v_a}$	$\bar{\sigma}_{v_b}$	$\bar{\sigma}_{v_c}$	E	C_3	C_3^2
σ_{v_b}	σ_{v_b}	$\bar{\sigma}_{v_c}$	σ_{v_a}	$\bar{C}_3^2$	$\bar{E}$	C_3	$\bar{\sigma}_{v_b}$	σ_{v_c}	$\bar{\sigma}_{v_a}$	C_3^2	E	$\bar{C}_3$
σ_{v_c}	σ_{v_c}	σ_{v_a}	$\bar{\sigma}_{v_b}$	$\bar{C}_3$	C_3^2	$\bar{E}$	$\bar{\sigma}_{v_c}$	$\bar{\sigma}_{v_a}$	σ_{v_b}	C_3	$\bar{C}_3^2$	E
$\bar{E}$	$\bar{E}$	$\bar{C}_3$	$\bar{C}_3^2$	$\bar{\sigma}_{v_a}$	$\bar{\sigma}_{v_b}$	$\bar{\sigma}_{v_c}$	E	C_3	C_3^2	σ_{v_a}	σ_{v_b}	σ_{v_c}
$\bar{C}_3$	$\bar{C}_3$	C_3^2	$\bar{E}$	$\bar{\sigma}_{v_c}$	$\bar{\sigma}_{v_a}$	σ_{v_b}	C_3	$\bar{C}_3^2$	E	σ_{v_c}	σ_{v_a}	$\bar{\sigma}_{v_b}$
$\bar{C}_3^2$	$\bar{C}_3^2$	$\bar{E}$	C_3	$\bar{\sigma}_{v_b}$	σ_{v_c}	$\bar{\sigma}_{v_a}$	C_3^2	E	$\bar{C}_3$	σ_{v_b}	$\bar{\sigma}_{v_c}$	σ_{v_a}
$\bar{\sigma}_{v_a}$	$\bar{\sigma}_{v_a}$	$\bar{\sigma}_{v_b}$	$\bar{\sigma}_{v_c}$	E	C_3	C_3^2	σ_{v_a}	σ_{v_b}	σ_{v_c}	$\bar{E}$	$\bar{C}_3$	$\bar{C}_3^2$
$\bar{\sigma}_{v_b}$	$\bar{\sigma}_{v_b}$	σ_{v_c}	$\bar{\sigma}_{v_a}$	C_3^2	E	$\bar{C}_3$	σ_{v_b}	$\bar{\sigma}_{v_c}$	σ_{v_a}	$\bar{C}_3^2$	$\bar{E}$	C_3
$\bar{\sigma}_{v_c}$	$\bar{\sigma}_{v_c}$	$\bar{\sigma}_{v_a}$	σ_{v_b}	C_3	$\bar{C}_3^2$	E	σ_{v_c}	σ_{v_a}	$\bar{\sigma}_{v_b}$	$\bar{C}_3$	C_3^2	$\bar{E}$

following features (see [18] and Appendix B):

(i) The double-group table is four times as large as the single-group table.

(ii) The product of an unbarred and a barred operator is the same as the product of the two corresponding unbarred operators except with a change from the unbarred to the barred or vice versa.

(iii) The product of barred and unbarred operators is the same as in (ii).

(iv) The product of two barred operators is the same as the product of the two corresponding unbarred operators.

Exercises

5.1 The rotation group O_2 is of infinite order covering all the operations of rotations of a circle about the z axis passing through its centre. If we denote the group elements by $R_z(\phi)$, where ϕ is a continuous parameter in the range $0 \leqslant \phi \leqslant 2$, find the matrix transformation for $R_z(\phi)$.

5.2 Find the law of composition for the group O_2 of exercise 5.1. What is (a) the identity element for this group and (b) the inverse element for $R_z(\phi)$?

5.3 Determine the matrix of transformation for $R_z(\phi)$ for the group O_3 defined in §5.1.

5.4 What is meant by a spinor group? How different is it compared with the corresponding related point group?

5.5 Find the multiplication table for the double group $^{d}C_2$.

5.6 Use the multiplication table of C_{4v} (see exercise 1.5) to obtain the multiplication table for the double group $^{d}C_{4v}$.

6

ILLUSTRATIVE EXAMPLES OF THE APPLICATIONS OF SYMMETRY PRINCIPLES

6.1 Introduction

Various applications can be discussed at this stage based on the knowledge gained so far [12–19, 26–28, 30–32, 35, 41–45, 49, 50]. However, for our purposes we have chosen four examples from the literature and presented detailed solutions in the hope that the procedures described here could be readily extended (with some effort) to more complicated problems of current interest. Furthermore readers are also strongly urged to refer to the original literature on each topic treated here. The original (classic) and the important references are cited alongside the topic.

6.2 Crystal field splittings (see [46, 29, 2])

Bethe's classic paper [46] entitled 'Splittings of terms in crystals' really marks the beginning of the applications of symmetry principles in crystal field theory. In this paper he showed how the degenerate energy levels of a free atom split when the atom is placed in a crystal field of a given symmetry.

Conventional crystal field theory usually begins by listing the kind of terms one expects in the Hamiltonian H for the electrons of a central paramagnetic ion residing in a crystalline electric field (figure 6.1), in the absence of any external electric or magnetic fields. In its simplest form

$$H = H_{\mathrm{F}} + V \tag{6.1}$$

where H_{F} is the Hamiltonian for the free ion and V is the electric potential provided by the symmetrical distribution of the negative point charges (ions) about the central positive ion. The positive and negative ions are assumed to be hard non-overlapping spheres. Thus (6.1) can be written as

$$H = -\frac{\hbar^2}{2m}\sum_i \nabla_i^2 - \sum_i \frac{Ze^2}{r_i} + \frac{1}{2}\sum_{i\neq j}\frac{e^2}{r_{ij}} + \lambda_{ij}\boldsymbol{l}_i\cdot\boldsymbol{s}_i - e_i\phi_{\mathrm{c}}(r_i). \tag{6.2}$$

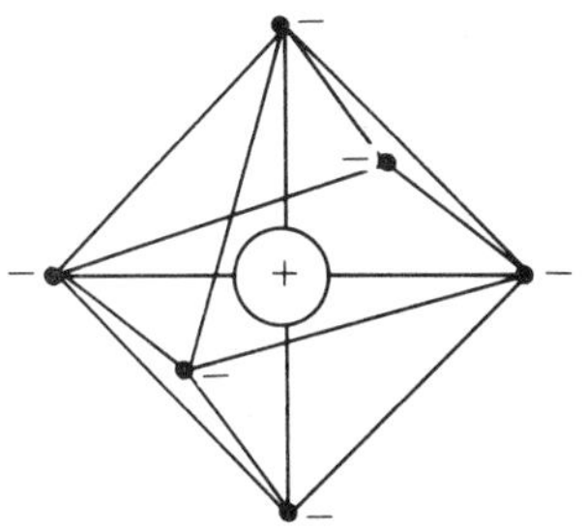

Figure 6.1 An example of a crystalline electric field of octahedral symmetry. Note that the central paramagnetic ion (positive ion) resides in a crystalline electric potential whose sources are negative point charges (ions) or point dipoles (e.g. H_2O molecules).

The first term on the right-hand side is the total kinetic energy of the electrons in the paramagnetic ion. The second term is the Coulomb attraction between the nucleus and these electrons. The third term is the electron–electron interaction term or the Coulomb repulsion between the electrons. The fourth term represents the influence of the spin–orbit coupling, the magnitude of which is determined by the coupling coefficient λ_{ij}; here i and j range over all electrons. The last term, which is the important one for us, represents the effect of the crystal (electric) field interaction with each electron e_i. The Hamiltonian (6.2) often contains other terms to account for: (i) fine structure splittings arising from axial symmetry; (ii) hyperfine structure splittings, arising from the interaction between electron and nuclear spins; and (iii) quadrupole splittings arising from the nuclear quadrupole moment (i.e. distortion of nucleus from spherical to elliptical symmetry) etc. In addition to these splittings (often called zero field splittings) further splittings can be produced by magnetic fields (called Zeeman splittings).

For our purposes we are not really concerned with all these aspects of crystal field studies. We are interested only in finding out, via symmetry principles, how the energy states of a free atom or ion split up owing to the perturbing influence of the crystal field potential $V = e_i\phi_c(\boldsymbol{r}_i)$. To make the problem as simple as possible we shall assume the atom to have a single spinless electron and consider for example the splitting of an f-level state in the free atom when it is subjected to a crystal field of cubic symmetry, O (say). For an f level, $l = 3$ and therefore the $(2l + 1) = 7$ eigenfunctions (which are degenerate in the free atom) will transform according to the representation $D^l \equiv D^3$ of the full rotation group O_3 (see Chapter 5). Our task is then to reduce the representation D^3 into its irreducible components.

Since the atom is placed in a cubic field of symmetry O, we first write down the character table of O as shown in table 6.1. Next, using equation (5.5), namely $\chi^l(\alpha) = \sin(l + \frac{1}{2})\alpha/\sin\frac{1}{2}\alpha$, we determine the characters generated

Table 6.1 Character table for the cubic group O. D^3 represents the reducible representation (see text).

	$\alpha = 0$	$\alpha = \pi$		$\alpha = 2\pi/3$	$\alpha = \pi/2$
O	E	$3C_4^2$	$6C_2$	$8C_3$	$6C_4$
Γ_1	1	1	1	1	1
Γ_2	1	1	-1	1	-1
Γ_3	2	2	0	-1	0
Γ_4	3	-1	-1	0	1
Γ_5	3	-1	1	0	-1
$D^l \equiv D^3$	7	-1	-1	1	-1

by the operations of O on the seven f-level functions as follows:

$$\chi(E) = (2l + 1) = 7$$

$$\chi(C_4^2) = \chi(\pi) = \frac{\sin(3 + \frac{1}{2})\pi}{\sin(\pi/2)} = -1$$

$$\chi(C_2) = \chi(\pi) = -1$$

$$\chi(C_3) = \chi\left(\frac{2\pi}{3}\right) = \frac{\sin[(3 + \frac{1}{2})2\pi/3]}{\sin(2\pi/6)} = 1$$

$$\chi(C_4) = \chi\left(\frac{\pi}{2}\right) = \frac{\sin(3 + \frac{1}{2})\pi/2}{\sin(\pi/4)} = 1.$$

These characters generated by the operations of O are listed at the bottom of the character table in the same order as in the table (see last row denoted by D^3 of table 6.1). In general, in determining these characters we observe that

$$\chi(E) = 2l + 1$$

$$\chi(C_2) = \chi(\pi) = (-1)^l$$

$$\chi(C_3) = \chi\left(\frac{2\pi}{3}\right) = \begin{cases} 1 & \text{if } l = 0, 3, \ldots \\ 0 & \text{if } l = 1, 4, \ldots \\ -1 & \text{if } l = 2, 5, \ldots \end{cases} \tag{6.3}$$

$$\chi(C_4) = \chi\left(\frac{\pi}{2}\right) = \begin{cases} 1 & \text{if } l = 0, 1, 4, 5 \\ -1 & \text{if } l = 2, 3, 6, 7. \end{cases}$$

The next step is to decompose this reducible representation D^3 into its irreducible components using (3.13). Thus

$$D^3 = \Gamma_2 + \Gamma_4 + \Gamma_5 \tag{6.4}$$

(usually one can check this from the table itself). Hence we see that the seven-fold degenerate level of a free atom (represented by D^3) splits into two three-fold degenerate levels Γ_4 and Γ_5 and one non-degenerate level Γ_2. This is shown schematically in figure 6.2. It should be noted that symmetry arguments will not indicate either the actual extent of the splittings or even the order of the energy levels. Thus it is not possible to say which will be the lowest energy level on the basis of symmetry alone.

6.2.1 *Additional splittings in fields of lower symmetry*

Using the above principles we can readily determine what further splittings could be obtained by reducing the symmetry of the lattice by, for example, straining the lattice uniformly along the (111) direction (i.e. along the cube diagonal) so as to reduce the symmetry to D_3. Then since D_3 is a subgroup of O we can readily determine how the representations Γ_2, Γ_4, Γ_5 of the cubic field O would split in the field of lower symmetry, that is D_3. The character table for the group D_3 is given in table 6.2 and below it are written the characters of the representations Γ_2, Γ_4, Γ_5 of the O group. This method is technically known as subducing the representations of O on to D_3.

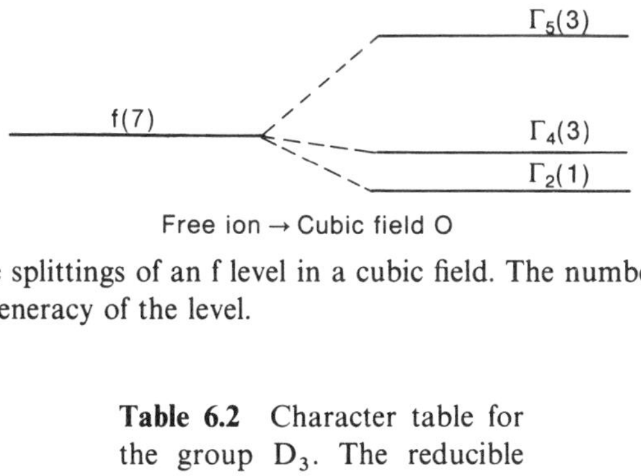

Figure 6.2 The splittings of an f level in a cubic field. The numbers in parentheses indicate the degeneracy of the level.

Table 6.2 Character table for the group D_3. The reducible representations from the cubic group O are written below the table.

D_3	E	$3C_2$	$2C_3$
Γ_1	1	1	1
Γ_2	1	-1	1
Γ_3	2	0	-1
Γ_2	1	-1	1
Γ_4	3	-1	0
Γ_5	3	1	0

Again using equation (3.13) one can readily see that the representations Γ_2, Γ_4, Γ_5 of the group O would break up as follows:

$$\begin{aligned}\Gamma_2 &\rightarrow \Gamma_2 \text{ of } D_3\\ \Gamma_4 &\rightarrow \Gamma_2 + \Gamma_3 \text{ of } D_3\\ \Gamma_5 &\rightarrow \Gamma_1 + \Gamma_3 \text{ of } D_3.\end{aligned} \tag{6.5}$$

Thus for a cubic field reduced to a lower symmetry field D_3 by a strain along the (111) direction, the splittings for an f level would be as shown in figure 6.3. It is perhaps worth emphasizing again that the order of the levels and the amount of splittings cannot be determined by symmetry alone but by crystal field calculations involving both the type and the strength of the crystal field potential.

6.2.2 *Introduction of spins and double groups*

We can now examine the effect of introducing spin into our discussion. As a specific example we consider an atom which has two 'inequivalent electrons' (i.e. quantum number n is different for the two electrons) labelled 2p3p and this is placed in a cubic field O. To solve this problem we make use of the following important theorem on the irreducible components of the direct product representations:

Theorem: Given two irreducible representations D^{l_1} and D^{l_2} of the full rotation group, the irreducible components of the direct product representation $D^{l_1} \otimes D^{l_2}$ are

$$D^{l_1+l_2}, D^{l_1+l_2-1}, D^{l_1+l_2-2}, \ldots, D^{|l_1-l_2|}.$$

In other words:

$$D^{l_1} \otimes D^{l_2} = \sum_{l=|l_1-l_2|}^{l_1+l_2} D^{l}. \tag{6.6}$$

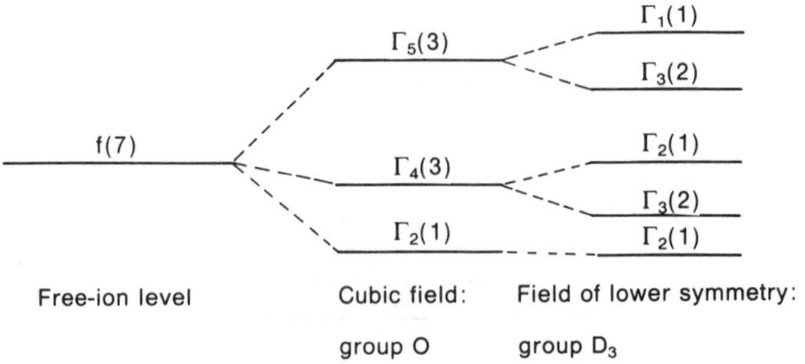

Figure 6.3 The splittings of an f level by a cubic field which is strained along the (111) direction. The numbers in parentheses indicate the degeneracy of the level.

Here we have used the symbol $\otimes$ to denote the direct product of two representations and the summation Σ is symbolically taken to mean what is technically called the 'direct sum' and not the matrix sum.

For our chosen example, $l_1 = 1$, $l_2 = 1$, so that

$$D^1 \otimes D^1 = D^2 + D^1 + D^0. \tag{6.7}$$

Equation (6.7) is of course a familiar result obtained by using the vector model of atoms in atomic physics. The general coupling of such l vectors and the states arising therefrom are given in table 6.3.

We can now obtain the effect of the cubic field O on these coupled electron states (6.7) by using equation (3.13) and the character table 6.2 for O to get the following results:

$$\begin{aligned} D^2 &\rightarrow \Gamma_3 + \Gamma_5 \\ D^1 &\rightarrow \Gamma_4 \\ D^0 &\rightarrow \Gamma_1. \end{aligned} \tag{6.8}$$

We observe that in a cubic field the D^2 representation for which $l = 2$ (i.e. a d state) splits up into a Γ_3 representation which is two-fold degenerate and a Γ_5 representation which is three-fold degenerate. Similarly the D^1 representation for which $l = 1$ (i.e. a p state) remains three-fold degenerate in a cubic field whilst the D^0 representation for which $l = 0$ (i.e. an s state) has no degeneracy.

In the above example we dealt with only the odd-dimensional representation D^l and not the even-dimensional representation D^j (see §5.3). The D^l representation is valid so long as we have integral values of spins, so that $j = l + s$ will be an integer. If, however, there is an odd number of electrons, then $j = l + s$ is half-integral and $(2j + 1)$ is even, so that here we must use the D^j or the double-group representations as explained in §5.3.

The method of dealing with double groups is very similar to the standard analysis that we have seen so far. In this case we first form the direct product

Table 6.3 The L values and term symbols for terms with different electron configurations. (Note: We are considering inequivalent electrons.)

Electron configuration	l_1	l_2	L	Term symbol
sp	0	1	1	P
pp	1	1	2, 1, 0	D, P, S
pd	1	2	3, 2, 1	F, D, P
dd	2	2	4, 3, 2, 1, 0	G, F, D, P, S

of the spin representation D^j with the space representation D^l and then decompose the resulting representations as before using equation (3.13).

As a concrete example we consider the splittings of a ^{4}F term in a cubic field O. Here $L = 3$ for an F state, and since multiplicity $2S + 1 = 4$, we have $S = 3/2$. We therefore have $D^j = D^{3/2}$ and $D^l = D^3$, so that we have to find the direct product of $D^{3/2}$ with the D^3 representations which we obtained earlier as $D^3 = \Gamma_2 + \Gamma_4 + \Gamma_5$ (see equation (6.4)). To obtain the transformation of the $D^{3/2}$ representation we must first write down the double-group character table of O, that is ^{d}O, as shown in table 6.4. It is then easy to see using equation (5.6) and remembering that $\alpha = 4\pi$ for E and $\alpha = 2\pi$ for $\bar{E}$ etc, that $D^{3/2}$ transforms like the Γ_8 representation (see table 6.5). Thus we now have to form the direct product of $D^3 = \Gamma_2 + \Gamma_4 + \Gamma_5$ with Γ_8 and then decompose it in the usual manner using equation (3.13). The result is as follows:

$$
\begin{aligned}
\Gamma_2 \otimes \Gamma_8 &= \Gamma_8(4) \\
\Gamma_4 \otimes \Gamma_8 &= \Gamma_6(2) + \Gamma_7(2) + \Gamma_8(4) + \Gamma_8(4) \\
\Gamma_5 \otimes \Gamma_8 &= \Gamma_6(2) + \Gamma_7(2) + \Gamma_8(4) + \Gamma_8(4).
\end{aligned}
\tag{6.9}
$$

The above results show that the Γ_2 level keeps its four-fold degeneracy ($M_s = \frac{3}{2}, \frac{1}{2}, -\frac{1}{2}, -\frac{3}{2}$) whereas the Γ_4 and Γ_5 levels each split into 4 and 4. The schematic representation of the splittings of a ^{4}F level in a cubic field is shown in figure 6.4. As mentioned earlier, the order and magnitude of the splittings cannot be predicted by symmetry theory alone.

This is as far as we can go for the present purposes. In our discussions we have not included the effects of the strengths of the crystalline field or the effects of spin–orbit couplings. But these are too extensive and are well treated in the list of references given earlier. If one understands the procedures that we have adopted so far, it is possible, with some effort, to extend the discussion to include the above effects.

Table 6.4 Character table for the double group ^{d}O (see Koster *et al* [10]).

^{d}O	E	$\bar{E}$	$8C_3$	$8\bar{C}_3$	$3C_2$ $3\bar{C}_2$	$6C'_2$ $6\bar{C}'_2$	$6C_4$	$6\bar{C}_4$
Γ_1	1	1	1	1	1	1	1	1
Γ_2	1	1	1	1	1	-1	-1	-1
Γ_3	2	2	-1	-1	2	0	0	0
Γ_4	3	3	0	0	-1	-1	1	1
Γ_5	3	3	0	0	-1	1	-1	-1
Γ_6	2	-2	1	-1	0	0	$\sqrt{2}$	$-\sqrt{2}$
Γ_7	2	-2	1	-1	0	0	$-\sqrt{2}$	$\sqrt{2}$
Γ_8	4	-4	-1	1	0	0	0	0

Table 6.5 Transformation properties of the D^j representations in the double group ${}^dO$. (Recall that $\chi^j(\alpha) = \sin(j+\frac{1}{2})\alpha/\sin(\alpha/2)$.)

${}^dO$	E	$\bar{E}$	$8C_3$	$8\bar{C}_3$		$3C_2$ $3\bar{C}_2$	$6C'_2$ $6\bar{C}'_2$	$6C_4$	$6\bar{C}_4$	
D^j	$(2j+1)$	$-(2j+1)$	1	-1	$(j=\frac{1}{2}, \frac{7}{2}, \ldots)$	0	0	$\sqrt{2}$	$-\sqrt{2}$	$(j=\frac{1}{2}, \frac{9}{2}, \ldots)$
			-1	1	$(j=\frac{3}{2}, \frac{9}{2}, \ldots)$			0	0	$(j=\frac{3}{2}, \frac{7}{2}, \frac{11}{2}, \ldots)$
			0	0	$(j=\frac{5}{2}, \frac{11}{2}, \ldots)$			$-\sqrt{2}$	$\sqrt{2}$	$(j=\frac{5}{2}, \frac{13}{2}, \ldots)$

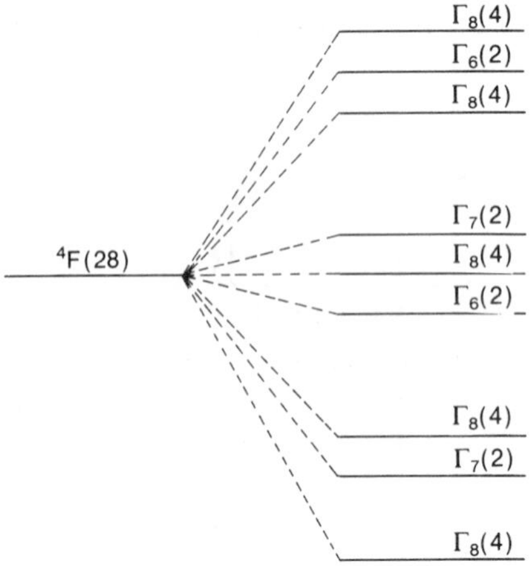

Figure 6.4 The splittings of a ^{4}F level by a cubic field O. The numbers in parentheses indicate the degeneracy of the level (see Tinkham [2] for further details).

6.3 Symmetry properties of electron wavefunctions in crystals (see [47, 8, 2, 9])

In this section we shall confine our attention to the symmetry classification of electronic states at various points in the Brillouin zone of the crystal space lattice. For simplicity, we shall consider in some detail only the BCC lattice and refer readers to the references given above for the symmetry classification in other lattices and for more details on the subject matter.

To enable us to explain the method we shall first present some important results from solid state physics and then develop the necessary group theoretical techniques in their simplest form.

6.3.1 Bloch's theorem

Since all crystals have translational (periodic) symmetry, Bloch has shown that for an electron moving in a periodic potential the solutions of the Schrödinger equation with a periodic potential are of the form (see Kittel [34])

$$\Psi_k(\boldsymbol{r}) = U_k(\boldsymbol{r})\exp(\mathrm{i}\boldsymbol{k}\cdot\boldsymbol{r})$$

where

$$U_k(\boldsymbol{r}) = U_k(\boldsymbol{r} + \boldsymbol{T}_n). \qquad (6.10)$$

In other words the solutions are plane waves ($\exp(\mathrm{i}\boldsymbol{k}\cdot\boldsymbol{r})$) modulated by the function $U_k(\boldsymbol{r})$ which has the same periodicity as the lattice. For our purposes, the wavevector $\boldsymbol{k}$ is chosen to be one of the allowed wavevectors in the first Brillouin zone of the crystal lattice and the translational group $\boldsymbol{T}_n$ is defined as before (see equation (4.2)) by

$$\boldsymbol{T}_n = n_1\boldsymbol{t}_1 + n_2\boldsymbol{t}_2 + n_3\boldsymbol{t}_3 \tag{6.11}$$

where $\boldsymbol{t}_1$, $\boldsymbol{t}_2$, $\boldsymbol{t}_3$ must of course be the primitive translation vectors of the crystal (direct) lattice because they are used for the construction of the Brillouin zone. We need to consider only the $\boldsymbol{k}$ vectors lying in or on the surface of the first Brillouin zone since it can be shown quite easily that any $\boldsymbol{k}$ vector lying outside this zone is related to one inside the zone by the addition or subtraction of a reciprocal lattice vector.

6.3.2 *The 'little group' and the 'small representation'*

Consider the Brillouin zone of the cubic lattices (see figure 4.9) which have a point-group symmetry O_h. If we consider the special wavevectors, as indicated by the symmetry symbols in figure 4.9, and carry out successively all the point-group operations R contained in O_h, we would generate what is known as the group of the $\boldsymbol{k}$ vector, that is a collection of all those operations in O_h which would leave the particular $\boldsymbol{k}$ invariant or transform it into $\boldsymbol{k}^* = \boldsymbol{k} + \boldsymbol{g}$ where $\boldsymbol{g}$ is any reciprocal lattice vector including zero. This collection of symmetry operations, which forms a subgroup of the main group (O_h in our example), is also called the *little group* of $\boldsymbol{k}$. For a $\boldsymbol{k}$ vector at the centre of the zone (i.e. at $\boldsymbol{k} = 0$), usually denoted by the symmetry symbol Γ, it is obvious that the $\boldsymbol{k}$ vector must be invariant under all the operations of the crystal point group. Thus, for example, in our case the little group at Γ ($\boldsymbol{k} = 0$) must be O_h. The little groups for all the important points and lines of symmetry in the Brillouin zone of a simple cubic lattice and a body-centred cubic lattice (see figure 4.9) are given in table 6.6.

If we now apply an operation of the little group of $\boldsymbol{k}$ to Ψ_k (equation (6.10)) then such an operation must either leave Ψ_k unchanged (apart from a phase factor perhaps) or else it should transform it into a new function Ψ'_k but with the same $\boldsymbol{k}$ vector. In this case, however, there will be more than one distinct $U_k(r)$ associated with $\exp(\mathrm{i}\boldsymbol{k}\cdot\boldsymbol{r})$. These various $U_k(\boldsymbol{r})$ will transform among themselves according to the irreducible representation of the little group of $\boldsymbol{k}$ which is called the *small representation.*

If a wavevector $\boldsymbol{k}$ in a Brillouin zone forms a little group then the wavefunction of an electron with the same value of $\boldsymbol{k}$ must also transform like the operations of the little group of $\boldsymbol{k}$. The representation for the rotational part R in the space-group operation $\{R|\boldsymbol{t}\}$ is given by the representations $\Gamma(R)$ obtained from the point-group table for the little group, whereas that for the translational part is given simply by $\exp(\mathrm{i}\boldsymbol{k}\cdot\boldsymbol{t})$. Thus

$$\mathbf{D}\{R|\boldsymbol{t}\} = \exp(\mathrm{i}\boldsymbol{k}\cdot\boldsymbol{t}) \times \Gamma(R). \tag{6.12}$$

Table 6.6 The little groups for all the important points and lines of symmetry for (*a*) simple cubic lattice (see figure 4.9(*a*)) and (*b*) body-centred cubic lattice (figure 4.9(*b*)). Note: A general point is indicated by the components (α, β, γ).

(*a*)

Symmetry symbol	$\boldsymbol{k}$	Little group	Order h of little group
Γ	$\frac{2\pi}{a}(0, 0, 0)$	O_h	48
Δ	$\frac{2\pi}{a}(\alpha, 0, 0)$	C_{4v}	8
X	$\frac{2\pi}{a}(\frac{1}{2}, 0, 0)$	D_{4h}	16
Λ	$\frac{2\pi}{a}(\alpha, \beta, \gamma)$	C_{3v}	6
R	$\frac{2\pi}{a}(\frac{1}{2}, \frac{1}{2}, \frac{1}{2})$	O_h	48
Σ	$\frac{2\pi}{a}(\alpha, \beta, 0)$	C_{2v}	4
M	$\frac{2\pi}{a}(\frac{1}{2}, \frac{1}{2}, 0)$	D_{4h}	16
T	$\frac{2\pi}{a}(\alpha, \frac{1}{2}, \frac{1}{2})$	C_{4v}	8
Z	$\frac{2\pi}{a}(\alpha, \frac{1}{2}, 0)$	C_{2v}	4
S	$\frac{2\pi}{a}(\alpha, \beta, \frac{1}{2})$	C_{2v}	4

If $\{R|\boldsymbol{t}\}$ and $\{R'|\boldsymbol{t}'\}$ are two space-group operations in the little group, their product is given by

$$\{R|\boldsymbol{t}\}\{R'|\boldsymbol{t}'\} = \{RR'|R\boldsymbol{t}' + \boldsymbol{t}\} \tag{6.13}$$

and their representations would multiply as

$$\mathbf{D}\{R|\boldsymbol{t}\}\mathbf{D}\{R'|\boldsymbol{t}'\} = \exp(\mathrm{i}\boldsymbol{k}\cdot\boldsymbol{t})\exp(\mathrm{i}\boldsymbol{k}\cdot\boldsymbol{t}')\Gamma(R)\Gamma(R') \tag{6.14}$$

$$= \exp[\mathrm{i}\boldsymbol{k}\cdot(\boldsymbol{t}+\boldsymbol{t}')]\Gamma(RR'). \tag{6.15}$$

Table 6.6 *continued.*

(*b*)

Symmetry symbol	k	Little group	Order h of little group
Γ	$\frac{2\pi}{a}(0, 0, 0)$	O_h	48
Δ	$\frac{2\pi}{a}(0, 0, \gamma)$	C_{4v}	8
H	$\frac{2\pi}{a}(0, 0, 1)$	O_h	48
Σ	$\frac{2\pi}{a}(\alpha, \beta, 0)$	C_{2v}	4
N	$\frac{2\pi}{a}(\frac{1}{2}, \frac{1}{2}, 0)$	D_{2h}	8
D	$\frac{2\pi}{a}(\frac{1}{2}, \frac{1}{2}, \gamma)$	C_{2v}	4
P	$\frac{2\pi}{a}(\frac{1}{2}, \frac{1}{2}, \frac{1}{2})$	T_d	24
Λ	$\frac{2\pi}{a}(\alpha, \beta, \gamma)$	C_{3v}	6
G	$\frac{2\pi}{a}(\alpha, \frac{1}{2}, 0)$	C_{2v}	4

6.3.3 Illustrative example

We shall explain the method of symmetry classification of electron states at various points in the Brillouin zone of a crystal by considering the Brillouin zone of a crystal whose space group is symmorphic (see §4.10). For the case of non-symmorphic space groups the method is somewhat complicated and we shall not describe it here (see, however, Koster [9]). To be more specific, we shall choose the Brillouin zone of a BCC lattice for illustrative purposes and urge readers to look up Tinkham [2] and Joshi [16] where a similar treatment is given using the Brillouin zone of a simple cubic lattice.

6.3.4 Method

As mentioned earlier, we begin by first establishing the crystal in direct lattice space; work out the reciprocal lattice vectors and then construct the appropriate Brillouin zone. We take this zone as our symmetrical unit cell, choose various vectors inside and on the surface of the zone, and determine the 'little group' for all the important points and lines of symmetry. Figure 4.9(*b*) shows the Brillouin zone of a BCC lattice with all the special $\boldsymbol{k}$ vectors labelled in the notation of Bouckaert *et al* [47]. The little group for all these special points is given in table 6.6(*b*).

The tables of the irreducible representations for these little groups can now be readily written using equation (6.12). $\Gamma(R)$ is given from the character tables and $\exp(i\boldsymbol{k}\cdot\boldsymbol{t})$ can be obtained by substituting the values of $\boldsymbol{k}$ and $\boldsymbol{t}$. In writing the character tables of the small representation it is customary to label the irreducible representations by writing subscripts on the special symmetry symbols. For example the representations of the little group at the point Δ are labelled as Δ_1, Δ_2, Δ_3, Δ_4, Δ_5. The character tables of the small representation for all the important symmetry points indicated in figure 4.9(*b*) are given in table 6.7.

6.3.5 *The Jones' symbol for the symmetry operation*

If the effect of a symmetry operation R on the components of a vector (x, y, z) is to produce a new vector with components (x', y', z') say, then in such cases it is convenient to express (x', y', z') in terms of the old components as shown below.

Example: Let

$$R = C_4 \equiv \begin{pmatrix} 0 & -1 & 0 \\ 1 & 0 & 0 \\ 0 & 0 & 1 \end{pmatrix}.$$

Then $(x', y', z') = C_4(x, y, z)$ can be written as

$$\begin{pmatrix} x' \\ y' \\ z' \end{pmatrix} = \begin{pmatrix} 0 & -1 & 0 \\ 1 & 0 & 0 \\ 0 & 0 & 1 \end{pmatrix} \begin{pmatrix} x \\ y \\ z \end{pmatrix} = (\bar{y}, x, z)$$

or $C_4(x, y, z) = (\bar{y}, x, z)$.

The symbol on the RHS expressing the effect of a symmetry operation on (x, y, z) is called the Jones' symbol. The Jones' symbols for the symmetry operations present in the little groups are given in table 6.7.

6.3.6 *Compatibility tables*

Looking at the tables for the little groups (table 6.7) it is clear that the order of groups (i.e. h is the number of elements in the group) is higher at the special (end) points of symmetry than on the lines of symmetry. For example, if we consider the line of symmetry ΓH (figure 4.9(*b*)), the symmetry group at the end points Γ and H is O_h with $h = 48$, whereas at the general point

Table 6.7 The character tables of the small representation at the special points of symmetry in the Brillouin zone of a BCC lattice (see figure 4.9(b)). The notations used for the representations are those of Bouckaert *et al* [47]. The Jones' symbols (see text) for the symmetry operations present in the little groups are also given below.

(a) The character table at symmetry points Γ and H. Little group = O_h.

Γ or H		E	$3C_4^2$	$6C_4$	$6C_2$	$8C_3$	I	$3\sigma_h$	$6S_4$	$6\sigma_d$	$8S_6$
Γ_1	H_1	1	1	1	1	1	1	1	1	1	1
Γ_2	H_2	1	1	−1	−1	1	1	1	−1	−1	1
Γ_{12}	H_{12}	2	2	0	0	−1	2	2	0	0	−1
Γ'_{15}	H'_{15}	3	−1	1	−1	0	3	−1	1	−1	0
Γ'_{25}	H'_{25}	3	−1	−1	1	0	3	−1	−1	1	0
Γ'_1	H'_1	1	1	1	1	1	−1	−1	−1	−1	−1
Γ'_2	H'_2	1	1	−1	−1	1	−1	−1	1	1	−1
Γ'_{12}	H'_{12}	2	2	0	0	−1	−2	−2	0	0	1
Γ_{15}	H_{15}	3	−1	1	−1	0	−3	1	−1	1	0
Γ_{25}	H_{25}	3	−1	−1	1	0	−3	1	1	−1	0

The Jones' symbol for the operations of O_h.

E	xyz
$3C_4^2 = 3C_2$	$\bar{x}\bar{y}z$; $x\bar{y}\bar{z}$; $\bar{x}y\bar{z}$
$6C_4$	$\bar{y}xz$; $y\bar{x}z$; $x\bar{z}y$; $xz\bar{y}$; $zy\bar{x}$; $\bar{z}yx$
$6C_2$	$yx\bar{z}$; $z\bar{y}x$; $\bar{x}zy$; $\bar{y}\bar{x}\bar{z}$; $\bar{z}\bar{y}\bar{x}$; $\bar{x}\bar{z}\bar{y}$
$8C_3$	zxy; yzx; $z\bar{x}\bar{y}$; $\bar{y}\bar{z}x$, $\bar{z}\bar{x}y$; $\bar{y}z\bar{x}$; $\bar{z}x\bar{y}$; $y\bar{z}\bar{x}$

(b) The character table at symmetry point Δ. Little group = C_{4v}.

Δ	E	$2C_4$	C_4^2	$2\sigma_v$	$2\sigma_d$
Δ_1	1	1	1	1	1
Δ_2	1	1	1	−1	−1
Δ_3	1	−1	1	1	−1
Δ_4	1	−1	1	−1	1
Δ_5	2	0	−2	0	0

The Jones' symbol for the operations of C_{4v}.

E	xyz
$2C_4$	$y\bar{x}z$; $\bar{y}xz$
C_4^2	$\bar{x}\bar{y}z$
$2\sigma_v$	$x\bar{y}z$; $\bar{x}yz$
$2\sigma_d$	$\bar{y}\bar{x}z$; yxz

Table 6.7 *continued.*

(*c*) The character table for Σ, D and G. Little group = C_{2v}.

Σ or D or G			E	C_2	$\sigma_{v_{xz}}$	$\sigma'_{v_{yz}}$
Σ_1	D_1	G_1	1	1	1	1
Σ_2	D_2	G_2	1	1	−1	−1
Σ_3	D_3	C_3	1	−1	1	−1
Σ_4	D_4	G_4	1	−1	−1	1

The Jones' symbol for the operations of C_{2v}.

E	xyz
C_2	$yx\bar{z}$
$\sigma_{v_{xz}}$	$xy\bar{z}$
$\sigma_{v_{yz}}$	yxz

(*d*) The character table for the symmetry point N. Little group = D_{2h}.

N	E	C_{2z}	C_{2y}	C_{2x}	I	σ_{xy}	σ_{xz}	σ_{yz}
N_1	1	1	1	1	1	1	1	1
N_2	1	1	−1	−1	1	1	−1	−1
N_3	1	−1	1	−1	1	−1	1	−1
N_4	1	−1	−1	1	1	−1	−1	1
N'_1	1	1	1	1	−1	−1	−1	−1
N'_2	1	1	−1	−1	−1	−1	1	1
N'_3	1	−1	1	−1	−1	1	−1	1
N'_4	1	−1	−1	1	−1	1	1	−1

(*e*) The character table for Λ. Little group = C_{3v}.

Λ	E	$2C_3$	$3\sigma_v$
Λ_1	1	1	1
Λ_2	1	1	−1
Λ_3	2	−1	0

The Jones' symbol for the operations of C_{3v}.

E	xyz
$2C_3$	zxy; yzx
$3\sigma_v$	yxz; zyx; xzy

(*f*) The character table for the symmetry point P. Little group = T_d.

P	E	$8C_3$	$3C_4^2$	$6S_4$	$6\sigma_d$
P_1	1	1	1	1	1
P_2	1	1	1	−1	−1
P_3	2	−1	2	0	0
P_4	3	0	−1	1	−1
P_5	3	0	−1	−1	1

The Jones' symbol for the operations of T_d.

E	xyz
$8C_3$	zxy; yzx; $z\bar{x}\bar{y}$; $\bar{y}\bar{z}x$; $\bar{z}\bar{x}y$; $\bar{y}z\bar{x}$; $\bar{z}x\bar{y}$; $y\bar{z}\bar{x}$
$3C_4^2 = 3C_2$	$\bar{x}\bar{y}z$; $x\bar{y}\bar{z}$; $\bar{x}y\bar{z}$
$6IC_4 = 6S_4$	$y\bar{x}\bar{z}$; $\bar{y}x\bar{z}$; $\bar{x}z\bar{y}$; $\bar{x}\bar{z}y$; $\bar{z}\bar{y}x$; $z\bar{y}\bar{x}$
$6IC_2 = 6\sigma_d$	$\bar{y}\bar{x}z$; $\bar{z}y\bar{x}$; $x\bar{z}\bar{y}$; yxz; zyx; xzy

Δ on the line of symmetry, the symmetry group is C_4, with $h = 8$ (see table 6.6(*b*)). We therefore expect more degeneracy of the electron energy bands at the points Γ and H than along the line ΓH. (Similarly if the group of the $\boldsymbol{k}$ vector has $h = 1$, that is only the identity element E, then we expect the degeneracy of all the bands to be completely lifted.) However, the representations at the general point Δ must be contained in the representations at the end points Γ and H. In other words the symmetry of any type along an axis must be contained in the symmetry of the type at the end points of the axis. The tables establishing such relationships between the representations at the special (end) points and those along the lines are called compatibility tables. They can be obtained quite simply by inspection of the character tables of the appropriate symmetry points or by the use of the decomposition formula (equation (3.13)). Such tables are very useful for the correct labelling of the energy bands. The compatibility relations between some important end points and lines of symmetry for a BCC lattice are given in table 6.8. Similar relations can be set up between the other end points and general points on lines of symmetry, for example between P_i, D_i, N_i, etc.

6.3.7 *Energy bands and symmetrized wavefunctions*

The energy eigenvalues are given by

$$E_t(\boldsymbol{k}) = \frac{\hbar^2}{2m} |\boldsymbol{k} + \boldsymbol{G}_n|^2 \tag{6.16}$$

Table 6.8 The compatibility relations for a BCC lattice.

(*a*) Between Γ and Δ.

Γ_1	Γ_2	Γ_{12}	Γ'_{15}	Γ'_{25}	Γ'_1	Γ'_2	Γ'_{12}	Γ_{15}	Γ_{25}
Δ_1	Δ_2	$\Delta_1\Delta_3$	$\Delta_2\Delta_5$	$\Delta_4\Delta_5$	Δ_2	Δ_4	$\Delta_2\Delta_4$	$\Delta_1\Delta_5$	$\Delta_3\Delta_5$

(*b*) Between Γ and Λ.

Γ_1	Γ_2	Γ_{12}	Γ'_{15}	Γ'_{25}	Γ'_1	Γ'_2	Γ'_{12}	Γ_{15}	Γ_{25}
Λ_1	Λ_2	Λ_3	$\Lambda_2\Lambda_3$	$\Lambda_1\Lambda_3$	Λ_2	Λ_1	Λ_3	$\Lambda_1\Lambda_3$	$\Lambda_2\Lambda_3$

(*c*) Between Γ and Σ.

Γ_1	Γ_2	Γ_{12}	Γ'_{15}	Γ'_{25}	Γ'_1	Γ'_2	Γ'_{12}	Γ_{15}	Γ_{25}
Σ_1	Σ_4	$\Sigma_1\Sigma_4$	$\Sigma_2\Sigma_3\Sigma_4$	$\Sigma_1\Sigma_2\Sigma_3$	Σ_2	Σ_3	$\Sigma_2\Sigma_3$	$\Sigma_1\Sigma_3\Sigma_4$	$\Sigma_1\Sigma_2\Sigma_4$

(*d*) Between Γ and Δ, Λ, Σ.

Γ_1	Γ_2	Γ_{12}	Γ'_{15}	Γ'_{25}	Γ'_1	Γ'_2	Γ'_{12}	Γ_{15}	Γ_{25}
Δ_1	Δ_2	$\Delta_1\Delta_3$	$\Delta_2\Delta_5$	$\Delta_4\Delta_5$	Δ_2	Δ_4	$\Delta_2\Delta_4$	$\Delta_1\Delta_5$	$\Delta_3\Delta_5$
Λ_1	Λ_2	Λ_3	$\Lambda_2\Lambda_3$	$\Lambda_1\Lambda_3$	Λ_2	Λ_1	Λ_3	$\Lambda_1\Lambda_3$	$\Lambda_2\Lambda_3$
Σ_1	Σ_4	$\Sigma_1\Sigma_4$	$\Sigma_2\Sigma_3\Sigma_4$	$\Sigma_1\Sigma_2\Sigma_3$	Σ_2	Σ_3	$\Sigma_2\Sigma_3$	$\Sigma_1\Sigma_3\Sigma_4$	$\Sigma_1\Sigma_2\Sigma_4$

where $\boldsymbol{G}_n$ is any reciprocal lattice vector. If we denote any reciprocal lattice point by $\boldsymbol{t}\mathbf{g}$ we have

$$E_t(\boldsymbol{k}) = \frac{\hbar^2}{2m}\,|\boldsymbol{k} + \boldsymbol{t}\cdot\mathbf{g}|^2. \tag{6.17}$$

The $\mathbf{g}$ matrix for the BCC lattice with cube edge a is given by

$$\mathbf{g} = \frac{2\pi}{a}\begin{pmatrix} 0 & 1 & 1 \\ 1 & 0 & 1 \\ 1 & 1 & 0 \end{pmatrix}$$

hence

$$\boldsymbol{t}\cdot\mathbf{g} = \frac{2\pi}{a}(t_1, t_2, t_3)\begin{pmatrix} 0 & 1 & 1 \\ 1 & 0 & 1 \\ 1 & 1 & 0 \end{pmatrix}$$

$$= \frac{2\pi}{a}\,[(t_2 + t_3), (t_1 + t_3), (t_1 + t_2)].$$

Put

$$t_2 + t_3 = n_1 \qquad t_1 + t_3 = n_2 \qquad t_1 + t_2 = n_3 \tag{6.18}$$

so that $t_1 = \frac{1}{2}(-n_1 + n_2 + n_3)$, $t_2 = \frac{1}{2}(n_1 - n_2 + n_3)$, $t_3 = \frac{1}{2}(n_1 + n_2 - n_3)$. Then

$$\boldsymbol{G}_n = \boldsymbol{t}\mathbf{g} = \frac{2\pi}{a}(n_1, n_2, n_3). \tag{6.19}$$

If we write the vector $\boldsymbol{k}$ in the form

$$\boldsymbol{k} = \frac{2\pi}{a}(\alpha, \beta, \gamma) \tag{6.20}$$

then equation (6.17) becomes

$$E_t(\boldsymbol{k}) = \frac{\hbar^2}{2m}\left(\frac{2\pi}{a}\right)^2 [(\alpha + n_1)^2 + (\beta + n_2)^2 + (\gamma + n_3)^2] \tag{6.21}$$

and defining

$$\lambda_E = \frac{2ma^2}{\hbar^2} E_t(\boldsymbol{k}) \tag{6.22}$$

we get

$$\lambda_E = \frac{2ma^2}{\hbar^2} E_t(\boldsymbol{k}) = (\alpha + n_1)^2 + (\beta + n_2)^2 + (\gamma + n_3)^2. \tag{6.23}$$

Here only those values of n are allowed which make $\boldsymbol{t}$ an integral vector. The wavefunctions are given by

$$\Psi_n(\boldsymbol{k}) = \exp\{(2\pi i/a)[(\alpha + n_1)x + (\beta + n_2)y + (\gamma + n_3)z]\}. \tag{6.24}$$

The $\boldsymbol{k}$ vectors for the various points and lines of symmetry are given in table 6.6(*b*).

(a) *Free-electron energy bands along the (001) or Δ axis.* The lowest energy lies at Γ and corresponds to $n = (0, 0, 0)$, that is

$$\lambda_{E_\Gamma} = 0.$$

The wavefunction is $\Psi = 1$ and is therefore the symmetry type Γ_1.

Starting from this point and moving along the Δ axis or the $(0, 0, 1)$ axis for which $\alpha = \beta = 0$, we find that the lowest band is characterized by $n_1 = n_2 = n_3 = 0$ and is just the parabola

$$\lambda_{E_{\Delta_1}} = \gamma^2.$$

The wavefunction is $\Psi = \exp[(2\pi i/a)\gamma z]$, $(0 < \gamma < 1)$, which is the Δ_1 state.

The next highest bands are those found by taking n_1, n_2, n_3 from the first reciprocal lattice. Taking n of type $(1, 0, 0)$ we find

$$t_1 = \tfrac{1}{2}(-n_1 + n_2 + n_3) = -\tfrac{1}{2}$$

which is not allowed since only those values of n are allowed which make $\boldsymbol{t}$ an integral vector. Taking n of type $(1, 1, 0)$ we have $t_1 = 0$, $t_2 = 0$, $t_3 = 1$. Thus the next highest bands are found by taking n of the type $(1, 1, 0)$. These give:

(i)
$$\lambda_{E_{\Lambda_2}} = (\gamma - 1)^2 + 1$$

for which n can take the following values: $n = (1, 0, \bar{1}); (\bar{1}, 0, \bar{1}); (0, 1, \bar{1}); (0, \bar{1}, \bar{1})$.

(ii)
$$\lambda_{E_{\Lambda_3}} = \gamma^2 + 2$$

for which n can take the following values: $n = (1, 1, 0); (\bar{1}, 1, 0); (1, \bar{1}, 0); (\bar{1}, \bar{1}, 0)$.

(iii)
$$\lambda_{E_{\Lambda_4}} = (\gamma + 2)^2 + 1$$

for which n can take the following values: $n = (1, 0, 1); (\bar{1}, 0, 1); (0, 1, 1); (0, \bar{1}, 1)$.

We note that each of these bands is four-fold degenerate since there are four possible choices of reciprocal lattice vectors which lead to the above expressions.

The next higher bands are obtained by taking n of the type $(2, 0, 0)$. These give:

$$\lambda_{E_{\Lambda_5}} = (\gamma - 2)^2$$

for which $n = (0, 0, \bar{2})$;

$$\lambda_{E_{\Lambda_6}} = 4 + \gamma^2$$

for which n can take the following values: $n = (2, 0, 0)$; $(\bar{2}, 0, 0)$; $(0, 2, 0)$; $(0, \bar{2}, 0)$;

$$\lambda_{E_{\Lambda_7}} = (\gamma + 2)^2$$

for which $n = (0, 0, 2)$.

Here we note that

$$\lambda_{E_{\Lambda_5}} \quad \text{and} \quad \lambda_{E_{\Lambda_7}}$$

are non-degenerate whereas

$$\lambda_{E_{\Lambda_6}}$$

is four-fold degenerate.

Proceeding in this manner we can get to higher and higher bands. Let us now see what happens at the end point H of the Δ axis for which $\alpha = \beta = 0$

and $\gamma = 1$. Here

$$\lambda_{E_H} = n_1^2 + n_2^2 + (1 + n_3)^2$$

and when n is of the type $(0, 0, 0)$ we get

$$\lambda_{E_{H_1}}.$$

When n is of the type $(1, 1, 0)$ we get

$$\lambda_{E_{H_2}} = 1$$

corresponding to $n = (1, 0, \bar{1}); (\bar{1}, 0, \bar{1}); (0, 1, \bar{1}); (0, \bar{1}, \bar{1})$. Similarly

$$\lambda_{E_{H_3}} = 3$$

for which n can take the following values: $n = (1, 1, 0); (\bar{1}, 1, 0); (1, \bar{1}, 0); (\bar{1}, \bar{1}, 0)$. Also

$$\lambda_{E_{H_4}} = 5$$

for which n can take the values: $n = (1, 0, 1); (\bar{1}, 0, 1); (0, 1, 1); (0, \bar{1}, 1)$. When n is of the type $(2, 0, 0)$ we get

$$\lambda_{E_{H_5}} = 1$$

corresponding to $n = (0, 0, \bar{2})$ and

$$\lambda_{E_{H_6}} = 5$$

for which n can take the following values: $n = (2, 0, 0); (\bar{2}, 0, 0); (0, 2, 0); (0, \bar{2}, 0)$.

$$\lambda_{E_{H_7}} = 9$$

corresponds to $n = (0, 0, 2)$.

In this way we can proceed to higher and higher bands. We see that for $\lambda_{E_H} = 1$ we have three bands:

$$\lambda_{E_{H_1}} \qquad \lambda_{E_{H_2}} \quad \text{and} \quad \lambda_{E_{H_5}}.$$

The first and the third are non-degenerate whereas the second is four-fold degenerate. Similarly for $\lambda_{E_H} = 3$ we have two bands,

$$\lambda_{E_{H_3}}$$

which is four-fold degenerate and

$$\lambda_{E_{H_8}}$$

which is also four-fold degenerate corresponding to n of the type: $(1, 1, \bar{2}); (\bar{1}, 1, \bar{2}); (1, \bar{1}, \bar{2}); (\bar{1}, \bar{1}, \bar{2})$.

Coming back to the point Γ, let us see what values λ_{E_Γ} takes for those values of n mentioned above. At Γ(0, 0, 0) we have

$$\lambda_{E_\Gamma} = n_1^2 + n_2^2 + n_3^2$$

and when $n_1 = n_2 = n_3 = 0$ we get

$$\lambda_{E_{\Gamma_1}} = 0$$

the lowest energy state. When n is of the type (1, 1, 0) we get

$$\lambda_{E_{\Gamma_2}} = 2$$

for which n can take the following values: $(1, 1, 0)$; $(\bar{1}, 1, 0)$; $(1, \bar{1}, 0)$; $(\bar{1}, \bar{1}, 0)$; $(1, 0, 1)$; $(\bar{1}, 0, 1)$; $(1, 0, \bar{1})$; $(\bar{1}, 0, \bar{1})$; $(0, 1, 1)$; $(0, \bar{1}, 1)$; $(0, 1, \bar{1})$; $(0, \bar{1}, \bar{1})$. That is,

$$\lambda_{E_{\Gamma_2}}$$

is 12-fold degenerate. When n is of the type (2, 0, 0) we get

$$\lambda_{E_{\Gamma_3}} = 4$$

for which n can take the following values: $n = (2, 0, 0)$; $(0, 2, 0)$; $(0, 0, 2)$; $(\bar{2}, 0, 0)$; $(0, \bar{2}, 0)$; $(0, 0, \bar{2})$. That is

$$\lambda_{E_{\Gamma_3}}$$

is six-fold degenerate.

Using the above information we can sketch the free-electron energy bands for a BCC lattice along the Δ axis as shown in figure 6.5.

(b) *Free-electron energy bands along the (111) or Λ axis.* We proceed in the same manner as for the Δ axis.

We have for the Γ point

$$\lambda_{E_\Gamma} = n_1^2 + n_2^2 + n_3^2.$$

For n of the type (0, 0, 0) we get

$$\lambda_{E_{\Gamma_1}} = 0$$

which is the lowest energy state. For n of the type (1, 1, 0) we get

$$\lambda_{E_{\Gamma_2}} = 2$$

for which n can take the following values: $n = (1, 1, 0)$; $(\bar{1}, 1, 0)$; $(1, \bar{1}, 0)$; $(\bar{1}, \bar{1}, 0)$; $(1, 0, 1)$; $(\bar{1}, 0, 1)$; $(1, 0, \bar{1})$; $(\bar{1}, 0, \bar{1})$; $(0, 1, 1)$; $(0, \bar{1}, 1)$; $(0, 1, \bar{1})$; $(0, \bar{1}, \bar{1})$. That is

$$\lambda_{E_{\Gamma_2}} = 2$$

is 12-fold degenerate.

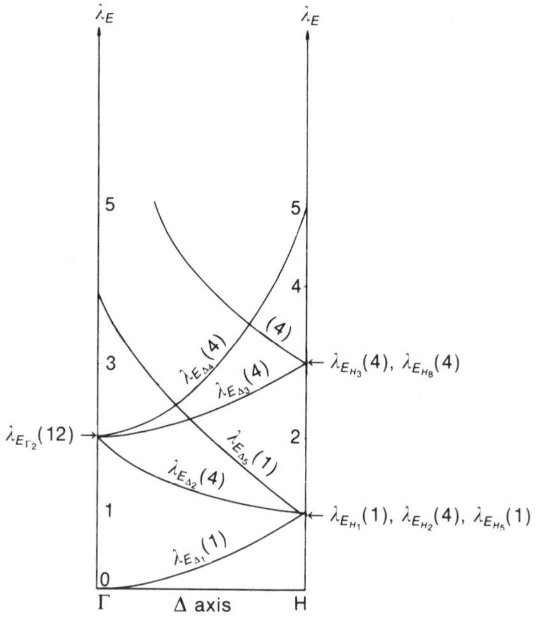

Figure 6.5 The free-electron energy bands for a BCC lattice along ΓH, i.e. the Δ axis. The degeneracy of each band is shown in parentheses.

In a similar manner it is easy to see that

$$\lambda_{E_{\Gamma_3}} = 4$$

corresponding to n of type $(2, 0, 0)$ is six-fold degenerate.

Now, moving along the Λ axis where $\alpha = \beta = \gamma \leqslant \frac{1}{2}$, we have

$$\lambda_{E_\Lambda} = (\gamma + n_1)^2 + (\gamma + n_2)^2 + (\gamma + n_3)^2.$$

For $n = (0, 0, 0)$ we get

$$\lambda_{E_{\Lambda_1}} = 3\gamma^2.$$

For n of the type $(1, 1, 0)$ we get

$$\lambda_{E_{\Lambda_2}} = 2(\gamma - 2)^2 + \gamma^2$$

corresponding to $n = (\bar{1}, \bar{1}, 0); (\bar{1}, 0, \bar{1}); (0, \bar{1}, \bar{1})$; that is, three-fold degenerate;

$$\lambda_{E_{\Lambda_3}} = 3\gamma^2 + 2$$

corresponding to $n = (\bar{1}, 1, 0); (1, \bar{1}, 0); (0, \bar{1}, 0); (0, 1, \bar{1}); (\bar{1}, 0, 1); (1, 0, \bar{1})$, that is, six-fold degenerate;

$$\lambda_{E_{\Lambda_4}} = 2(\gamma + 1)^2 + \gamma^2$$

corresponding to $n = (1, 1, 0); (1, 0, 1); (0, 1, 1)$; that is, three-fold degenerate.

Similarly for n of type $(2, 0, 0)$ it is easy to see that we get:

$$\lambda_{E_{\Lambda_5}} = (\gamma - 2)^2 + (2\gamma^2)$$

which is three-fold degenerate and

$$\lambda_{E_{\Lambda_6}} = (\gamma + 2)^2 + (2\gamma^2)$$

which is also three-fold degenerate.

Finally at the point P where $\alpha = \beta = \gamma = \frac{1}{2}$ and

$$\lambda_{E_P} = (\tfrac{1}{2} + n_1)^2 + (\tfrac{1}{2} + n_2)^2 + (\tfrac{1}{2} + n_3)^2$$

we have the following bands:

$$\lambda_{E_{P_1}} = \tfrac{3}{4}$$

corresponding to $n = (0, 0, 0)$;

$$\lambda_{E_{P_2}} = \tfrac{3}{4}$$

corresponding to $n = (\bar{1}, \bar{1}, 0); (\bar{1}, 0, \bar{1}); (0, \bar{1}, \bar{1})$; that is, three-fold degenerate;

$$\lambda_{E_{P_3}} = 2\tfrac{3}{4}$$

corresponding to $n = (\bar{1}, 1, 0); (1, \bar{1}, 0); (0, \bar{1}, 1); (0, 1, \bar{1}); (\bar{1}, 0, 1); (1, 0, \bar{1})$; that is, six-fold degenerate;

$$\lambda_{E_{P_4}} = 4\tfrac{3}{4}$$

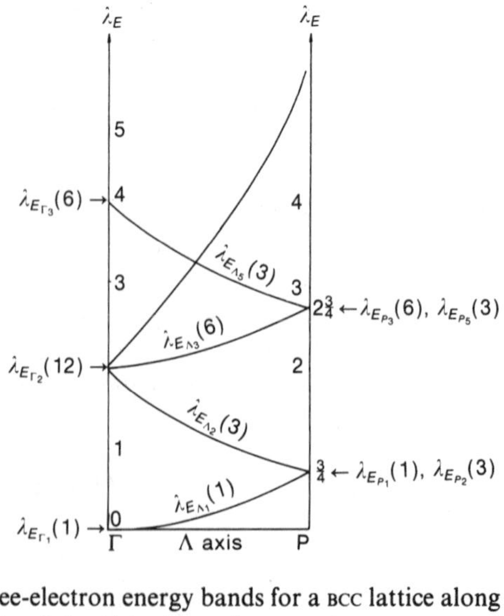

Figure 6.6 The free-electron energy bands for a BCC lattice along the (111) or Λ axis. The degeneracy of each band is shown in parentheses.

corresponding to $n = (1, 1, 0); (1, 0, 1); (0, 1, 1)$; that is, three-fold degenerate;

$$\lambda_{E_{P_5}} = 2\tfrac{3}{4}$$

corresponding to $n = (\bar{2}, 0, 0); (0, \bar{2}, 0); (0, 0, \bar{2})$; that is, three-fold degenerate;

$$\lambda_{E_{P_6}} = 6\tfrac{1}{4}$$

corresponding to $n = (2, 0, 0); (0, 2, 0); (0, 0, 2)$; that is, three-fold degenerate.

Using the above information we can sketch the free-electron energy bands along the Λ axis as shown in figure 6.6. It is left as an exercise for readers to determine the free-electron energy bands along the $(1, 1, 0)$ or Σ axis.

This is as far as we can go in this section. It should be emphasized that in our treatment of labelling the energy bands we have assumed throughout that the crystal potential $V(\boldsymbol{r}) = 0$. This is known as the free-electron approximation. When $V(\boldsymbol{r}) \neq 0$, some modifications have to be made in going from the case $V(\boldsymbol{r}) = 0$ to $V(\boldsymbol{r}) \neq 0$. We shall not discuss these changes here but refer readers to the references given earlier.

6.4 Symmetry and lattice vibration spectra (see [13, 1, 55])

We have seen earlier (Chapter 4) that attached to every lattice point in a crystal there was a basis of atoms (or ions or molecules or radicals). At any temperature these bases at the respective lattice sites are not at rest but vibrate about their mean position. In the simplest case, if the basis atoms were to be considered as perfectly free, then the vibrations of the atoms arising from thermal energy could be described by a simple harmonic type of vibration. The atoms would vibrate freely in the x or y or z direction and hence would have three degrees of freedom. The atoms in a solid are, however, not free but bound to their neighbours by the crystal lattice. In the study of lattice vibrations it is common practice to depict the lattice by elastic springs as shown in figure 6.7.

This means that the thermal vibrations of the atoms are more complex, being rather like coupled simple harmonic oscillators since each displaced atom would pull its neighbour along. Consequently, neighbouring atoms would have a complex displacement pattern, either in the plane of the lattice or perpendicular to it. For a solid composed of N (similar) atoms we therefore expect $3N$ degrees of freedom and we say that the solid could vibrate in $3N$

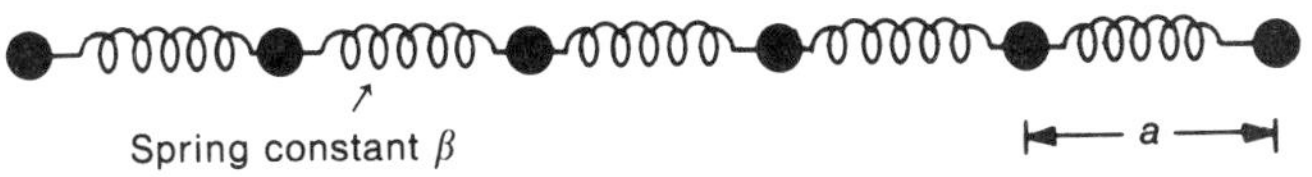

Figure 6.7 Schematic representation of a linear monatomic lattice. The basis atoms (●) are joined to each other by elastic springs. Here a is the lattice constant.

independent or normal modes of vibration. A normal mode is defined as consisting of a vibration of all the atoms in the crystal, each with its own amplitude, direction and phase but all with the same frequency.

For a linear monatomic lattice of the type shown in figure 6.7 the normal modes of vibration of the system are given by (see Kittel [34])

$$\omega = \pm\left(\frac{4\beta}{m}\right)^{1/2} \sin(\tfrac{1}{2}ka)$$

where β is the spring constant, m the mass of atom, a the lattice constant (i.e. separation between the atoms) and $\boldsymbol{k}$ the wavevector of the lattice wave with magnitude $2\pi/\lambda$. The functional relation between the frequency ω (or energy $E = h\omega$) and $\boldsymbol{k}$ is called the dispersion relation, and for the model considered this is shown in figure 6.8(a). This mode of vibration is termed the longitudinal acoustic mode since the lattice wave which is propagated on the line of atoms is in the horizontal direction. Furthermore, the vibrations are of relatively low frequency, very much like those acoustic vibrations one comes across in the theory of sound. For a 3D case (e.g. for a cubic crystal with one atom/unit cell) there will be three degrees of freedom, so that in addition to the longitudinal mode there will be two transverse doubly degenerate acoustic modes, in which the displacements of the atoms are perpendicular to the wave propagation (figure 6.8(b)). It is worthwhile mentioning at this point that for crystals the wavevector $\pm\boldsymbol{k}$ is restricted to the first Brillouin zone.

So far we have looked at only the low-frequency acoustic modes. As we go to higher and higher frequencies the mode of vibrations begins to get more complex and in fact when the wavelength of the modes become comparable to the interatomic spacings the departure is very great and the mode is termed 'optic mode', being similar to the motion of ions being excited by electromagnetic waves lying in the visible and infrared regions. This case

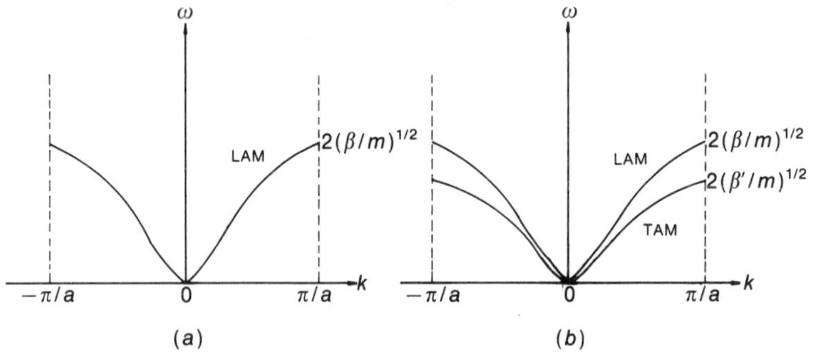

Figure 6.8 The vibration spectrum for modes propagating in a [100] direction for (a) a linear monatomic lattice and (b) for a simple cubic crystal. The transverse mode is doubly degenerate. LAM = longitudinal acoustic mode, TAM = transverse acoustic mode.

is illustrated by a linear diatomic lattice where the dispersion relation takes the form (see Kittel [34])

$$\omega^2 = \beta\left(\frac{1}{m}+\frac{1}{M}\right) \pm \beta\left[\left(\frac{1}{m}+\frac{1}{M}\right)^2 - \frac{4\sin^2 ka}{mM}\right]^{1/2}.$$

Here m and M are the masses of the two atoms.

Figure 6.9 shows a plot of this relation and we note that there are now two branches corresponding to a single value of k. These two branches are labelled as acoustic (low-frequency) and optic (high-frequency) modes.

The energy of the lattice modes or lattice waves, measured in quantized units, is known as phonons—just as we define photons as quantized electromagnetic waves—so that figures 6.8 and 6.9 are also known as phonon dispersion curves.

It is not our intention to discuss the theory of lattice vibrations here. Readers who are interested will find ample discussions on this topic in [13, 40, 55]. In this section we shall discuss only the applications of symmetry principles in determining the infrared and optical (Raman) lattice spectra of insulating crystals.

6.4.1 *Symmetry principles involved in infrared and Raman optical transition processes*

As explained earlier, in the theory of lattice vibrations, each lattice wave or mode of vibration is assigned a wavevector $\boldsymbol{k}$ and that for crystals $\boldsymbol{k}$ is restricted to the first Brillouin zone. Thus a wavevector $\boldsymbol{k}_q$ indicates the direction of propagation (q denotes direction (100), (110), (111), etc) of a lattice mode having a wavelength λ_q such that $|\boldsymbol{k}_q| = 2\pi/\lambda_q$. It follows, therefore, that for a solid there will be a number of energy levels (phonon states) arising from different values of $\boldsymbol{k}$. The actual values of these phonon energy states have of course to be calculated by writing down and solving the quantum mechanical equations of motion [55]. Once this has been done

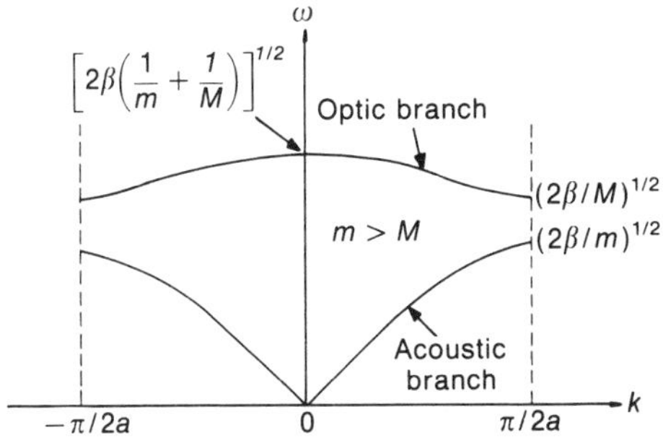

Figure 6.9 Dispersion relation for a linear diatomic lattice.

and the various phonon states determined, the problem which remains to be solved is to determine the restrictions imposed, via symmetry analysis, on the infrared and Raman active transitions between the various available (calculated) levels. The results obtained are particularly useful in the interpretation of optical spectra of crystals and also in studying the actual spacings between two vibrational states of the crystal.

We shall illustrate the procedure by considering the symmetries of the various vibrational modes of a crystal having a simple cubic lattice. The Brillouin zone for such a crystal is shown in figure 4.9(*a*). The symbols Γ, X, R, M and Σ, Δ, Λ, T, Z, S corresponds to $\boldsymbol{k}$ vectors which possess special symmetry (see §6.3). The point with $\boldsymbol{k} = 0$ (i.e. the Γ point) possesses the full symmetry of the cubic group. We take this group to be O_h. Its character table is given in table 6.9. The last column indicates the transformation properties of the (basis) functions x, y, z or x^2, y^2, z^2, etc.

The normal modes of vibration at $\boldsymbol{k} = 0$ will then be labelled by one or other of the irreducible representations of the group O_h, that is Γ_1, Γ_2, Γ_{12}, Γ'_{15}, Γ'_{25}, etc, in the notation of Bouckaert *et al* [47]. For simplicity let us assume a set of three allowed modes of vibration designated by the symbols Γ'_2, Γ'_{15} and Γ'_{25} (figure 6.10). We now wish to determine the transitions between these levels in the presence of electromagnetic radiation that will give rise to Raman or infrared (absorption or emission) spectra.

The rule stating which transitions are allowed and which are forbidden is determined by finding the transition matrix element corresponding to the particular process in question. If $\hat{O}$ is the operator representing the transition process (infrared or Raman) between two levels with wavefunctions denoted by Ψ_a and Ψ_b, then quantum mechanics tells us that if the matrix element of $\hat{O}$

$$\langle \Psi_b | \hat{O} | \Psi_a \rangle = 0 \quad \text{the transition is forbidden} \tag{6.25a}$$

but if

$$\langle \Psi_b | \hat{O} | \Psi_a \rangle \neq 0 \quad \text{the transition is allowed.} \tag{6.25b}$$

Translated into group theoretical terms, a transition is allowed if $\langle \Psi_b | \hat{O} | \Psi_a \rangle$ contains the totally symmetrical representation of the group, where Ψ_a and Ψ_b now denote the initial and final states labelled by the irreducible representation of the symmetry group of the crystal.

The operator $\hat{O}$ corresponding to the infrared transition process is the dipole moment operator $e\boldsymbol{r}$ which transforms like the basis functions x, y, z in the group, while the operator corresponding to Raman transitions is the symmetric polarizability tensor α_{ij} which transforms as the symmetric product of $\boldsymbol{r}$ with itself, that is it transforms like the basis functions x^2, y^2, z^2, xy, yz, zx or their linear combinations.

Let us now be more specific and consider whether an infrared transition is possible from the state labelled Γ'_{25} to the lower state labelled Γ'_2 (figure

Dirac $\Gamma_{15} \times \Gamma'_{25}$ 9 1 −1 −1 0 9 1 −1 −1 0 $= \Gamma'_2 + \Gamma'_{12} + \Gamma_{25} + \Gamma_{15}$

Table 6.9 The character table and basis functions at $\boldsymbol{k} = 0$ for the cubic group O_h in the notation of Bouckaert *et al* [47].

Γ	E	$3C_4^2$	$6C_4$	$6C_2$	$8C_3$	I	$3\sigma_h$	$6S_4$	$6\sigma_d$	$8S_6$	Basis functions
Γ_1	1	1	1	1	1	1	1	1	1	1	$x^2 + y^2 + z^2$
Γ_2	1	1	−1	−1	1	1	1	−1	−1	1	$(2z^2 - x^2 - y^2, x^2 - y^2)$
Γ_{12}	2	2	0	0	−1	2	2	0	0	−1	
Γ'_{15}	3	−1	1	−1	0	3	−1	1	−1	0	
Γ'_{25}	3	−1	−1	1	0	3	−1	−1	1	0	(xz, yz, xy)
Γ'_1	1	1	1	1	1	−1	−1	−1	−1	−1	
Γ'_2	1	1	−1	−1	1	−1	−1	1	1	−1	
Γ'_{12}	2	2	0	0	−1	−2	−2	0	0	1	
Γ_{15}	3	−1	1	−1	0	−3	1	−1	1	0	(x, y, z)
Γ_{25}	3	−1	−1	1	0	−3	1	1	−1	0	

Raman $\Gamma'_{25} \otimes \Gamma_2$ 3 −1 1 −1 0 3 −1 1 −1 0 $= \Gamma_{12}$

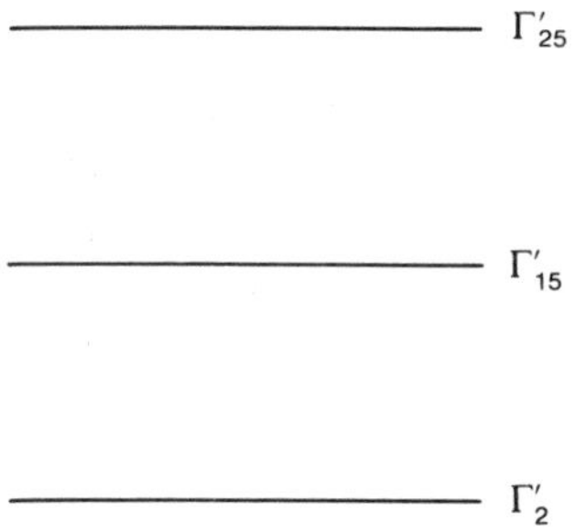

Figure 6.10 A set of allowed modes of vibration.

6.10). This means determining the matrix element

$$\langle \Gamma_2' | e\boldsymbol{r} | \Gamma_{25}' \rangle \quad \text{or} \quad \langle \Gamma_2' | \boldsymbol{r} | \Gamma_{25}' \rangle.$$

Looking at table 6.9, we see that the operator $\boldsymbol{r}$ transforms as the representation Γ_{15}' in the group so that $\boldsymbol{r}|\Gamma_{25}'\rangle$ transforms as

$$\Gamma_{15} \otimes \Gamma_{25}' = \Gamma_2' + \Gamma_{12}' + \Gamma_{25} + \Gamma_{15}. \tag{6.26}$$

Here we have used equation (3.13). Since the RHS of (6.26) contains Γ_2' it follows immediately that such a transition would be allowed. Note, however, that a transition from the state labelled Γ_{25}' to the state labelled Γ_{15}' is forbidden for infrared transition since Γ_{15}' is not contained in the RHS of (6.26).

The same principle outlined above is applicable in determining whether a transition between two levels is Raman active. Here one uses equations (6.25) but now $\hat{O}$ transforms like the functions x^2, y^2, z^2, xy, yz, zx and so on, as mentioned earlier, so that if Ψ_b in (6.25) transforms like any of these functions or any linear combinations of them, the transition is allowed Raman active, otherwise it is forbidden.

Readers should note that a full discussion on infrared and Raman active spectra using symmetry principles would provide a good framework for understanding experimental results. The treatment given here is of necessity rather oversimplified, so that when dealing with actual crystal space groups and their representations, where other influences come into play, the applications of the symmetry principles go beyond what has been described, but this is really something we cannot deal with at present and readers who are interested should consult the references given earlier.

6.5 Factorization of secular determinants in band structure calculations: a practical example

(See [48, 51–54, 56, 17] and Harisson W A 1965 *Pseudopotential in the Theory of Metals* (New York: Benjamin), and Bassani F 1966 *Semiconductors and Semimetals* vol. 1 (New York: Academic Press).)

In this section we shall show by means of examples the usefulness of symmetry analysis in energy band calculations. In particular we shall refer to the literature on the calculation of energy band structures of group IV and III–V semiconductors and show, by using the symmetry group of the crystal, how to factorize the $n \times n$ secular determinants formed by using the data given by Cohen and Bergstresser [48], thereby permitting easy numerical solution.

6.5.1 *Basic principles*

In all energy band calculations there are two main problems [52], namely determination of a suitable crystal potential and numerical solution of the crystal wave equations. In recent work the local pseudopotential model has been found to provide an adequate description of the band structure of several solids (see references given above). The success of the pseudopotential approximation lies in the fact that very few parameters are needed to obtain a reasonable fit to experimental data. In fact Cohen and Bergstresser [48] have shown that within the framework of the local pseudopotential model it should be possible to obtain adequate descriptions of the band structures of various semiconductors from secular determinants of order as low as 20×20. In what follows, we shall first present the essential mathematical features of the pseudopotential method that are required to build up the $n \times n$ secular determinants for group IV and III–V semiconductors, and then consider the method of factorizing the secular determinants formed by using the pseudopotential form factors of Cohen and Bergstresser.

6.5.2 *Pseudopotential formalism of band structure*

The local pseudopotential method has been treated in great detail by Harrison (see above). The pseudopotential Schrödinger equation for an electron in a crystal can be written as

$$\left(-\frac{\hbar^2}{2m}\nabla^2 + V_\mathrm{p}\right)\phi_{nk}(\boldsymbol{r}) = E_n(\boldsymbol{k})\phi_{nk}(\boldsymbol{r}). \tag{6.27}$$

Here the pseudopotential

$$V_\mathrm{p} = V(\boldsymbol{r}) + V_\mathrm{R} \tag{6.28}$$

where $V(\boldsymbol{r})$ is the periodic crystal potential and V_R has the character of a repulsive potential which cancels in part the large attractive Coulomb potential such that in the region of the core the higher Fourier coefficients of the pseudopotential V_p are small enough to be neglected in the first-order approximation. The functions $\phi_{nk}(\boldsymbol{r})$ are smooth pseudowavefunctions and are free of the convergence difficulties of the corresponding Bloch functions

$$\Psi_{nk}(r) = \exp(\mathrm{i}\boldsymbol{k}\cdot\boldsymbol{r}) \tag{6.29}$$

which are obtained for the Schrödinger equation with a perfectly periodic potential $V(\boldsymbol{r})$. The matrix elements of V_p can be written as a product of a structure factor $S(\boldsymbol{K})$ and a pseudopotential form factor $V_{\boldsymbol{K}_i}$, where the $\boldsymbol{K}$ are reciprocal lattice vectors [48]. Since the potential for the zincblende-type crystals of the III–V group is not invariant under inversion (with $\boldsymbol{r} \to -\boldsymbol{r}$), V_p is generally expressed as the sum of two potentials, one of which is symmetric and the other antisymmetric with respect to inversion.

We thus have

$$V_p = \sum_{|\boldsymbol{K}_i|} [S^s(\boldsymbol{K}_i) V^s_{\boldsymbol{K}_i} + iS^a(\boldsymbol{K}_i) V^a_{\boldsymbol{K}_i}] \exp[i(\boldsymbol{K}_i \cdot \boldsymbol{r})]. \tag{6.30}$$

If we choose the origin of the coordinates to be midway between the two atoms of the diamond or zincblende unit cell, then

$$S^s(\boldsymbol{K}) = \cos \boldsymbol{K} \cdot \boldsymbol{\tau} \qquad S^a(\boldsymbol{K}) = \sin \boldsymbol{K} \cdot \boldsymbol{\tau} \tag{6.31}$$

where $\boldsymbol{\tau} = a(\frac{1}{8}, \frac{1}{8}, \frac{1}{8})$ is the glide translation of the crystal space group and a is the cube edge. The atomic potentials $V^s_{\boldsymbol{K}}$ and $V^a_{\boldsymbol{K}}$ are given in table 2 of Cohen and Bergstresser [48] for various group IV and III–V semiconductors. The first three reciprocal lattice vectors with non-zero structure factors are $\langle 111 \rangle$, $\langle 220 \rangle$ and $\langle 311 \rangle$ with $|\boldsymbol{K}|^2 = 3$, 8 and 11 respectively. We can now solve the Schrödinger equation (6.27) by determining the roots of the secular equation [54]

$$\det |[(\boldsymbol{K} - \boldsymbol{k})^2 - E]\delta_{\boldsymbol{K}\boldsymbol{K}'} + V_{\boldsymbol{K}-\boldsymbol{K}'}| = 0 \tag{6.32}$$

where the $\boldsymbol{K}$ are measured in units of $2\pi/a$ and the $\boldsymbol{k}$ are the reduced wavevectors lying in the first Brillouin zone. The $V_{\boldsymbol{K}-\boldsymbol{K}'}$ are non-zero only for $|\boldsymbol{K} - \boldsymbol{K}'|^2 = 3$, 8 or 11. In the next section we shall show that, for wavevectors $\boldsymbol{k}$ of high symmetry, equation (6.32) can be readily factorized permitting easy numerical solutions.

6.5.3 Symmetry properties of the energy bands of diamond and zincblende lattices

The group IV and III–V semiconductors have the diamond and zincblende structures respectively. The symmetry properties of these lattices have been discussed in great detail in [9, 18, 47].

The diamond structure belongs to the crystallographic space group O_h^7 and consists of two interpenetrating FCC lattices displaced along the body diagonal by one-fourth of its length. The zincblende structure belongs to the crystallographic space group T_d^2 and the crystal lattice is similar to diamond except that two different kinds of atom form the two FCC sublattices. The zincblende lattice, therefore, does not possess a centre of inversion symmetry. The point group T_d of the zincblende lattice has 24 symmetry elements and is a subgroup of the diamond group which has, in addition to the 24 symmetry elements of T_d, their product with the inversion element I which gives the full octahedral group O_h consisting of 48 elements. As the two structures have the FCC translation symmetry the Brillouin zone is the same for the two cases. This zone is shown in figure 4.19(c). The points and lines of symmetry are given in table 6.10.

If the symmetry group $G(\boldsymbol{k})$ of the wavevector $\boldsymbol{k}$ is known it is possible to predict not only the form of the various crystal wavefunctions but also the overall form of the various energy bands in the reduced zone. However, the relative order of the eigenstates at a given point can be determined only by solving the crystal wave equation numerically. For high symmetry points this can be done quite easily by factorizing the secular determinant (6.32) by means of orthogonal matrices as shown below.

Table 6.10 The important points and lines of symmetry for the Brillouin zone of an FCC lattice (see [1, 9]).

Symmetry symbol	$\boldsymbol{k}$
Γ	$\frac{2\pi}{a}(0, 0, 0)$
Λ	$\frac{2\pi}{a}(\alpha, \alpha, \alpha)$
L	$\frac{2\pi}{a}(\frac{1}{2}, \frac{1}{2}, \frac{1}{2})$
Δ	$\frac{2\pi}{a}(0, \beta, 0)$
X	$\frac{2\pi}{a}(0, 1, 0)$
Σ	$\frac{2\pi}{a}(\alpha, \alpha, 0)$
W	$\frac{2\pi}{a}(\frac{1}{2}, 1, 0)$
K	$\frac{2\pi}{a}(\frac{3}{4}, \frac{3}{4}, 0)$
Z	$\frac{2\pi}{a}(\alpha, 1, 0)$
S	$\frac{2\pi}{a}(\alpha, 1, \alpha)$
U	$\frac{2\pi}{a}(\frac{1}{2}, 1, \frac{1}{2})$
Q	$\frac{2\pi}{a}(\frac{1}{2}, \frac{1}{2} + \alpha, \frac{1}{2} - \alpha)$

6.5.4 *Method of constructing orthogonal matrices and factorization of secular determinants*

In the following treatment, $\boldsymbol{K} \equiv (K_1K_2K_3)$, such as (111), (200), etc, will be used to represent the plane wave

$$\exp[(2\pi i/a)(K_1x + K_2y + K_3z)] \tag{6.33}$$

with

$$|\boldsymbol{K}|^2 = K_1^2 + K_2^2 + K_3^2 \tag{6.34}$$

while $\langle K_1K_2K_3 \rangle$ will represent an ensemble of waves $(K_1K_2K_3)$. For example $\langle 111 \rangle$ will denote the ensemble of eight waves (111), $(1\bar{1}\bar{1})$, $(\bar{1}1\bar{1})$, $(\bar{1}\bar{1}1)$, $(\bar{1}\bar{1}\bar{1})$, $(\bar{1}11)$, $(1\bar{1}1)$ and $(11\bar{1})$, with $|\boldsymbol{K}|^2 = 3$. The physical significance of $|\boldsymbol{K}|^2$ is that for an empty lattice it represents the energy to within a factor of $4\pi^2$; in fact

$$E = \left(\frac{2\pi}{a}\right)^2 (K_1^2 + K_2^2 + K_3^2). \tag{6.35}$$

The bases states used to form the secular determinant (6.32) consist of plane waves with wavevector $(\boldsymbol{K} + \boldsymbol{k})$. In the first approximation all vectors $\boldsymbol{K}$, such that $(\boldsymbol{K} + \boldsymbol{k})^2 \leqslant E_1$ with $E_1 = 7$, form the bases [48]. The second-order contribution is given by $E_1 < (\boldsymbol{K} + \boldsymbol{k})^2 \leqslant E_2$ with $E_2 = 21$. The method of selecting $\boldsymbol{K}$ is as follows.

Let plane waves be represented by

$$E = \exp[(2\pi i/a)(\boldsymbol{K} \cdot \boldsymbol{r})].$$

Now operate on this with all the operators of the diamond or zincblende lattice. If it is transformed into another real wave then this set forms a basis. If, however, it is transformed into an imaginary wave, then this set has to be rejected because the crystal potential is developed in terms of real and not imaginary waves.

To illustrate the general method of constructing orthogonal matrices let us consider a concrete example for the diamond and zincblende structures. For the symmetry point Γ it is easily seen using the above method of selecting $\boldsymbol{K}$ that only the sets $\langle 000 \rangle$, $\langle 111 \rangle$ and $\langle 200 \rangle$ are permitted to form the bases in the first-order approximation (i.e. $E_1 \leqslant 7$). Now the number of plane waves corresponding to the set $\langle 000 \rangle$, $\langle 111 \rangle$ and $\langle 200 \rangle$ are 1, 8 and 6 respectively. Thus in the first order we should be involved with a 15×15 matrix. The irreducible representations for the plane waves $\langle 000 \rangle$, $\langle 111 \rangle$ and $\langle 200 \rangle$ are given in table 6.11 for the diamond and zincblende structures.

Using the method described in §3.4 it is easy to see that the ensemble of 15 waves for the diamond structure should form the bases of the reduced representation

$$\Gamma = 2\Gamma_1 + 2\Gamma_2' + 2\Gamma_{25}' + \Gamma_{15} + \Gamma_{12}'. \tag{6.36}$$

Table 6.11 Irreducible representations for plane waves at symmetry point Γ in diamond and zincblende structures.

$\langle \boldsymbol{K} \rangle$	$\lvert \boldsymbol{K} \rvert^2$	No. of plane waves	Diamond Γ_1	Diamond Γ_2'	Diamond Γ_{12}'	Diamond Γ_{15}	Diamond Γ_{25}'	Zincblende Γ_1	Zincblende Γ_{12}	Zincblende Γ_{15}
$\langle 000 \rangle$	0	1	1	–	–	–	–	1	–	–
$\langle 111 \rangle$	3	8	1	1	–	1	1	2	–	2
$\langle 200 \rangle$	4	6	–	1	1	–	1	1	1	1
Degeneracy of IR			1	1	2	3	3	1	2	3

The 15 × 15 secular determinant $D_d(\Gamma)$ will thus factorize for this structure at the symmetry point Γ as follows:

$$
\begin{array}{cccccccccc}
\Gamma_1 & \Gamma_2' & \Gamma_{25}' & \Gamma_{25}' & \Gamma_{25}' & \Gamma_{15} & \Gamma_{15} & \Gamma_{15} & \Gamma_{12}' & \Gamma_{12}'
\end{array}
$$
$$
\begin{vmatrix}
\begin{vmatrix} a_{11} & a_{12} \\ a_{21} & a_{22} \end{vmatrix} &&&&&&&&& \\
& \begin{vmatrix} b_{11} & b_{12} \\ b_{21} & b_{22} \end{vmatrix} &&&&&&&& \\
&& \begin{vmatrix} c_{11} & c_{12} \\ c_{21} & c_{22} \end{vmatrix} &&&&&&& \\
&&& \begin{vmatrix} c_{11} & c_{12} \\ c_{21} & c_{22} \end{vmatrix} &&&&&& \\
&&&& \begin{vmatrix} c_{11} & c_{12} \\ c_{21} & c_{22} \end{vmatrix} &&&&& \\
&&&&& \lvert d_{11} \rvert &&&& \\
&&&&&& \lvert d_{11} \rvert &&& \\
&&&&&&& \lvert d_{11} \rvert && \\
&&&&&&&& \lvert e_{11} \rvert & \\
&&&&&&&&& \lvert e_{11} \rvert
\end{vmatrix}. \tag{6.37}
$$

That is, we should now break the large 15 × 15 determinant into:

one second-order determinant (Γ_1 with elements a_{mn}),
one second-order determinant (Γ_2' with elements b_{mn}),
three equal second-order determinants (Γ_{25}' with elements c_{mn}),
three equal first-order determinants (Γ_{15} with elements d_{11}), and
two equal first-order determinants (Γ_{12}' with elements e_{11}).

Similarly, for the zincblende structure the ensemble of 15 waves forms the bases of the reduced representation

$$\Gamma = 4\Gamma_1 + 3\Gamma_{15} + \Gamma_{12} \tag{6.38}$$

and as such the 15×15 determinant $D_{zb}(\Gamma)$ in this case will factorize as follows:

$$\begin{vmatrix}
\overbrace{\begin{vmatrix} f_{11} & f_{12} & f_{13} & f_{14} \\ f_{21} & f_{22} & f_{23} & f_{24} \\ f_{31} & f_{32} & f_{33} & f_{34} \\ f_{41} & f_{42} & f_{43} & f_{44} \end{vmatrix}}^{\Gamma_1} & & & & & \\
& \overbrace{\begin{vmatrix} g_{11} & g_{12} & g_{13} \\ g_{21} & g_{22} & g_{23} \\ g_{31} & g_{32} & g_{33} \end{vmatrix}}^{\Gamma_{15}} & & & & \\
& & \overbrace{\begin{vmatrix} g_{11} & g_{12} & g_{13} \\ g_{21} & g_{22} & g_{23} \\ g_{31} & g_{32} & g_{33} \end{vmatrix}}^{\Gamma_{15}} & & & \\
& & & \overbrace{\begin{vmatrix} g_{11} & g_{12} & g_{13} \\ g_{21} & g_{22} & g_{23} \\ g_{31} & g_{32} & g_{33} \end{vmatrix}}^{\Gamma_{15}} & & \\
& & & & \overbrace{|h_{11}|}^{\Gamma_{12}} & \\
& & & & & \overbrace{|h_{11}|}^{\Gamma_{12}}
\end{vmatrix} . \tag{6.39}$$

That is, we should now break the 15×15 determinant into:

one fourth-order determinant (Γ_1 with elements f_{mn}),
three equal third-order determinants (Γ_{15} with elements g_{mn}), and
two equal first-order determinants (Γ_{12} with elements h_{11}).

The above method is in fact quite straightforward and general, and exactly similar analyses can be performed at other symmetry points in figure 4.19(c).

For example, at the symmetry point X of the diamond structure the ensemble of 14 plane waves (100), $(\bar{1}00)$, (011), $(0\bar{1}\bar{1})$, $(0\bar{1}1)$, $(01\bar{1})$, (120), $(1\bar{2}0)$, $(\bar{1}20)$, $(\bar{1}\bar{2}0)$, (102), $(10\bar{2})$, $(\bar{1}02)$, $(\bar{1}0\bar{2})$ should form the bases of the reduced representation

$$X^{(14)} = 3X_1^{(2)} + X_2^{(2)} + X_3^{(2)} + 2X_4^{(2)} \tag{6.40}$$

where the superscripts indicate the degeneracy of a representation. Thus the secular determinant $D_{\mathrm{d}}(X)$ should factorize as follows:

$$\begin{vmatrix}
\overbrace{\begin{vmatrix} i_{11} & i_{12} & i_{13} \\ i_{21} & i_{22} & i_{23} \\ i_{31} & i_{32} & i_{33} \end{vmatrix}}^{X_1} & & & & & & & \\
& \overbrace{\begin{vmatrix} i_{11} & i_{12} & i_{13} \\ i_{21} & i_{22} & i_{23} \\ i_{31} & i_{32} & i_{33} \end{vmatrix}}^{X_1} & & & & & & \\
& & \overbrace{|j_{11}|}^{X_2} & & & & & \\
& & & \overbrace{|j_{11}|}^{X_2} & & & & \\
& & & & \overbrace{|k_{11}|}^{X_3} & & & \\
& & & & & \overbrace{|k_{11}|}^{X_3} & & \\
& & & & & & \overbrace{\begin{vmatrix} l_{11} & l_{12} \\ l_{21} & l_{22} \end{vmatrix}}^{X_4} & \\
& & & & & & & \overbrace{\begin{vmatrix} l_{11} & l_{12} \\ l_{21} & l_{22} \end{vmatrix}}^{X_4}
\end{vmatrix}. \quad (6.41)$$

That is, we should have

two equal third-order determinants (X_1 with elements i_{mn}),
two equal first-order determinants (X_2 with elements j_{11}),
two equal first-order determinants (X_3 with elements k_{11}), and
two equal second-order determinants (X_4 with elements l_{mn}).

As the formalism is exactly the same for the diamond and zincblende structures at all the various symmetry points, we shall consider only the Γ point in constructing the orthogonal matrices and using them to factorize the secular determinants.

The orthogonal matrices are constructed by using a basis formed from symmetrized combinations of plane waves (SCPW). These can be found in a straightforward way [53] and since tables of SCPW for the diamond and zincblende lattices are readily available in the literature ([51], Bassani (see above) and references therein) we shall simply quote from these tables. Let us represent the ensemble of 15 waves at the Γ point for the two structures as

$$\begin{array}{lllll}
\alpha = (100) & \phi_1 = (111) & \phi_2 = (1\bar{1}\bar{1}) & \phi_3 = (\bar{1}1\bar{1}) & \phi_4 = (\bar{1}\bar{1}1) \\
\phi_5 = (\bar{1}\bar{1}\bar{1}) & \phi_6 = (\bar{1}11) & \phi_7 = (1\bar{1}1) & \phi_8 = (11\bar{1}) & \psi_1 = (200) \\
\psi_2 = (020) & \psi_3 = (002) & \psi_4 = (\bar{2}00) & \psi_5 = (0\bar{2}0) & \psi_6 = (00\bar{2}).
\end{array}$$

Since the SCPW for the diamond structure are

$$W_{11}(\Gamma_1) = \alpha$$

$$W_{11}(\Gamma_1) = \sqrt{\tfrac{1}{8}}(\phi_1 - \phi_2 - \phi_3 - \phi_4 + \phi_5 - \phi_6 - \phi_7 - \phi_8)$$

$$W_{11}(\Gamma'_2) = \sqrt{\tfrac{1}{8}}(\phi_1 - \phi_2 - \phi_3 - \phi_4 - \phi_5 + \phi_6 + \phi_7 + \phi_8)$$

$$W_{11}(\Gamma'_{25}) = \sqrt{\tfrac{1}{8}}(\phi_1 + \phi_2 + \phi_3 - \phi_4 + \phi_5 + \phi_6 + \phi_7 - \phi_8)$$

$$W_{12}(\Gamma'_{25}) = \sqrt{\tfrac{1}{8}}(\phi_1 + \phi_2 - \phi_3 + \phi_4 + \phi_5 + \phi_6 - \phi_7 + \phi_8)$$

$$W_{13}(\Gamma'_{25}) = \sqrt{\tfrac{1}{8}}(\phi_1 - \phi_2 + \phi_3 + \phi_4 + \phi_5 - \phi_6 + \phi_7 + \phi_8)$$

$$W_{11}(\Gamma_{15}) = \sqrt{\tfrac{1}{8}}(\phi_1 + \phi_2 + \phi_3 - \phi_4 - \phi_5 - \phi_6 - \phi_7 + \phi_8)$$

$$W_{12}(\Gamma_{15}) = \sqrt{\tfrac{1}{8}}(\phi_1 + \phi_2 - \phi_3 + \phi_4 - \phi_5 - \phi_6 + \phi_7 - \phi_8)$$

$$W_{13}(\Gamma_{15}) = \sqrt{\tfrac{1}{8}}(\phi_1 - \phi_2 + \phi_3 + \phi_4 - \phi_5 + \phi_6 - \phi_7 - \phi_8)$$

$$W_{11}(\Gamma'_2) = 2\sqrt{\tfrac{1}{6}}(\psi_1 + \psi_2 + \psi_3 - \psi_4 - \psi_5 - \psi_6)$$

$$W_{11}(\Gamma'_{12}) = 2\sqrt{\tfrac{1}{4}}(\psi_2 - \psi_3 - \psi_5 + \psi_6)$$

$$W_{12}(\Gamma'_{12}) = 2\sqrt{\tfrac{1}{12}}(-2\psi_1 + \psi_2 + \psi_3 + 2\psi_4 - \psi_5 - \psi_6)$$

$$W_{11}(\Gamma'_{25}) = 2\sqrt{\tfrac{1}{2}}(\psi_3 + \psi_6)$$

$$W_{12}(\Gamma'_{25}) = 2\sqrt{\tfrac{1}{2}}(\psi_2 + \psi_5)$$

$$W_{13}(\Gamma'_{25}) = 2\sqrt{\tfrac{1}{2}}(\psi_1 + \psi_4),$$

it follows that the 15×15 orthogonal matrix $\mathbf{O}_{\mathrm{d}}(\Gamma)$ for this symmetry point is

$$\begin{bmatrix}
1 & 0 & 0 & 0 & 0 & 0 & 0 & 0 & 0 & 0 & 0 & 0 & 0 & 0 & 0 \\
0 & \sqrt{\frac{1}{8}} & -\sqrt{\frac{1}{8}} & -\sqrt{\frac{1}{8}} & -\sqrt{\frac{1}{8}} & \sqrt{\frac{1}{8}} & -\sqrt{\frac{1}{8}} & -\sqrt{\frac{1}{8}} & -\sqrt{\frac{1}{8}} & 0 & 0 & 0 & 0 & 0 & 0 \\
0 & \sqrt{\frac{1}{8}} & -\sqrt{\frac{1}{8}} & -\sqrt{\frac{1}{8}} & -\sqrt{\frac{1}{8}} & -\sqrt{\frac{1}{8}} & \sqrt{\frac{1}{8}} & \sqrt{\frac{1}{8}} & \sqrt{\frac{1}{8}} & 0 & 0 & 0 & 0 & 0 & 0 \\
0 & 0 & 0 & 0 & 0 & 0 & 0 & 0 & 0 & \sqrt{\frac{1}{6}} & \sqrt{\frac{1}{6}} & \sqrt{\frac{1}{6}} & -\sqrt{\frac{1}{6}} & -\sqrt{\frac{1}{6}} & -\sqrt{\frac{1}{6}} \\
0 & \sqrt{\frac{1}{8}} & \sqrt{\frac{1}{8}} & \sqrt{\frac{1}{8}} & -\sqrt{\frac{1}{8}} & \sqrt{\frac{1}{8}} & \sqrt{\frac{1}{8}} & \sqrt{\frac{1}{8}} & -\sqrt{\frac{1}{8}} & 0 & 0 & 0 & 0 & 0 & 0 \\
0 & 0 & 0 & 0 & 0 & 0 & 0 & 0 & 0 & 0 & 0 & \sqrt{\frac{1}{2}} & 0 & 0 & \sqrt{\frac{1}{2}} \\
0 & \sqrt{\frac{1}{8}} & \sqrt{\frac{1}{8}} & -\sqrt{\frac{1}{8}} & \sqrt{\frac{1}{8}} & \sqrt{\frac{1}{8}} & \sqrt{\frac{1}{8}} & -\sqrt{\frac{1}{8}} & \sqrt{\frac{1}{8}} & 0 & 0 & 0 & 0 & 0 & 0 \\
0 & 0 & 0 & 0 & 0 & 0 & 0 & 0 & 0 & 0 & \sqrt{\frac{1}{2}} & 0 & 0 & \sqrt{\frac{1}{2}} & 0 \\
0 & \sqrt{\frac{1}{8}} & -\sqrt{\frac{1}{8}} & \sqrt{\frac{1}{8}} & \sqrt{\frac{1}{8}} & \sqrt{\frac{1}{8}} & -\sqrt{\frac{1}{8}} & \sqrt{\frac{1}{8}} & \sqrt{\frac{1}{8}} & 0 & 0 & 0 & 0 & 0 & 0 \\
0 & 0 & 0 & 0 & 0 & 0 & 0 & 0 & 0 & \sqrt{\frac{1}{2}} & 0 & 0 & \sqrt{\frac{1}{2}} & 0 & 0 \\
0 & \sqrt{\frac{1}{8}} & \sqrt{\frac{1}{8}} & \sqrt{\frac{1}{8}} & -\sqrt{\frac{1}{8}} & -\sqrt{\frac{1}{8}} & -\sqrt{\frac{1}{8}} & -\sqrt{\frac{1}{8}} & \sqrt{\frac{1}{8}} & 0 & 0 & 0 & 0 & 0 & 0 \\
0 & \sqrt{\frac{1}{8}} & \sqrt{\frac{1}{8}} & -\sqrt{\frac{1}{8}} & \sqrt{\frac{1}{8}} & -\sqrt{\frac{1}{8}} & -\sqrt{\frac{1}{8}} & \sqrt{\frac{1}{8}} & -\sqrt{\frac{1}{8}} & 0 & 0 & 0 & 0 & 0 & 0 \\
0 & \sqrt{\frac{1}{8}} & -\sqrt{\frac{1}{8}} & \sqrt{\frac{1}{8}} & \sqrt{\frac{1}{8}} & -\sqrt{\frac{1}{8}} & \sqrt{\frac{1}{8}} & -\sqrt{\frac{1}{8}} & \sqrt{\frac{1}{8}} & 0 & 0 & 0 & 0 & 0 & 0 \\
0 & 0 & 0 & 0 & 0 & 0 & 0 & 0 & 0 & 0 & \frac{1}{2} & -\frac{1}{2} & 0 & -\frac{1}{2} & \frac{1}{2} \\
0 & 0 & 0 & 0 & 0 & 0 & 0 & 0 & 0 & -\sqrt{\frac{1}{3}} & \frac{1}{2}\sqrt{\frac{1}{3}} & \frac{1}{2}\sqrt{\frac{1}{3}} & \sqrt{\frac{1}{3}} & -\frac{1}{2}\sqrt{\frac{1}{3}} & -\frac{1}{2}\sqrt{\frac{1}{3}}
\end{bmatrix}$$

Similarly for the Γ point of the zincblende structure the SCPW are

$$
\begin{aligned}
W_{11}(\Gamma_1) &= \alpha \\
W_{11}(\Gamma_1) &= \tfrac{1}{2}(\phi_1 - \phi_2 - \phi_3 - \phi_4) \\
W_{21}(\Gamma_1) &= \tfrac{1}{2}(\phi_5 - \phi_6 - \phi_7 - \phi_8) \\
W_{11}(\Gamma_{15}) &= \tfrac{1}{2}(\phi_1 + \phi_2 + \phi_3 - \phi_4) \\
W_{12}(\Gamma_{15}) &= \tfrac{1}{2}(\phi_1 + \phi_2 - \phi_3 + \phi_4) \\
W_{13}(\Gamma_{15}) &= \tfrac{1}{2}(\phi_1 - \phi_2 + \phi_3 + \phi_4) \\
W_{21}(\Gamma_{15}) &= \tfrac{1}{2}(-\phi_5 - \phi_6 - \phi_7 + \phi_8) \\
W_{22}(\Gamma_{15}) &= \tfrac{1}{2}(-\phi_5 - \phi_6 + \phi_7 - \phi_8) \\
W_{23}(\Gamma_{15}) &= \tfrac{1}{2}(-\phi_5 + \phi_6 - \phi_7 - \phi_8) \\
W_{11}(\Gamma_1) &= 2\sqrt{\tfrac{1}{6}}(\psi_1 + \psi_2 + \psi_3 - \psi_4 - \psi_5 - \psi_6) \\
W_{11}(\Gamma_{15}) &= 2\sqrt{\tfrac{1}{2}}(\psi_3 + \psi_6) \\
W_{12}(\Gamma_{15}) &= 2\sqrt{\tfrac{1}{2}}(\psi_2 + \psi_5) \\
W_{13}(\Gamma_{15}) &= 2\sqrt{\tfrac{1}{2}}(\psi_1 + \psi_4) \\
W_{11}(\Gamma_{12}) &= 2\sqrt{\tfrac{1}{4}}(\psi_2 - \psi_3 - \psi_5 + \psi_6) \\
W_{12}(\Gamma_{12}) &= 2\sqrt{\tfrac{1}{12}}(-2\psi_1 + \psi_2 + \psi_3 + 2\psi_4 - \psi_5 - \psi_6)
\end{aligned}
$$

and the corresponding matrix $\mathbf{O}_{zb}(\Gamma)$ is

$$
\begin{bmatrix}
1 & 0 & 0 & 0 & 0 & 0 & 0 & 0 & 0 & 0 & 0 & 0 & 0 & 0 & 0 \\
0 & \frac{1}{2} & -\frac{1}{2} & -\frac{1}{2} & -\frac{1}{2} & 0 & 0 & 0 & 0 & 0 & 0 & 0 & 0 & 0 & 0 \\
0 & 0 & 0 & 0 & 0 & 0 & 0 & 0 & 0 & \sqrt{\frac{1}{6}} & \sqrt{\frac{1}{6}} & \sqrt{\frac{1}{6}} & -\sqrt{\frac{1}{6}} & -\sqrt{\frac{1}{6}} & -\sqrt{\frac{1}{6}} \\
0 & 0 & 0 & 0 & 0 & \frac{1}{2} & -\frac{1}{2} & -\frac{1}{2} & -\frac{1}{2} & 0 & 0 & 0 & 0 & 0 & 0 \\
0 & \frac{1}{2} & \frac{1}{2} & \frac{1}{2} & -\frac{1}{2} & 0 & 0 & 0 & 0 & 0 & 0 & 0 & 0 & 0 & 0 \\
0 & 0 & 0 & 0 & 0 & 0 & 0 & 0 & 0 & 0 & 0 & \sqrt{\frac{1}{2}} & 0 & 0 & \sqrt{\frac{1}{2}} \\
0 & 0 & 0 & 0 & 0 & -\frac{1}{2} & -\frac{1}{2} & -\frac{1}{2} & \frac{1}{2} & 0 & 0 & 0 & 0 & 0 & 0 \\
0 & \frac{1}{2} & \frac{1}{2} & -\frac{1}{2} & \frac{1}{2} & 0 & 0 & 0 & 0 & 0 & 0 & 0 & 0 & 0 & 0 \\
0 & 0 & 0 & 0 & 0 & 0 & 0 & 0 & 0 & 0 & \sqrt{\frac{1}{2}} & 0 & 0 & \sqrt{\frac{1}{2}} & 0 \\
0 & 0 & 0 & 0 & 0 & -\frac{1}{2} & -\frac{1}{2} & \frac{1}{2} & -\frac{1}{2} & 0 & 0 & 0 & 0 & 0 & 0 \\
0 & \frac{1}{2} & -\frac{1}{2} & \frac{1}{2} & \frac{1}{2} & 0 & 0 & 0 & 0 & 0 & 0 & 0 & 0 & 0 & 0 \\
0 & 0 & 0 & 0 & 0 & 0 & 0 & 0 & 0 & \sqrt{\frac{1}{2}} & 0 & 0 & \sqrt{\frac{1}{2}} & 0 & 0 \\
0 & 0 & 0 & 0 & 0 & -\frac{1}{2} & \frac{1}{2} & -\frac{1}{2} & -\frac{1}{2} & 0 & 0 & 0 & 0 & 0 & 0 \\
0 & 0 & 0 & 0 & 0 & 0 & 0 & 0 & 0 & 0 & \frac{1}{2} & -\frac{1}{2} & 0 & -\frac{1}{2} & \frac{1}{2} \\
0 & 0 & 0 & 0 & 0 & 0 & 0 & 0 & 0 & -\sqrt{\frac{1}{3}} & \frac{1}{2}\sqrt{\frac{1}{3}} & \frac{1}{2}\sqrt{\frac{1}{3}} & \sqrt{\frac{1}{3}} & -\frac{1}{2}\sqrt{\frac{1}{3}} & -\frac{1}{2}\sqrt{\frac{1}{3}}
\end{bmatrix}.
$$

When the orthogonal matrix **O** so formed is applied to the original matrix **D** such that

$$\bar{\mathbf{D}} = \mathbf{O}\mathbf{D}\mathbf{O}^{-1}$$

the determinant of $\bar{\mathbf{D}}$ will be automatically factorized in the form (6.37) or (6.39). Thus, for example, the 15 × 15 secular determinants $D(\Gamma)$ formed by using the pseudopotentials V_3, V_4^{a}, V_8 and V_{11} of Cohen and Bergstresser [48] factorize as follows for the diamond and zincblende structures at the symmetry point Γ.

(1) *Diamond structure*

$$\Gamma_1 = \begin{vmatrix} 0 & -2V_3^{\mathrm{s}} \\ -2V_3^{\mathrm{s}} & 3(2\pi/a)^2 + 3V_8^{\mathrm{s}} \end{vmatrix}$$

$$\Gamma_2' = \begin{vmatrix} 3(2\pi/a)^2 + 3V_8^{\mathrm{s}} & \sqrt{6}(V_3^{\mathrm{s}} + V_{11}^{\mathrm{s}}) \\ \sqrt{6}(V_3^{\mathrm{s}} + V_{11}^{\mathrm{s}}) & 4(2\pi/a)^2 + 4V_8^{\mathrm{s}} \end{vmatrix}$$

$$\Gamma_{25}' = \begin{vmatrix} 3(2\pi/a)^2 - V_8^{\mathrm{s}} & \sqrt{2}(V_3^{\mathrm{s}} - V_{11}^{\mathrm{s}}) \\ \sqrt{2}(V_3^{\mathrm{s}} - V_{11}^{\mathrm{s}}) & 4(2\pi/a)^2 \end{vmatrix}$$

$$\Gamma_{15} = |3(2\pi/a)^2 - V_8^{\mathrm{s}}|$$

$$\Gamma_{12}' = |4(2\pi/a)^2 - 2V_8^{\mathrm{s}}|.$$

(2) *Zincblende structure*

$$\Gamma_1 = \begin{vmatrix} 0 & \sqrt{2}(V_3^{\mathrm{s}} + \mathrm{i}V_3^{\mathrm{a}}) & \sqrt{2}(V_3^{\mathrm{s}} - \mathrm{i}V_3^{\mathrm{a}}) & -\mathrm{i}\sqrt{6}V_4^{\mathrm{a}} \\ \sqrt{2}(V_3^{\mathrm{s}} - \mathrm{i}V_3^{\mathrm{a}}) & 3(2\pi/a)^2 + 3V_8^{\mathrm{s}} & -3\mathrm{i}V_4^{\mathrm{a}} & -\sqrt{3}(V_{\mathrm{s}}^{\mathrm{s}} + \mathrm{i}V_{\mathrm{s}}^{\mathrm{a}}) \\ \sqrt{2}(V_3^{\mathrm{s}} + \mathrm{i}V_3^{\mathrm{a}}) & 3\mathrm{i}V_4^{\mathrm{a}} & 3(2\pi/a)^2 + 3V_8^{\mathrm{s}} & \sqrt{3}(V_{\mathrm{s}}^{\mathrm{s}} - \mathrm{i}V_{\mathrm{s}}^{\mathrm{a}}) \\ \mathrm{i}\sqrt{6}V_4^{\mathrm{a}} & -\sqrt{3}(V_{\mathrm{s}}^{\mathrm{s}} - \mathrm{i}V_{\mathrm{s}}^{\mathrm{a}}) & \sqrt{3}(V_{\mathrm{s}}^{\mathrm{s}} + \mathrm{i}V_{\mathrm{s}}^{\mathrm{a}}) & 4(2\pi/a)^2 + 4V_8^{\mathrm{s}} \end{vmatrix}$$

$$\Gamma_{15} = \begin{vmatrix} 3(2\pi/a)^2 - V_8^{\mathrm{s}} & -\mathrm{i}V_4^{\mathrm{a}} & -V_{\mathrm{D}}^{\mathrm{s}} - \mathrm{i}V_{\mathrm{D}}^{\mathrm{a}} \\ \mathrm{i}V_4^{\mathrm{a}} & 3(2\pi/a)^2 - V_8^{\mathrm{s}} & V_{\mathrm{D}}^{\mathrm{s}} - \mathrm{i}V_{\mathrm{D}}^{\mathrm{a}} \\ -V_{\mathrm{D}}^{\mathrm{s}} + \mathrm{i}V_{\mathrm{D}}^{\mathrm{a}} & V_{\mathrm{D}}^{\mathrm{s}} + \mathrm{i}V_{\mathrm{D}}^{\mathrm{a}} & 4(2\pi/a)^2 \end{vmatrix}$$

$$\Gamma_{12} = |4(2\pi/a)^2 - 2V_8^{\mathrm{s}}|$$

where

$$V_{\mathrm{s}}^{\mathrm{s}} = V_3^{\mathrm{s}} + V_{11}^{\mathrm{s}} \qquad V_{\mathrm{s}}^{\mathrm{a}} = V_3^{\mathrm{a}} + V_{11}^{\mathrm{a}}$$

$$V_{\mathrm{D}}^{\mathrm{s}} = V_3^{\mathrm{s}} - V_{11}^{\mathrm{s}} \qquad V_{\mathrm{D}}^{\mathrm{a}} = V_3^{\mathrm{a}} - V_{11}^{\mathrm{a}}.$$

In a similar way the secular determinants formed from the Cohen and Bergstresser pseudopotentials for other symmetry points in figure 4.9(*c*) can be easily factorized using the readily available SCPW tables.

This is as far as we can go in emphasizing the usefulness of symmetry analysis in energy band calculations. As mentioned earlier, symmetry arguments not only greatly simplify the calculations but also often give much insight into the physical situation. It is therefore always advisable to carry out a symmetry analysis before undertaking any quantum mechanical calculations of problems in crystal physics. The present example has shown how to eliminate a number of structure factors and enable the easy factorization of higher-order determinants at high symmetry points.

Exercises

6.1 Write down the Hamiltonian for a many-electron system and explain the meaning of all the terms used.

6.2 Explain the nature of the crystal field $V = e_i\phi_c(\boldsymbol{r}_i)$ (see equation (6.2)) in the following three cases and give examples of compounds where such fields exist.

(a) $V < \lambda_{ij}\boldsymbol{l}_i \cdot \boldsymbol{s}_i$

(b) $\lambda_{ij}\boldsymbol{l}_i \cdot \boldsymbol{s}_i < V < e^2/r_{ij}$

(c) $V > e^2/r_{ij}$.

6.3 A transition metal ion with a ^{3}G free-ion level is placed in a crystal with site symmetry D_{4h}. Use the character table given by Koster *et al* [10] and obtain the number and symmetries of the levels derived from crystal field splitting.

6.4 Ti^{3+} has a $3d^1$ configuration. If this ion is placed in a crystal field with a cubic site symmetry, calculate the number and symmetries of the levels derived from this crystal splitting. Would the levels split further if a tetragonal distortion were introduced, for example by small symmetrical displacements of the charges on the z axis? Draw rough sketches depicting all the splittings of the free-ion level.

6.5 Find the free-ion energy level ground state for electrons in the d shell, that is d^n ($n = 1$–9).

6.6 Show schematically how an ion in a ^{4}F ground state would split when it is placed successively in cubic crystal fields with increasing field strengths compared with spin–orbit coupling.

6.7 Construct the Brillouin zone of a simple square lattice with lattice constant a and label all the special points and lines of symmetry within it. Write down the wavevectors of the special points and lines and hence determine their little groups.

6.8 Draw the free-electron energy bands for the square lattice of exercise 6.7 along the ΓX and ΓM directions.

6.9 Consider the Brillouin zone of a simple cubic lattice with lattice constant a (see figure 4.9(a)). Write down the wavevectors for all the points and lines of symmetry indicated and verify that the little groups for all these special points are as given in table 6.6(a).

6.10 Write down explicitly the character tables for the following points and lines of symmetry in the Brillouin zone of a simple cubic lattice (figure 4.9(a)): Γ, Δ, X, Λ and R.

Determine the compatibility relations between Γ, Δ, X and Γ, Λ, R and hence sketch the free-electron energy bands along the Δ axis (001) and Λ axis (111).

6.11 Write down the compatibility relations between the Γ points for the double-valued representations of the cubic groups O_h and O.

6.12 Obtain the free-electron energy bands for an FCC lattice.

6.13 In group theoretical terms state the selection rules governing transitions between states leading to infrared and Raman spectra.

6.14 Determine the point-group symmetry of the NH_3 molecule. Use the selection rules given in exercise 6.13 and consider transitions involving Γ_1 (the totally symmetric normal mode) at one end and deduce which normal modes are active in infrared and Raman active transitions (use tables given by Koster *et al* [10]).

6.15 The CH_4 (methane) molecule has a point-group symmetry belonging to the tetrahedral group T_d. Write down the various representations for this group and assign functions of the type x, y, z, and x^2, y^2, z^2, xy, yz, zx, etc to these representations. Hence determine, as in exercise 6.14, which modes are infrared active and which are Raman active.

6.16 Write down the character table for the point group of the water molecule H_2O. By assigning functions to the operators corresponding to the Raman active transitions, determine the modes which can be excited from the ground state by Raman transitions.

6.17 Using the method described in §6.5, determine the number of plane waves corresponding to the set ⟨000⟩, ⟨100⟩, ⟨110⟩.

6.18 Operate on the wave (100) by the operator $\{C_{4z}|O\}$ of the diamond space group O_h^7 (as explained in §6.5) and show that

$$E = \exp\left(\frac{2\pi i}{a}(\boldsymbol{K}\cdot\boldsymbol{r})\right) = i.$$

6.19 Use the information given in §6.5 to draw the free-electron energy bands for Ge along the ΓX direction.

6.20 Calculate the value of the determinant Γ'_{25} and Γ'_{2} for Ge using the following data: $V_3^s = -0.23$ Ryd, $V_8^s = 0.01$ Ryd, $V_{11}^s = 0.06$ Ryd. Take the lattice constant $a = 5.66$ Å; 1 atomic unit $= 0.528$ Å.

6.21 Determine the orthogonal matrices which will factorize the determinant formed by the following ensemble of plane waves of diamond (O_h^7):
(a) $\langle 111 \rangle$; (b) $\langle 200 \rangle$.

PART II

SYMMETRY AND THE MAGNETIC STATE

7

INTRODUCTION AND SCOPE OF PART II

In Part I of this book we presented an elementary introduction to symmetry principles and its applications to the physical properties of crystals closely linked with structures. Such applications are now very well established and a number of books are available on the subject (see [1–8]). In recent years, however, the use of symmetry principles in studying the symmetry of the Fermi surface and the symmetry properties of wavefunctions in magnetic and non-magnetic crystals, symmetry of phonons, magnons and other quasi-particles etc, have been widely developed (see [9–15]). One of the most recent developments in the use of symmetry principles in solid state physics has been the study of the symmetry properties of strongly magnetic crystals (ferromagnetic, antiferromagnetic and ferrimagnetic structures) and these are now being discussed extensively in the literature (see [12–31]).

The study of the antiferromagnetic transition metal difluorides MnF_2, FeF_2, CoF_2, NiF_2 and the transition metal oxides MnO, FeO, CoO and NiO have been of considerable experimental and theoretical interest in recent years, both in order to explain the remarkable magnetic properties possessed by these electrical insulators and also because they provide a great deal of insight into the fundamental mechanisms of exchange interactions. The theories applied to these compounds are in fact applicable to a wide range of magnetically ordered crystals and as such in this part we have presented a comprehensive account of some important symmetry properties of a transition metal difluoride (NiF_2) and a transition metal oxide (NiO). These examples should prove to be of considerable interest to theoretical and experimental physicists investigating the symmetry properties of closely related magnetic materials.

The scope for this part is as follows. Chapter 8 deals briefly with the broad classification of magnetic materials and the rest is concerned mainly with the physical properties of magnetic crystals which are closely linked to their symmetries. Thus, studies on susceptibilities and magnetization measurements, theories of magnetism and the thermodynamic properties etc have been omitted. For a more general coverage on a wide range of magnetic materials readers are referred to the recent review article by De Jonge and Miedema [45].

Chapters 9 and 10 deal with a brief introduction to magnetic groups and corepresentation theory. Chapter 10 has been written with the intention of acquainting those readers who are already familiar with the principles and applications of the ordinary crystallographic point groups and space groups (Federov groups) to the general extensions of these groups to magnetic point groups and space groups (Shubnikov groups) which are utilized in the study of magnetic crystals.

Chapter 11 is a detailed account of some symmetry properties of NiF_2 and NiO and Chapter 12 gives a brief account of magnetic space-group selection rules with applications to NiF_2 and NiO. It is perhaps worth adding that, even though the emphasis in this part is on the recently gained knowledge of the symmetry properties of antiferromagnetic NiF_2 and NiO, the principles are in fact quite general and the techniques can be shown 'at work' in many magnetic crystals [40].

8

CLASSIFICATION OF MAGNETIC MATERIALS

8.1 Introduction

Many features of the magnetic properties of materials are formally similar to the electrical properties (dielectrics, ferroelectrics, etc). For example, magnetic polarization characterized by $\boldsymbol{M}$ is analogous to induced dielectric polarization characterized by $\boldsymbol{P}$. Atoms and molecules with permanent magnetic dipoles exist just as do those with permanent electric dipoles. Some materials possess spontaneous magnetization (ferromagnets etc) just as some possess spontaneous electric polarization (ferroelectrics etc).

There is, however, one important difference between the magnetic and electric properties of matter. Individual electric charges of one sign do exist, that is we have isolated positive and negative charges. A corresponding magnetic monopole does not occur (see, however, [46]) but this, as we shall see, is entirely unnecessary in describing the magnetic properties of materials. Just as the main topics of the electrical properties of matter can be described in terms of $\boldsymbol{P}$ so also can the main topic of magnetization be described by $\boldsymbol{M}$.

The connecting relations between the $\boldsymbol{B}$ (magnetic induction), $\boldsymbol{H}$ (magnetic field) and $\boldsymbol{M}$ (magnetization) vectors are exactly similar to the connecting relations between the $\boldsymbol{D}$ (electric induction or displacement vector), $\boldsymbol{E}$ (electric field) and $\boldsymbol{P}$ (polarization) vectors. For instance, exactly analogous to the $\boldsymbol{D}$, $\boldsymbol{E}$, $\boldsymbol{P}$ vectors we may write

$$\boldsymbol{B} = \mu \boldsymbol{H} \tag{8.1}$$

where

$$\mu = \mu_r \mu_0 \tag{8.2}$$

and for media in which $\boldsymbol{M}$ (magnetization) exists we have

$$\boldsymbol{B} = \mu_0 \boldsymbol{H} + \mu_0 \boldsymbol{M} \tag{8.3}$$

or using (8.1) and (8.2) in (8.3) we get

$$\boldsymbol{M} = (\mu_r - 1)\boldsymbol{H} \tag{8.4}$$

or

$$\boldsymbol{M} = \chi_m \boldsymbol{H} \tag{8.5}$$

where

$$\chi_m = (\mu_r - 1). \tag{8.6}$$

χ_m is defined as the magnetic susceptibility and is similar to the electric susceptibility χ_e given by

$$\chi_e = (\varepsilon_r - 1). \tag{8.7}$$

The most natural way to classify magnetic materials is by studying the response of the material to an externally applied field $\boldsymbol{H}$. The response is characterized by the susceptibility relation

$$\boldsymbol{M} = \chi \boldsymbol{H} \tag{8.8}$$

(here we have dropped the subscript m in χ_m since it is not going to create any confusion because we are dealing only with magnetic materials).

In general χ is a function of both $\boldsymbol{H}$ and temperature T. For isotropic materials $\boldsymbol{M}$ and $\boldsymbol{H}$ are parallel and χ is a scalar (i.e. a tensor of rank 0; see [2]). For anisotropic materials $\boldsymbol{M}$ and $\boldsymbol{H}$ need not be parallel and χ is then a tensor of the second rank.

The magnetic properties of all materials depend upon (a) whether the substance has unpaired electron spins (i.e. permanent magnetic dipoles) and (b) if it does have unpaired electron spins, how are these spins oriented with respect to each other?

Based on these criteria magnetic materials are generally classified into five (or six) categories as follows:

(i) diamagnetic;
(ii) paramagnetic;
(iii) ferromagnetic;
(iv) antiferromagnetic;
(v) ferrimagnetic;
(vi) complicated magnetic ordering structures like helical, canted, spiral, umbrella, etc.

In this part we are primarily concerned with the strongly magnetic solids, namely ferromagnetics, ferrimagnetics and antiferromagnetics. However, for the sake of completeness, we give in the following sections a brief summary of the magnetic properties of the various types of materials listed above.

8.2 Diamagnetics

A diamagnetic material possesses no permanent magnetic dipoles and hence its magnetic effects are very weak. Diamagnetism is analogous to the induced dielectric effect. When placed in a magnetic field $\boldsymbol{H}$ the magnetic susceptibility χ is negative and, moreover, is quite independent of $\boldsymbol{H}$ and temperature T. The origin of diamagnetism is explained classically using Faraday's and

Lenz's laws of electromagnetic induction. When $\boldsymbol{H}$ is applied the angular velocity ω of the electrons is altered so that the orbital (quantum) states of the electrons in the material are modified by the magnetic force. The modified motion of electrons in turn produces a local magnetic moment that opposes the applied field $\boldsymbol{H}$. $\boldsymbol{M}$ is therefore negative (in accordance with Lenz's law) and since $\chi = \boldsymbol{M}/\boldsymbol{H}$ it follows that, for diamagnetics, χ is negative.

8.3 Paramagnetics

These substances possess unpaired electron spins. In the absence of an external $\boldsymbol{H}$ these spins are randomly oriented and hence magnetization is zero (figure 8.1(*a*)). At a finite temperature T each atom has a thermal energy approximately equal to kT. To align the spins in a given direction we need to apply an external $\boldsymbol{H}$ which is strong enough to overcome the force due to thermal agitation. Thus the net magnetization $\boldsymbol{M}$ in paramagnets is due to the alignment of spins by $\boldsymbol{H}$ (figure 8.1(*b*)). $\boldsymbol{M}$ is therefore positive and hence χ is positive; moreover, χ is highly dependent on $\boldsymbol{H}$ and T.

The temperature dependence of χ is given by the well known Curie law, namely $\chi = C/T$ (figure 8.2).

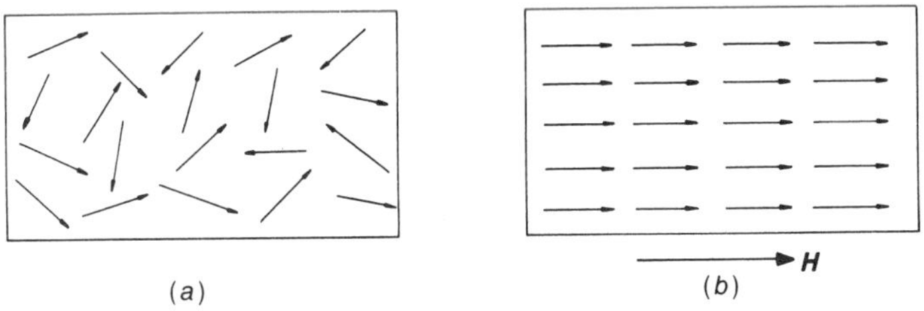

Figure 8.1 Electron spins in paramagnetic materials (schematic): (*a*) random orientation; (*b*) alignment of spins by means of an external magnetic field $\boldsymbol{H}$.

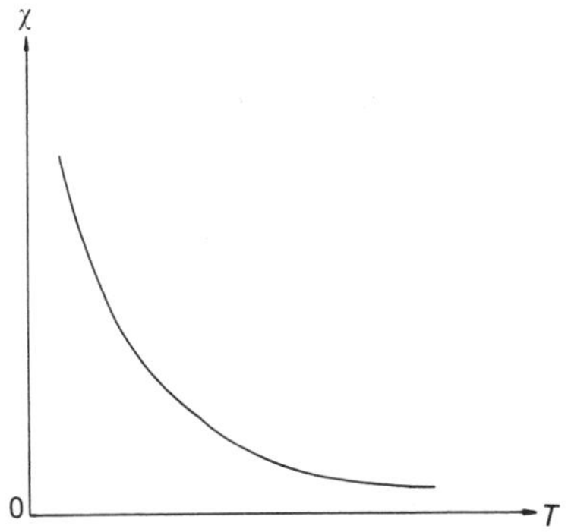

Figure 8.2 Temperature dependence of paramagnetic susceptibility (schematic).

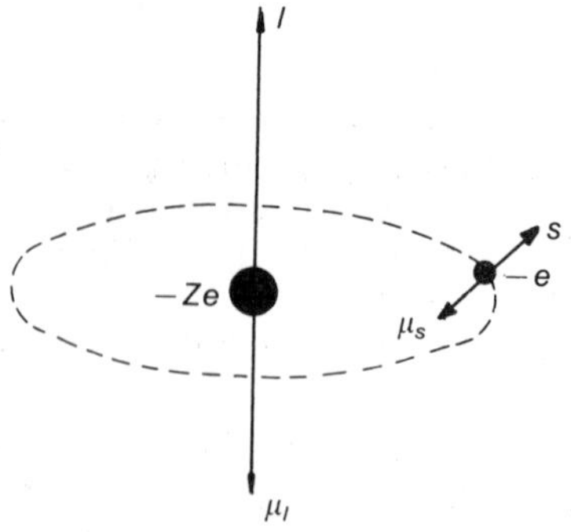

Figure 8.3 Diagram showing the original of orbital (l) and spin (s) angular momenta and their associated magnetic moments μ_l and μ_s respectively.

The magnetic moment that we have been referring to comes of course from the orbital (μ_l) and spin (μ_s) motions of the unpaired electron(s) (figure 8.3), and has a value $\mu = \mu_l + \mu_s$, where μ_l and μ_s are given by the Bohr magneton μ_B. For an unpaired s electron, however, $l = 0$ so that the total magnetic moment is $\mu = \mu_s = \mu_B$.

8.4 Strongly magnetic materials

In this part we shall no longer be concerned with diamagnetics and paramagnetics. Instead we shall be concerned with those materials which show magnetic ordering of unpaired electron spins below a certain critical temperature even in the absence of an externally applied $\boldsymbol{H}$. These materials are the ferromagnetics, ferrimagnetics and antiferromagnetics. Furthermore we shall only be considering magnetic materials which are insulators. This means that we shall be considering materials in which the spins (magnetic moments) are localized at the respective lattice sites.

Weiss was the first to postulate quite empirically that the ordering of spins in these substances was due to large internal fields, commonly referred to as the Weiss molecular field. This molecular field, as we now know, is quite complicated in its origin and arises from various types of interaction between the neighbouring magnetic ions. Furthermore this interaction is of quantum mechanical origin and has no classical counterpart. We shall say a little more about this type of interaction when we deal with the energies of spin waves and dispersion relations. For the moment we shall summarize some important and well known physical properties of ferromagnetic, ferrimagnetic and antiferromagnetic solids.

8.5 Ferromagnetics

Ferromagnetic solids are quite different from the solids we considered earlier—namely diamagnetics and paramagnetics—in that:

(1) they exhibit spontaneous magnetization (see, however, [85–87]) below a certain critical temperature;
(2) the permeability attains a very much higher value, as high as 10^5 in some solids; and
(3) a magnetization remains after the field is removed and the familiar hysteresis loop occurs in the relation between $\boldsymbol{M}$ and $\boldsymbol{H}$ (figure 8.4). $\boldsymbol{M}$ is thus a multi-valued function of $\boldsymbol{H}$ and is not given by the simple relation (8.8).

In these solids the nature of the internal field is such that below a certain critical temperature T_c, called the Curie temperature, the individual spins are aligned parallel to each other (figure 8.5) resulting in a large spontaneous bulk magnetization.

Above T_c, thermal energy becomes stronger than the internal (molecular) field and the spins get randomly oriented. At absolute zero the spins are all perfectly aligned and the spontaneous magnetization has its maximum value $M(0)$. As the temperature is increased the spins begin to deviate from their perfect alignment as a result of thermal agitation and consequently the spontaneous magnetization decreases with increasing temperature until at $T = T_c$ the bulk magnetization becomes zero. Figure 8.6 shows a plot of spontaneous magnetization as a function of temperature in a ferromagnetic material.

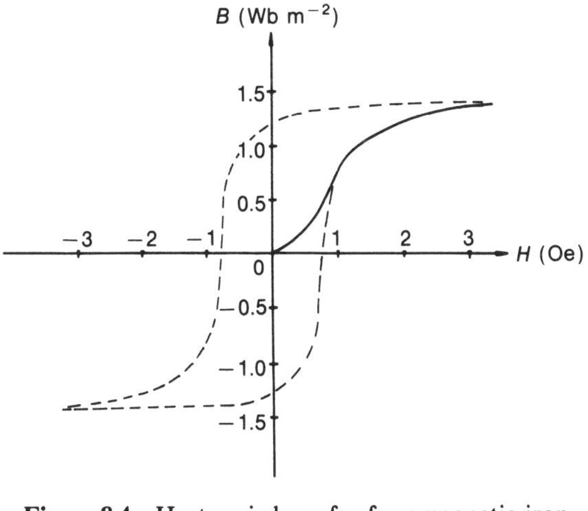

Figure 8.4 Hysteresis loop for ferromagnetic iron.

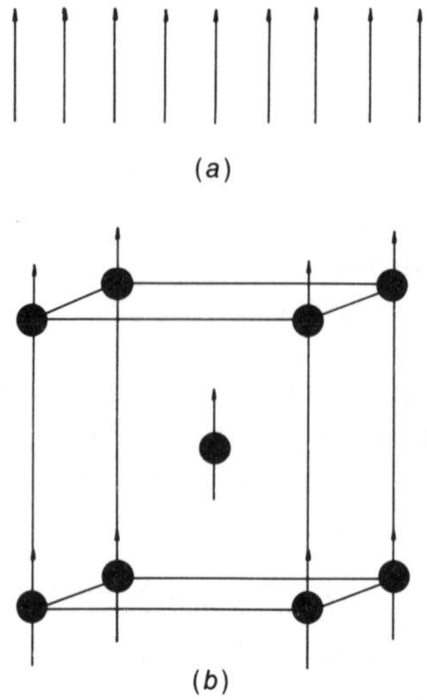

Figure 8.5 Ordering of spins in a ferromagnetic material: (*a*) parallel arrangement of spins; (*b*) parallel ordering in a body-centred cubic lattice.

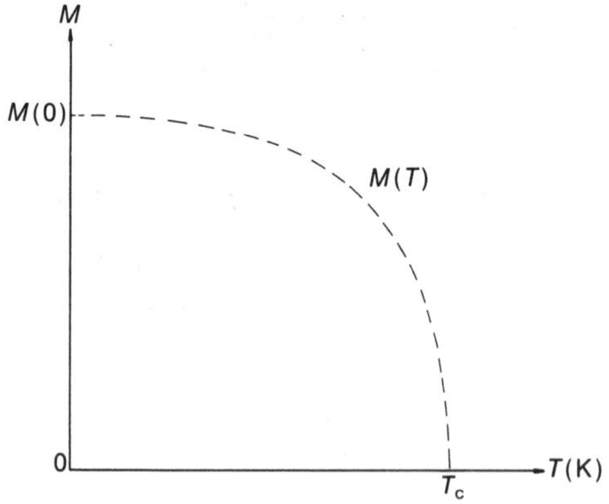

Figure 8.6 Temperature dependence of magnetization for ferromagnets.

In the very-low-temperature region (i.e. $T \to 0\,\mathrm{K}$) the temperature dependence of magnetization has the form given by

$$M(T) = M(0)(1 - a_1 T^{3/2} - a_2 T^{5/2} \ldots) \tag{8.9}$$

where a_1, $a_2 \ldots$ etc are constants and $M(0) = Ng\mu_B S$, N being the number of magnetic atoms, S the spin and g not greatly different from the 'spin-only' value 2. At T close to T_c the dependence of $M(T)$ is given approximately by

$$M(T) \simeq M(0)(1 - T/T_c)^{1/2}. \tag{8.10}$$

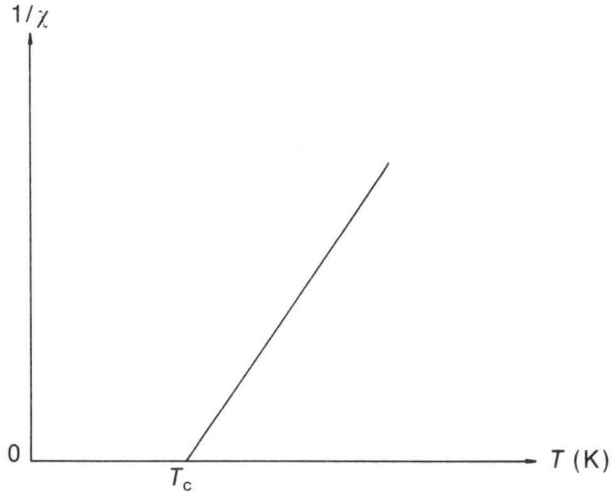

Figure 8.7 Temperature dependence of reciprocal susceptibility for ferromagnets (schematic).

At T_c the transition from the ferromagnetic phase to the paramagnetic phase occurs and this transition is generally a second-order phase transition, the symmetry aspects of which are briefly discussed in Chapter 12.

The temperature dependence of susceptibility is generally given by the Curie–Weiss law, namely

$$\chi = \frac{C}{T - T_c} \tag{8.11}$$

where C is the Curie constant. As for paramagnetic materials, the most common way of displaying experimental high-temperature susceptibility data is by plotting $1/\chi$ against T. This would give a straight line of slope $1/C$ and the intercept on the T axis would correspond to T_c (figure 8.7).

8.6 Antiferromagnetics

In antiferromagnetic crystals the nature of the internal field is such that below a certain critical temperature, called the Néel temperature T_N, all the spins are aligned in an antiparallel arrangement (figure 8.8). In the simplest case the crystal lattice can be divided into two similar sublattices such that below T_N the spins of one sublattice are all parallel to each other and pointing in one direction, whilst the spins on the other sublattice are also parallel to each other but are pointing in the opposite direction.

In the absence of an external field $\boldsymbol{H}$ the crystal possesses no net magnetization. However, each sublattice has a spontaneous magnetization whose temperature dependence is similar to ferromagnets (figure 8.9).

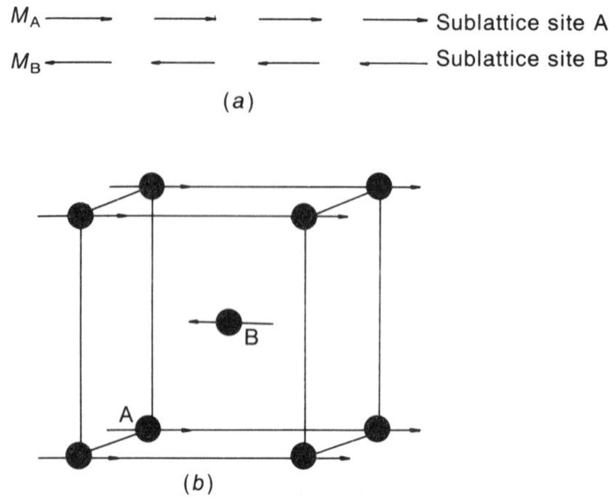

Figure 8.8 Ordering of spins in an antiferromagnetic material. (*a*) Antiparallel arrangement of spins (schematic): M_A, magnetization on sublattice A; M_B, magnetization on sublattice B; $M_A = M_B$. (*b*) Antiparallel ordering in a body-centred cubic lattice. A (corner sites) and B (body-centred sites) denote two similar sublattices.

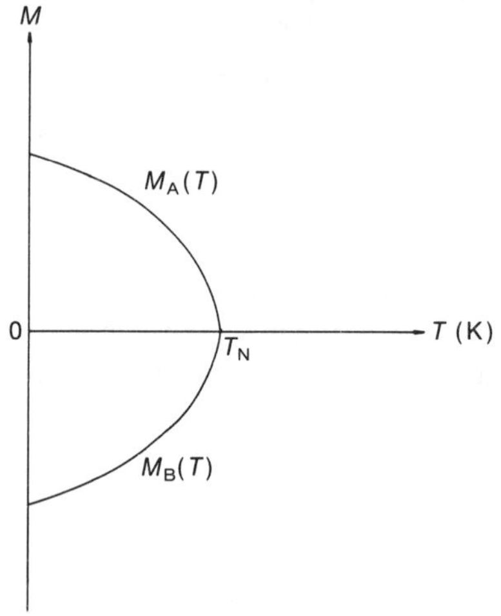

Figure 8.9 Temperature dependence of sublattice magnetization in antiferromagnets (schematic).

As in ferromagnets there is also an internal anisotropy field present whose role is to determine the direction of easy magnetization, that is the direction along which the bulk magnetization vector $\boldsymbol{M}$ points on the two sublattices.

The temperature dependence of the antiferromagnetic susceptibility is shown in figure 8.10. The susceptibility below T_N depends on whether $\boldsymbol{H}$ is applied parallel or perpendicular to the direction of magnetization. When

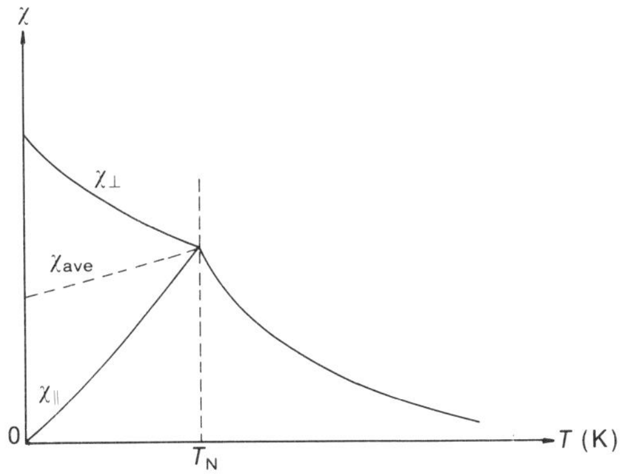

Figure 8.10 Temperature dependence of susceptibility in antiferromagnets. (Highly schematic. Many antiferromagnets show wide variations in susceptibility as a function of T below T_N.)

H is applied parallel to ***m***, $\chi_\parallel$ goes to zero at $T = 0$. This is because the moments on the spin-down sublattice cannot turn over owing to the strong internal field. When ***H*** is perpendicular to ***M***, however, the sublattices respond feebly and $\chi_\perp$ has more or less a constant value from $T = 0$ to T_N, that is the temperature variation is usually small (figure 8.10). Above T_N, χ obeys the usual paramagnetic law, namely $\chi = C/(T + T_N)$, when only intersublattice interactions are considered.

8.7 Ferrimagnetics

In ferrimagnetic crystals the internal field is such that the individual spins are ordered in an antiparallel manner as for antiferromagnets, but unlike the antiferromagnetic case there is no longer complete cancellation between the magnetic moments of the two (or more) sublattices and the crystal would therefore exhibit a net spontaneous magnetization (figure 8.11). This would happen, for example, in materials which contain more than one kind of magnetic atom, for example ferrites ($MnFe_2O_4$, $NiFe_2O_4$, etc) so that $M_A - M_B \neq 0$, resulting in a temperature variation of sublattice magnetization as shown schematically in figure 8.12.

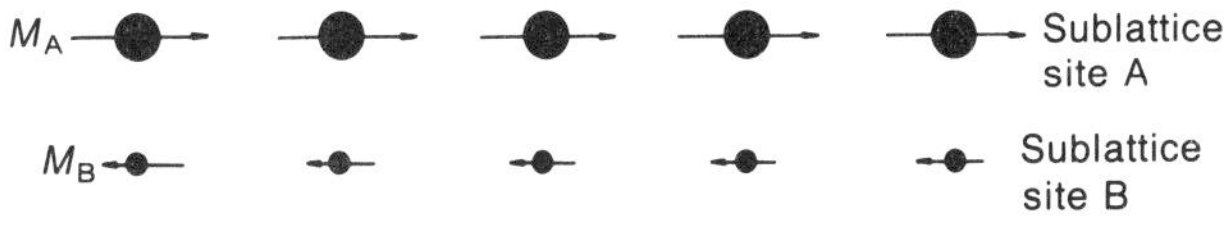

Figure 8.11 Ordering of spins in ferrimagnets (schematic). $M_A \neq M_B$.

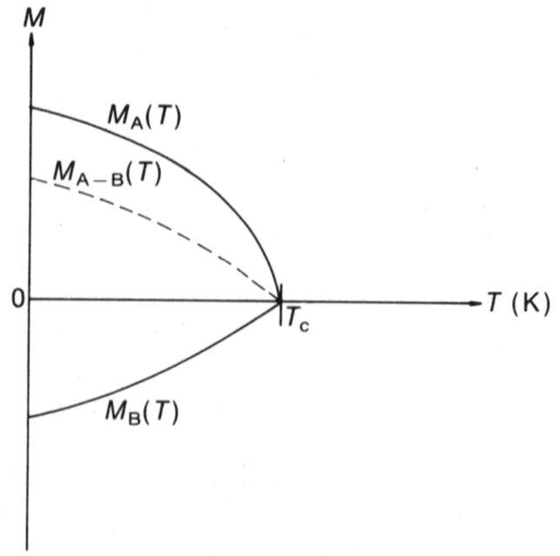

Figure 8.12 Temperature variation of sublattice and resultant magnetization in ferromagnets (schematic).

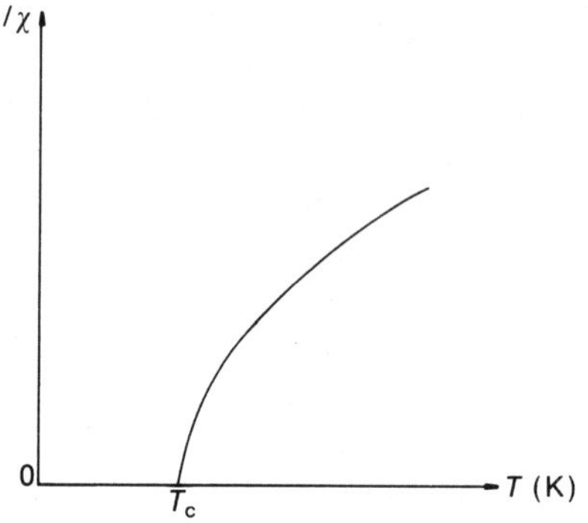

Figure 8.13 Néel curve for ferrimagnets showing the variation of inverse susceptibility with temperature (schematic).

The temperature variation of susceptibility in ferrimagnets is generally described by the Curie–Néel law (figure 8.13)

$$\frac{1}{\chi} = \frac{1}{\chi_0} + \frac{T}{C} - \frac{D}{T - T_c}$$

where χ_0, C and D are constants. Above T_c ferrimagnets exhibit the usual paramagnetic behaviour.

9

CONSTRUCTION OF MAGNETIC GROUPS

9.1 Magnetic point groups

We are all no doubt familiar with the 14 Bravais lattices (also called 'uncoloured' Bravais lattices), the seven crystal systems, the 32 crystal classes based on the 32 crystallographic point groups and the 230 space groups that are obtained by combining the 32 point groups with the translation group of the 14 Bravais lattices (see [32–36] and Chapter 4 of this book). This information is summarized in table 9.1.

The ordinary point groups merely describe the possible point symmetry of the mean charge density function $\rho(\boldsymbol{r})$ of the crystal in the equilibrium state. In strongly magnetic crystals (ferromagnetics, ferrimagnetics, antiferromagnetics, etc), however, besides $\rho(\boldsymbol{r})$ there may also be present a non-vanishing time-averaged distribution of current density $J(\boldsymbol{r})$ and spin density $S(\boldsymbol{r})$, or in other words a total magnetic moment density $\mu(\boldsymbol{r}) = J(\boldsymbol{r}) + S(\boldsymbol{r})$. Now the symmetry of $\mu(\boldsymbol{r})$ is characterized by a special transformation which involves the reversal of the vector direction (see [12–14, 37, 38]). This specific operation of vector reversal, which is not present in the ordinary crystallographic point groups, is incorporated in magnetic groups by means of a new antisymmetry operator R which simply reverses the sign of magnetic moment at each point in space but does not act on the space coordinates. Shubnikov [39] introduced the idea of antisymmetry by studying the symmetry groups of the polyhedra with coloured faces and derived 122 coloured groups. As an illustration of the derivation let us consider the symmetry operations of a square [40]. As shown in figure 9.1(a) there are in all eight symmetry operations present which we identify as follows:

E,	identify operation
C_{4z}^{+},	clockwise rotation through 90° about 0
C_{2z},	rotation through 180° about 0
C_{4z}^{-},	clockwise rotation through 270°, or an anticlockwise rotation through 90°, about 0
σ_{d_1}, σ_{d_2}, σ_{v_x}, σ_{v_y},	reflection operations in vertical planes as indicated in figure 9.1(a).

These eight operations make up the group C_{4v} (Schoenflies notation) or 4 mm (International notation).

Table 9.1 Summary of the 14 Bravais lattices, seven crystal systems, 32 crystal classes and 230 crystal space groups.

Crystal system No.	Name	Definition of unit cell	Lattices Symbol	No.	Highest symmetry symbol	Point-group symmetry of crystal Symbol	No.	No. of space group
1	Triclinic	$\alpha \neq \beta \neq \gamma$ $a \neq b \neq c$	P	1	C_i	C_i; C_1	2	2
2	Monoclinic	$\alpha = \beta = 90^\circ \neq \gamma$ $a \neq b \neq c$	P, C	2	C_{2h}	C_{2h}; C_{1h}; C_2	3	13
3	Orthorhombic	$\alpha = \beta = \gamma = 90^\circ$ $a \neq b \neq c$	P, C, I, F	4	D_{2h}	D_{2h}; D_2; C_{2v}	3	59
4	Tetragonal	$\alpha = \beta = \gamma = 90^\circ$ $a = b \neq c$	P, I	2	D_{4h}	D_{4h}; C_{4h}; D_{2d}; C_{4v}; D_4; S_4; C_4	7	68
5	Trigonal	$\alpha = \beta = \gamma \neq 90^\circ < 120^\circ$ $a = b = c$	R	1	D_{3d}	D_{3d}; D_3; C_{3v}; S_6; C_3	5	52
6	Hexagonal	$\alpha = \beta = 90^\circ$ $\gamma = 120^\circ$ $a = b \neq c$	P	1	D_{6h}	D_{6h}; C_{6h}; D_{3h}; C_{6v}; D_6; C_{3h}; C_6	7	
7	Cubic	$\alpha = \beta = \gamma = 90^\circ$ $a = b = c$	P, I, F	3	O_h	O_h; T_h; T_d O; T	5	36
			Total	14			32	230

Bravais lattice symbols:
P–primitive
I–body centred
F–face centred
C–side centred
R–rhombohedral

Crystal axes:

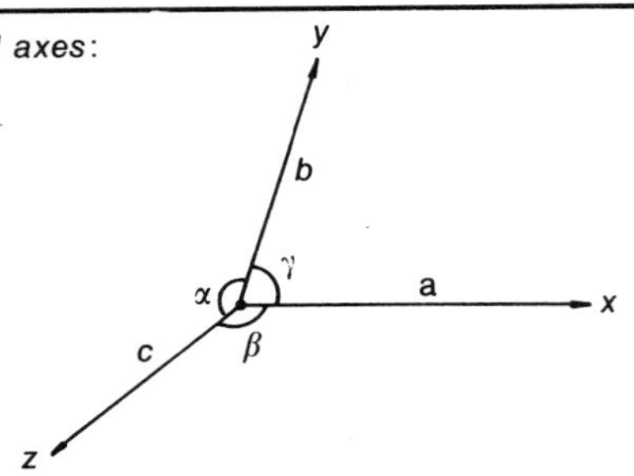

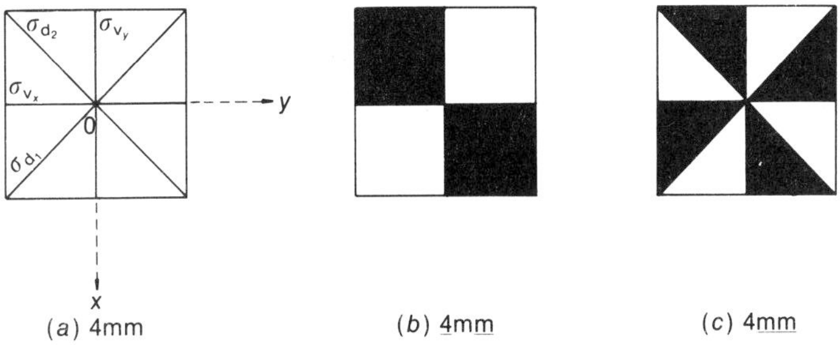

Figure 9.1 Point-group symmetry of uncoloured (a) and coloured (b, c) squares.

Suppose now we colour part of the square black and part white, as shown in figures 9.1(b) and 9.1(c), and introduce a new colour-changing operator R which acts only on the colour, that is turns black to white or white to black. We can then readily see that the symmetry operations of 9.1(b) for example must be E, C_{2z}, σ_{d_1}, σ_{d_2}, RC_{4z}^{+}, RC_{4z}^{-}, $R\sigma_{v_x}$, $R\sigma_{v_y}$, giving a new colour group $\underline{4}\,m\underline{m}$. In the International notation one takes the uncoloured group and underlines all those elements to which R is attached to give the resulting coloured group.

Similarly, if we considered the symmetry operations of figure 9.1(c) we could get the following elements: E, C_{4z}^{+}, C_{4z}^{-}, C_{2z}, $R\sigma_{v_x}$, $R\sigma_{v_y}$, $R\sigma_{d_1}$, $R\sigma_{d_2}$, giving rise to the new group designated $4\,\underline{mm}$. Thus, from the simple (uncoloured) group $4\,mm$ we have derived two coloured groups, $\underline{4}\,m\underline{m}$ and $4\,\underline{mm}$. It is interesting to note that, in each of the new colour groups derived, half of the operations are coloured operations, that is R is attached. This is in fact not just coincidence (as indicated later) and forms a necessary and sufficient condition in the construction of the new colour groups [3].

The colour-changing operation is of course analogous to the reversal of spin direction, that is R could be considered an operator which, when applied to a particle with spin pointing up, (↑) say, would change it to a particle with spin pointing down (↓) and vice versa. The operation R thus reverses a magnetic moment. Alternatively we can think of R as reversing the direction of an electric current because an orbiting charge constitutes a current, that is

Since $i = \mathrm{d}q/\mathrm{d}t$ we may write $-i = \mathrm{d}q/\mathrm{d}(-t)$ and therefore we could also associate R as a *time inversion* operator.

Let us now consider briefly the effect of R on a wavefunction ψ in a magnetic crystal. We may write the time-dependent Schrödinger equation as

$$H\psi = \mathrm{i}\hbar\,\frac{\partial \psi}{\partial t}.$$

H is real so that taking the complex conjugate we obtain

$$H\psi^* = -i\hbar\,\frac{\partial\psi^*}{\partial t}$$

or

$$H\psi^* = i\hbar\,\frac{\partial\psi^*}{\partial(-t)}.$$

Thus by reversing the sign of t and changing ψ to ψ^*, Schrödinger's equation is still satisfied so that we can make the connection

$$R\psi = \psi^*.$$

R is thus called an *antiunitary* operator and the colour groups which contain R are called non-unitary groups.

It is worth noting that the colour groups which we had derived earlier contain both unitary and antiunitary operators as follows:

Ordinary group G	Colour group M	Unitary elements forming subgroup H	Antiunitary elements
4 mm	$\underline{4}\,m\underline{m}$	E, C_{2z}, σ_{d_1}, σ_{d_2} (2 mm)	RC_{4z}^{+}, RC_{4z}^{-}, $R\sigma_{v_x}$, $R\sigma_{v_y}$
4 mm	$4\,\underline{mm}$	E, C_{4z}^{+}, C_{4z}^{-}, C_{2z} (4)	$R\sigma_{v_x}$, $R\sigma_{v_y}$, $R\sigma_{d_1}$, $R\sigma_{d_2}$

An operator U is said to be unitary if

$$\langle U\psi | U\phi\rangle = \langle\psi|\phi\rangle$$

where ψ and ϕ are state functions of the system. An operator A is said to be antiunitary if

$$\langle A\psi | A\phi\rangle = \langle\psi|\phi\rangle^* = \langle\phi|\psi\rangle.$$

If Φ is expanded in the form

$$\Phi = \sum_\alpha C_\alpha \phi_\alpha$$

then as shown by Wigner [38] U is linear since

$$U\Phi = \sum_\alpha C_\alpha U\Phi_\alpha.$$

The colour groups which we may now identify as magnetic groups contain equal numbers of unitary and antiunitary elements.

Having considered an example of the derivation of two colour groups ($\underline{4}\,m\underline{m}$ and $4\,\underline{mm}$) from an uncoloured group (4 mm) we may now list the entire colour groups (magnetic groups). These are usually classified under three types as follows:

type I: 32 uncoloured groups (these are the 32 ordinary crystallographic point groups);
type II: 32 *grey* point groups;
type III: 58 *black and white* point groups.

The procedure for constructing these 122 colour groups is as follows.

If G is an ordinary point group (type I) then a type II point group M is given by

$$M = G + RG.$$

A type III point group is given by

$$M = H + R(G - H)$$

where H is the unitary halving subgroup of G as explained in the above table. The halving subgroup H for each of the type III groups has been worked out by several authors (see [3, 13, 34, 37]) and are listed in Appendix C in both the Schoenflies and International notations.

It is worthwhile considering the construction of these magnetic groups by yet another and more elegant method using the theory of representation analysis [29].

The magnetic groups are derived quite easily by analysing the character tables for the 32 ordinary point groups. We use the character tables as given by Koster *et al* [35]. The method of construction of the magnetic groups is as follows:

(i) The totally symmetric representation Γ_1 or Γ_1^+ is considered as being synonymous with the ordinary point group. This therefore gives us the 32 type I groups.
(ii) The remaining real one-dimensional representations are used to construct the type III magnetic groups.
(iii) If Γ_i is a real one-dimensional representation, a magnetic group is constructed by taking all those elements with character +1 to be in the unitary subgroup H and all those elements with character −1 to be in (G − H) and therefore to be associated with R.
(iv) The complex one-dimensional representations and the degenerate representations are of no relevance in this context.

As an illustrative example we consider table D.5 of Appendix D for the group C_{2h} (Schoenflies notation) or 2/M (International notation). We see that Γ_1^+ is identified as the type I group. For the remaining real one-dimensional representations Γ_2^+, Γ_1^- and Γ_2^-, the invariant unitary subgroup H is formed from those elements having character +1 and thus corresponds to subgroups C_i, C_2 and C_{1h}. Hence the three magnetic groups derived from the ordinary group C_{2h} are, in the Schoenflies notation, $C_{2h}(C_i)$, $C_{2h}(C_2)$ and $C_{2h}(C_{1h})$ or, in the International notation, $2/\underline{M}$, $\underline{2}/\underline{M}$ and $\underline{2}/M$ respectively. In the International notation one underlines all the elements to which *R* is attached. The complete derivation of all the magnetic groups and the type of magnetic structure associated with each of them is given in tables D.1–D.32 of Appendix D. It is to be noted that in deriving these type III groups only distinct groups are counted. For example, table D.8 of Appendix D gives us only three distinct type III groups and not seven, as might appear at first sight.

Table 9.2 Symbols used in drawing the stereograms of the magnetic point groups.

Symbol	Description
○	Inversion
	Anti-inversion
⬮	Rotation-diad
	Antirotation-diad
▲	Rotation-triad
■	Rotation-tetrad
	Antirotation-tetrad
⬣	Rotation-hexad
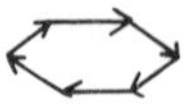	Antirotation-hexad
——	Reflection
══	Antireflection
×	General point (neglecting spin) above plane of paper
□	General point (neglecting spin) below plane of paper
●	General point with spin up above plane of paper
○	General point with spin down above plane of paper
	General point with spin up below plane of paper
	General point with spin down below plane of paper

The graphical symbols used in drawing the stereographic projections of the magnetic groups constructed above are summarized in table 9.2. The symbols are self-explanatory and are simple extensions of the standard notations. As can readily be seen, the character tables are a useful aid in drawing the stereograms since all symmetry elements with character -1 are represented with the corresponding symbols of antioperation. The stereographic projections of the 58 type III magnetic point groups are given in Appendix E.

9.2 Magnetic (coloured) lattices

In non-magnetic crystals there are 14 distinct translation groups (the Bravais lattices) which combine with the 32 ordinary crystallographic point groups to give the 230 space groups. When dealing with magnetic crystals, however, we have to combine the 90 coloured point groups with coloured (black and white) Bravais lattices. As an illustration for deriving these coloured lattices we again turn to the tetragonal system 4mm. If we consider the primitive tetragonal Bravais lattice (figure 9.2(a)) defined by the three basic vectors $\boldsymbol{t}_1, \boldsymbol{t}_2, \boldsymbol{t}_3$ we see that it is possible to add to these lattices of black points an identical set of white points (figure 9.2(b)) at sites defined by a non-primitive translation vector $\boldsymbol{\tau} = \frac{1}{2}(\boldsymbol{t}_1 + \boldsymbol{t}_2 + \boldsymbol{t}_3)$. $\boldsymbol{\tau}$ thus connects a black (white) lattice site to a white (black) lattice site. The black and white lattice sites are of course analogous to spin-up (↑) and spin-down (↓) sites in magnetic crystals. In this simple example we have in fact constructed a black and white coloured lattice of the tetragonal system which is commonly designated as $\mathrm{P_I}$. If we were to consider all the 14 Bravais lattices (uncoloured) which we listed in table 9.1 we would find that it is possible to derive 22 black and white Bravais lattices. These coloured lattices are shown in Appendix F.

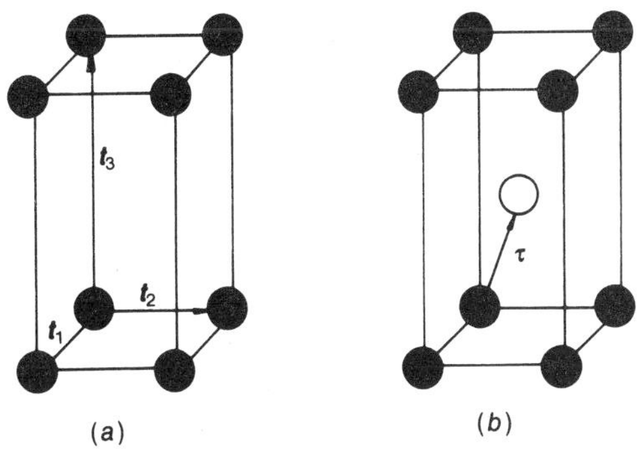

Figure 9.2 Uncoloured (a) and coloured (b) tetragonal Bravais lattices: (a) P, (b) $\mathrm{P_I}$.

9.3 Magnetic space groups

The full symmetry group of a magnetic crystal is of course the magnetic space group. Magnetic space groups are obtained by combining magnetic point groups derived in §9.1 with the translation group of the coloured lattices considered in §9.2. When this is done a total of 1651 colour (magnetic) groups are obtained (see Appendix G) and these can be classed as follows:

(i) *230 colourless groups* (M). These are the 230 ordinary (Federov) groups for non-magnetic crystals which do not involve the time reversal operator R. If G is any Federov group then the corresponding colourless (non-magnetic) group is given by

$$\mathrm{M} = \mathrm{G}.$$

(ii) *230 grey groups* ($\mathrm{M_g}$). These are obtained by associating R with each element of the Federov group G, that is

$$\mathrm{M_g} = \mathrm{G} + R\mathrm{G}.$$

(iii) *674 black and white groups associated with the 14 uncoloured or ordinary Bravais lattices* ($\mathrm{M_{bwO_L}}$). A black and white magnetic space group belonging to this class is defined by

$$\mathrm{M_{bwO_L}} = \mathrm{H} + R(\mathrm{G} - \mathrm{H}).$$

Here H is the halving subgroup of the unitary spatial Federov group G.

(iv) *517 black and white groups associated with the 22 coloured Bravais lattices* ($\mathrm{M_{bwC_L}}$). A group belonging to this class is defined by

$$\mathrm{M_{bwC_L}} = \mathrm{G} + \tau R\mathrm{G}$$

where τ is the extra translation operation introduced when dealing with coloured lattices (see figure 9.2(b)).

10

COREPRESENTATION THEORY FOR MAGNETIC GROUPS

Since the operator R is antiunitary (Chapter 9) the magnetic groups contain both unitary and antiunitary operators and are thus non-unitary groups. The conventional theory of representations valid for the unitary groups is therefore not directly relevant to magnetic groups and has to be replaced by the theory of corepresentations developed by Wigner [38].

The irreducible representations of non-unitary groups are termed corepresentations (corep for short). The distinction is made because the corep matrices do not multiply in the same way as the operators themselves. This is a direct result of the fact that the antiunitary operators of the magnetic groups are antilinear (Chapter 7). The multiplication scheme for the corep matrices of the magnetic groups is as follows:

$$\begin{aligned} \mathbf{D}^i(U_1)\mathbf{D}^i(U_2) &= \mathbf{D}^i(U_1U_2) \\ \mathbf{D}^i(U)\mathbf{D}^i(A) &= \mathbf{D}^i(UA) \\ \mathbf{D}^i(A)\mathbf{D}^i(U)^* &= \mathbf{D}^i(AU) \\ \mathbf{D}^i(A_1)\mathbf{D}^i(A_2)^* &= \mathbf{D}^i(A_1A_2). \end{aligned} \tag{10.1}$$

Wigner's results have shown that the coreps $\mathbf{D}^i$ for the non-unitary magnetic groups are obtained from the representations Δ^i of the unitary subgroup in one of three ways.

If U is a member of the unitary subgroup H and A is one of the antiunitary operators in the magnetic group, and if $\Delta^i(U)$ is a representation of H, the type of corep $\mathbf{D}^i$ depends on $\Delta^i(U)$ and $\bar{\Delta}^i(U)$, where $\bar{\Delta}^i(U) = \Delta^i(A_0^{-1}UA_0)^*$. Here A_0 is any fixed antiunitary operator in the magnetic group. If $\Delta^i(U)$ and $\bar{\Delta}^i(U)$ are equivalent, that is $\chi\{\Delta^i(U)\} = \chi\{\bar{\Delta}^i(U)\}$, it is possible to express $\bar{\Delta}^i(U)$ in the form $\beta^{-1}\Delta^i(U)\beta$, and in this case two possibilities arise depending on $\beta\beta^*$ (Wigner [38], Dimmock and Wheeler [12, 13]). If

(a) $$\beta\beta^* = \Delta^i(A_0^2) \tag{10.2}$$

then

$$\begin{aligned} \mathbf{D}^i(U) &= \Delta^i(U) \\ \mathbf{D}^i(A) &= \Delta^i(AA_0^{-1})\beta \end{aligned} \tag{10.3}$$

but if

(b)
$$\beta\beta^* = -\Delta^i(A_0^2) \tag{10.4}$$

then

$$\mathbf{D}^i(U) = \begin{pmatrix} \Delta^i(U) & 0 \\ 0 & \Delta^i(U) \end{pmatrix}$$
$$\mathbf{D}^i(A) = \begin{pmatrix} 0 & \Delta^i(AA_0^{-1})\beta \\ -\Delta^i(AA_0^{-1})\beta & 0 \end{pmatrix}. \tag{10.5}$$

(c) If $\Delta^i(U)$ and $\bar{\Delta}^i(U)$ are not equivalent representations, then the corep of the magnetic group is

$$\mathbf{D}^i(U) = \begin{pmatrix} \Delta^i(U) & 0 \\ 0 & \bar{\Delta}^i(U) \end{pmatrix}$$
$$\mathbf{D}^i(A) = \begin{pmatrix} 0 & \Delta^i(AA_0) \\ \Delta^i(A_0^{-1}A)^* & 0 \end{pmatrix}. \tag{10.6}$$

Wigner also pointed out that it does not matter which of the antiunitary operators is taken to be A_0.

The following statements concern the dimensionality of the coreps as obtained from the representations of Δ^i.

Case (a). No new degeneracy is introduced. The irreducible corep $\mathbf{D}^i$ of the magnetic group M corresponds to a single irreducible representation $\Delta^i(U)$ of H and has the same dimension.

Case (b). The degeneracy of $\Delta^i(U)$ is doubled. The irreducible corep $\mathbf{D}^i$ of M corresponds again to a single irreducible representation of H but has twice its dimension.

Case (c). The representation $\Delta^i(U)$ is degenerate with the representation $\Delta^j(U) = \Delta^i(A_0^{-1}UA_0)^*$ $(i \neq j)$. The irreducible corep $\mathbf{D}^i$ corresponds to two inequivalent irreducible representations of H, namely $\Delta^i(U)$ and $\Delta^j(U)$ such that in this case the antiunitary operators cause $\Delta^i(U)$ and $\Delta^j(U)$ to become degenerate.

The general results stated above will be applied in later sections to determine the corep of the magnetic space groups for NiF_2 and NiO.

The irreducible representations of H are obtained readily since H is unitary. It therefore remains to obtain a method by which one can decide between the three cases, given the group M and an irreducible representation of H. In order to do this we outline the method of Dimmock and Wheeler [12, 13]. We assume that β and $\Delta^i(U)$ are unitary and that the $\Delta^i(U)$ matrices satisfy the usual orthogonality relation

$$\sum_k \Delta^i(U_k)_{rl}\Delta^j(U_k)^*_{sm} = \frac{n}{l_i}\delta_{ij}\delta_{rs}\delta_{lm} \tag{10.7}$$

where the sum is over all the n elements in H and l_i is the dimension of the irreducible representation $\Delta^i(U)$. Adopting the usual convention of an understood summation over repeated indices, we can write

$$\sum_k \Delta^i(A_k^2)_{rr} = \sum_k \Delta^i(U_k A_0 U_k A_0)_{rr}$$

$$= \sum_k \Delta^i(U_k)_{rl} \Delta^i(A_0^2)_{lt} \Delta^i(A_0^{-1} U_k A_0)_{tr}. \qquad (10.8)$$

In cases (a) and (b) this yields, using the equations

$$\Delta^i(A_0^{-1} U_k A_0)^*_{tr} = \beta^*_{ts} \Delta^i(U_k)^*_{sm} \beta^*_{mr}$$

and (10.2) and (10.4),

$$\sum_k \Delta^i(A_k^2)_{rr} = \sum_k \Delta^i(U_k)_{rl} \Delta^i(A_0^2)_{lt} \beta_{ts}^{-1*} \Delta^i(U_k)^*_{sm} \beta^*_{mr}$$

$$= \frac{n}{l_i} \delta_{rs} \delta_{lm} \Delta^i(A_0^2)_{lt} \beta_{st} \beta^*_{mr}$$

$$= \frac{n}{l_i} \Delta^i(A_0^2)_{lt} \beta_{rt} \beta^*_{lr} \quad \text{for } r = s \text{ and } l = m$$

$$= \pm \frac{n}{l_i} \Delta^i(A_0^2)_{lt} \Delta^i(A_0^2)^*_{lt} \quad \text{using } \beta\beta^* = \pm \Delta^i(A_0^2)$$

$$= \pm \frac{n}{l_i} \Delta^i(A_0^2)_{lt} \Delta^i(A_0^2)^{-1}_{tl}$$

$$= \pm \frac{n}{l_i} \Delta^i(E)_{ll}$$

$$= \pm n$$

where the plus sign holds for case (a) and the minus sign for case (b). In case (c) $\Delta^i(A_0^{-1} U A_0)^*$ is not equivalent to $\Delta^i(U)$ so that, from (10.7),

$$\sum_k \Delta^i(A_k^2)_{rr} = \sum_k \Delta^i(U_k)_{rl} \Delta^i(A_0^2)_{lt} \Delta^i(A_0^{-1} U_k A_0)_{tr}$$

$$= \sum_k \Delta^i(U_k)_{rl} \Delta^i(A_0^2)_{lt} \Delta^j(U_k)^*_{tr}$$

$$= 0.$$

Collecting these results we find that

$$\begin{aligned}\sum_k \chi^i(A_k^2) &= +n \quad \text{case (a)} \\ &= -n \quad \text{case (b)} \\ &= 0 \quad \text{case (c)}\end{aligned} \tag{10.9}$$

where $\chi^i(A_k^2) = \Delta^i(A_k^2)_{rr}$ is the character of the representation matrix for the operator A_k^2. We recall from the properties of non-unitary groups that A_k^2 is a member of the unitary subgroup. We have written above $A_k = U_k A_0$ and it is still more convenient to write $A_0 = V_0 R$, where V_0 is unitary, such that, by using the fact that $R^2 = \omega E$ ($\omega = +1$ for single group representations and $\omega = -1$ for double-group representations) and that R commutes with spatial operators, equation (10.9) becomes

$$\begin{aligned}\sum_k \chi^i(V_0 U_k V_0 U_k) &= +\omega n \quad \text{case (a)} \\ &= -\omega n \quad \text{case (b)} \\ &= 0 \quad \text{case (c).}\end{aligned} \tag{10.10}$$

If the group M contains the time reversal operator itself (as for paramagnetic crystals) we can choose $V_0 = E$, the identity operator, and equation (10.10) reduces to

$$\begin{aligned}\sum_k \chi^i(U_k^2) &= +\omega n \quad \text{case (a)} \\ &= -\omega n \quad \text{case (b)} \\ &= 0 \quad \text{case (c).}\end{aligned} \tag{10.11}$$

The above results (10.9)–(10.11) are used in the next chapter to find the irreducible coreps of the magnetic little group of NiF_2 and NiO.

11

SYMMETRY PROPERTIES OF WAVEFUNCTIONS IN MAGNETIC CRYSTALS

11.1 Introduction

The use of group theory in determining the eigenstates of a crystal by the classification of these states according to the irreducible representations of crystal space groups was first introduced by Bouckaert *et al* [41]. Since their work there have been many papers considering the same thing for various space groups. The most important of these have been the employment of time reversal symmetry by Herring [43] and the procedure for finding the representations of space groups by Koster [42]. The major foundation relating to the present work was laid by the idea of antisymmetry, first introduced by Shubnikov [39]. (Actually Heesch [44] was the first to suggest the idea of antisymmetry but its importance was not realized at the time and the credit has passed on to Shubnikov.) This idea of antisymmetry as we have seen in Chapter 9 leads to the theory of magnetic groups or Shubnikov groups.

The symmetry properties of the wavefunctions in a magnetic crystal are determined by finding the irreducible corepresentations of the groups of the wavevector $\boldsymbol{k}$ for each of the points and lines of symmetry of the first Brillouin zone of the magnetic and non-magnetic lattices (see [5, 7, 10, 40, 41] for further discussion on groups of the wavevector $\boldsymbol{k}$, often called the little group). The first Brillouin zone is chosen because it not only contains all the irreducible representations of the pure translation on or inside its surface, but also has the further advantage of possessing the full symmetry of the lattice in $\boldsymbol{k}$ space. We therefore choose this to be the unit cell in $\boldsymbol{k}$ space. Above the critical temperature T_c, if H is the space group of the crystal then, because in this state the crystal possesses a vanishing time-averaged magnetic moment density, it follows that its potential will also be invariant under the time reversal operator R. The full space group of the crystal will then be given by $\mathrm{G} = \mathrm{H} + \mathrm{H}R$. Below T_c, however, there would be a reduction in symmetry brought about by the magnetic ordering of the lattice and the consequent introduction of a non-vanishing time-averaged magnetic moment density. In this case R will not leave the magnetic crystal unchanged so that the crystal potential will now be invariant under another group of unitary operators H′. Thus the full magnetic space group below T_c will be given by G′ = H′ + H′R. The groups G and G′ contain both unitary and antiunitary operations. The groups H and H′ are unitary. The subgroup relations between these four groups which characterize the crystal in its magnetic and non-magnetic (paramagnetic) states are as follows:

Lattice	Non-unitary		Unitary
Paramagnetic ($T > T_c$)	G	→	H
	↓		↓
Magnetic ($T < T_c$)	G′	→	H′

The symmetry properties are therefore determined by finding the irreducible corepresentations of the groups of the wavevector H_k and H'_k for all points and lines of symmetry in the paramagnetic and magnetic Brillouin zones. The effect of ordering is seen by forming the compatibility relations between the coreps of H_k and H'_k for the corresponding points in the appropriate Brillouin zones.

If the magnetic and non-magnetic unit cells are the same, that is if the translational symmetry of the crystal is unchanged by the magnetic ordering, the Brillouin zones below and above T_c will be the same. This is the case for the 3D transition metal difluorides MnF_2, FeF_2, CoF_2 and NiF_2 which are all antiferromagnets below their respective Néel temperatures. However, for the 3D transition metal oxides (MnO, FeO, CoO, NiO) for example, the magnetic unit cell is found experimentally to be twice as large as the non-magnetic unit cell so that the Brillouin zone for the two lattices is different. As an illustration of the two cases we give below a comprehensive survey of the symmetry properties of the wavefunctions in the antiferromagnets NiF_2 and NiO.

11.2 Symmetry properties of NiF_2

NiF_2, like the other isomorphic antiferromagnetic difluorides MnF_2, FeF_2 and CoF_2, exhibits a rutile structure as shown in figure 11.1. The disposition of the metallic ions (Ni^{2+}) which are denoted by open circles is indicated relative to a tetragonal unit cell with centre at 0. The F^- ions, denoted by full circles, are all at the same distance from the nearest metallic ion. Classically the crystal possesses a four-fold axis, four two-fold axes and a centre of symmetry 0 [14]. The appropriate non-magnetic point group is thus 4/mmm (D_{4h}). Below the Néel temperature the spins of the Ni^{2+} ions are perpendicular to the z axis and the antiferromagnetic structure is shown in figure 11.2. The spins are very nearly parallel and antiparallel to the y axis, but are canted by a small angle $\phi \approx 0.39°$ towards the x axis and consequently there is a weak resultant ferromagnetic moment in the x direction. However, the

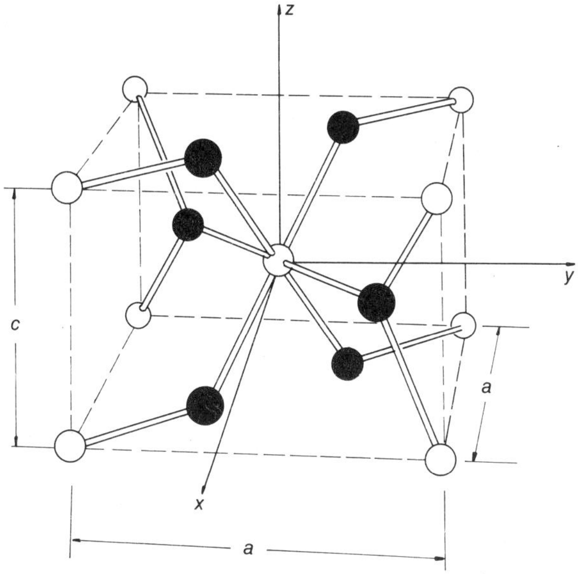

Figure 11.1 The unit cell of NiF_2: open circles represent Ni^{2+} ions and full circles represent F^- ions. $a = 4.710$ Å, $c = 3.118$ Å.

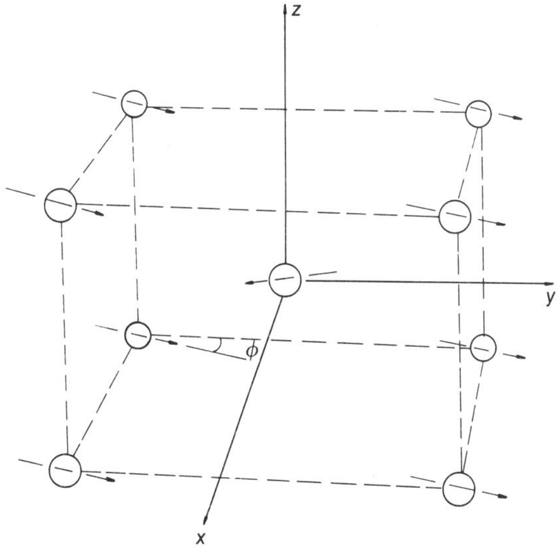

Figure 11.2 The orientation of the spins of the Ni^{2+} ions in antiferromagnetic NiF_2; the spins are in the x–y plane at ϕ to y axis.

symmetry of the crystal structure is exactly the same as it would be if one neglected the canting. This is true because it is possible to rotate the spins, for example from parallelism with $0y$ towards $0x$, giving rise to a small net magnetic moment. Weak ferromagnetism is prohibited for the other difluorides

(MnF_2, FeF_2, CoF_2) because any rotations of the spins away from parallelism or antiparallelism with the tetragonal axis $0z$ lowers the symmetry of the crystal since the direction is no longer a four-fold axis of symmetry. The appropriate magnetic point group for NiF_2 is m$\underline{mm}$ (International notation) or $D_{2h}(C_{2h})$ (Schoenflies notation) with symmetry elements E, I, C_{2x}, σ_{v_x}, $C_{2z} . R$, $C_{2y} . R$, $\sigma_h . R$, $\sigma_{v_y} . R$. We can now use the principal results of Chapter 10 to determine the irreducible coreps of the space group of the NiF_2 structure.

11.2.1 Paramagnetic NiF_2 ($T > T_N$)

The space group for NiF_2 above T_N is the same as that of the difluorides MnF_2, FeF_2 and CoF_2. This structure is described by the ordinary Fedorov space group $P4_2/mnm$ or, if time reversal is included by the grey Shubnikov space group, $P4_2/mnm1'$. We use the notation given by Dimmock and Wheeler [12] for the symmetry operations of the rutile structure. The basic translations of the primitive tetragonal lattice are

$$t_1 = a\hat{i} \qquad t_2 = a\hat{j} \qquad t_3 = c\hat{k}$$

where $\hat{i}, \hat{j}$ and $\hat{k}$ are unit vectors in the x, y and z directions respectively. The non-primitive translation is taken to be

$$\tau = \tfrac{1}{2}(t_1 + t_2 + t_3).$$

The reciprocal lattice is also primitive tetragonal and its basic vectors are

$$g_1 = \frac{2\pi}{a}\hat{i} \qquad g_2 = \frac{2\pi}{a}\hat{j} \qquad g_3 = \frac{2\pi}{a}\hat{k}.$$

The Brillouin zone for the primitive tetragonal lattice is illustrated in figure 11.3 with the points and lines of symmetry marked as in Dimmock and Wheeler [12].

In table 11.1 we give a summary for the description of the groups of the wavevector for $P4_2/mnm$.

Since the irreducible representation of $P4_2/mnm$ and the irreducible corepresentations of $P4_2/mnm1'$ have been tabulated by Dimmock and Wheeler [12, tables 1–3] we shall not repeat it here. The last column in these tables of Dimmock and Wheeler indicates to which of the three cases (a), (b) or (c) (see Chapter 10) a specific representation $\Delta^i(u)$ of the unitary group $P4_2/mnm$ belongs when it is considered as an invariant unitary subgroup of the non-unitary magnetic group $P4_2/mnm1'$.

11.2.2 Antiferromagnetic NiF_2 ($T < T_N$)

The unit cell of the antiferromagnetic structure of NiF_2 is the same as the unit cell of the paramagnetic structure and consequently the Shubnikov space group or magnetic space group M of the antiferromagnetic structure is a subgroup of $P4_2/mnm1'$. The magnetic space group M contains an equal number of unitary and antiunitary elements, and they are $\{E|0\}$, $\{C_{2x}|\tau\}$, $\{I|0\}$, $\{\sigma_{v_x}|\tau\}$ and $R\{C_2|0\}$, $R\{C_{2y}|\tau\}$, $R\{\sigma_h|0\}$, $R\{\sigma_{v_y}|\tau\}$.

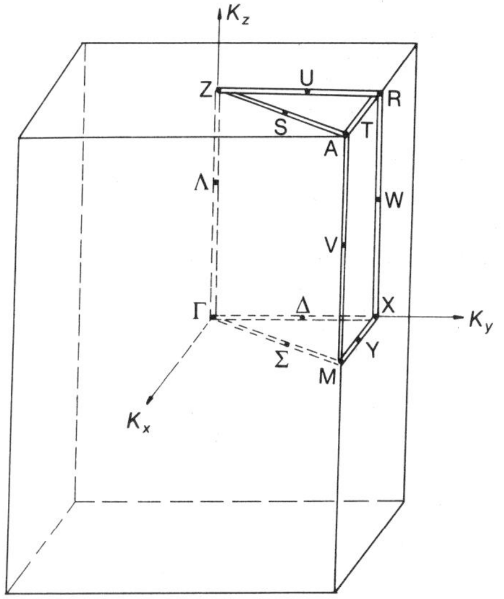

Figure 11.3 The Brillouin zone for the primitive tetragonal lattice of paramagnetic NiF_2 with the points and lines of symmetry indicated.

Table 11.1 Points and lines of symmetry in the Brillouin zone of paramagnetic NiF_2.

Point or line	*k*	Group of the wavevector: Group	Elements of the group
Γ	(0, 0, 0)	D_{4h}	$E, C_4, C_4^{-1}, C_2,$
Z	$(0, 0, \frac{1}{2})$		$C_{2x}, C_{2y}, C_{2a}, C_{2b},$
M	$(\frac{1}{2}, \frac{1}{2}, 0)$		$I, S_4^{-1}, S_4, \sigma_h,$
A	$(\frac{1}{2}, \frac{1}{2}, \frac{1}{2})$		$\sigma_{v_x}, \sigma_{v_y}, \sigma_{d_a}, \sigma_{d_b}$
X	$(0, \frac{1}{2}, 0)$	D_{2h}	$E, C_{2x}, C_{2y}, C_2,$
R	$(0, \frac{1}{2}, \frac{1}{2})$		$I, \sigma_{v_x}, \sigma_{v_y}, \sigma_h$
Λ	$(0, 0, \alpha)$	C_{4v}	E, C_4, C_4^{-1}, C_2
V	$(\frac{1}{2}, \frac{1}{2}, \alpha)$		
Δ	$(0, \alpha, 0)$	C_{2v}	$E, C_{2y}, \sigma_h, \sigma_{v_x}$
U	$(0, \alpha, \frac{1}{2})$		
Σ	$(\alpha, \alpha, 0)$	C_{2v}	$E, C_{2a}, \sigma_h, \sigma_{d_a}$
S	$(\alpha, \alpha, \frac{1}{2})$		
Y	$(\alpha, \frac{1}{2}, 0)$	C_{2v}	$E, C_{2x}, \sigma_h, \sigma_{v_y}$
T	$(\alpha, \frac{1}{2}, \frac{1}{2})$		
W	$(0, \frac{1}{2}, \alpha)$	C_{2v}	$E, C_2, \sigma_{v_x}, \sigma_{v_y}$

The subgroup of the unitary elements makes up the space group $P2_1/b$ (C_{2h}^5) and the complete magnetic space group is p*nn*′m′ in the notation of Shubnikov and Belov [11]. M is thus a primitive orthorhombic magnetic space group with basic vectors

$$\boldsymbol{t}_1 = a\hat{\boldsymbol{i}} \qquad \boldsymbol{t}_2 = b\hat{\boldsymbol{j}} \qquad \boldsymbol{t}_3 = c\hat{\boldsymbol{k}}$$

and reciprocal lattice vectors

$$\boldsymbol{g}_1 = (2\pi/a)\hat{\boldsymbol{i}} \qquad \boldsymbol{g}_2 = (2\pi/b)\hat{\boldsymbol{j}} \qquad \boldsymbol{g}_3 = (2\pi/c)\hat{\boldsymbol{k}}$$

whereas $P4_2/mnm1'$ was a primitive tetragonal space group. If there is no distortion of the lattice when NiF_2 becomes ordered, a and b are accidentally equal. In fact, a very-high-precision (1 part in 10^5) x-ray diffraction measurement of the lattice constants of NiF_2 below the Néel temperature has detected a small difference between the two lattice constants a and b (see [47]). The actual results were

$$a = (4.648\,44 \pm 0.000\,04)\,\text{Å} \qquad b = (4.647\,19 \pm 0.000\,04)\,\text{Å}$$

where a is the direction of the weak ferromagnetic moment, that is in the x direction. Although the unitary subgroup $P2_1/b$ is, strictly speaking, monoclinic it does have special values of the angle β and the axial ratios determined by the primitive orthorhombic Bravais lattice of the orthorhombic magnetic space group $Pnn'm'$. The Brillouin zone of the primitive orthorhombic Bravais lattice is shown in figure 11.4. It will be noted that the representative fraction of the Brillouin zone is larger than it was for the paramagnetic state; a similar situation arises in MnF_2 [48]. The points and lines of symmetry that were

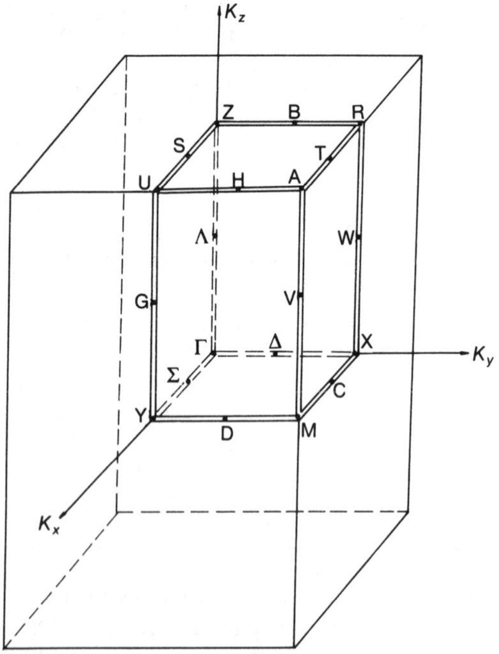

Figure 11.4 The Brillouin zone for the primitive orthorhombic lattice of antiferromagnetic NiF_2 with the points and lines of symmetry indicated.

present in the Brillouin zone of the paramagnetic NiF_2 are nearly all still present in the orthorhombic Brillouin zone of the antiferromagnetic NiF_2 (see table 11.1). $(\alpha, \alpha, 0)$ and $(\alpha, \alpha, \pi/c)$, which were lines of symmetry in the paramagnetic structure, are no longer lines of symmetry in the antiferromagnetic structure. However, there are two additional points of symmetry

Table 11.2 Points and lines of symmetry in the Brillouin zone of antiferromagnetic NiF_2.

Point or line	$\boldsymbol{k}$	Magnetic little group Unitary elements	Antiunitary elements
Γ	(0, 0, 0)	$E, C_{2x}, I, \sigma_{v_x}$	$C_2, C_{2y}, \sigma_h, \sigma_{v_y}$
X	$(0, \frac{1}{2}, 0)$		
Y	$(\frac{1}{2}, 0, 0)$		
Z	$(0, 0, \frac{1}{2})$		
U	$(\frac{1}{2}, 0, \frac{1}{2})$		
R	$(0, \frac{1}{2}, \frac{1}{2})$		
M	$(\frac{1}{2}, \frac{1}{2}, 0)$		
A	$(\frac{1}{2}, \frac{1}{2}, \frac{1}{2})$		
Δ(ΓX)	$(0, \alpha, 0)$	E, σ_{v_x}	C_2, σ_{v_y}
D(YM)	$(\frac{1}{2}, \alpha, 0)$		
H(UA)	$(\frac{1}{2}, \alpha, \frac{1}{2})$		
B(ZR)	$(0, \alpha, \frac{1}{2})$		
Σ(ΓY)	$(\alpha, 0, 0)$	E, C_{2x}	C_2, C_{2y}
C(XM)	$(\alpha, \frac{1}{2}, 0)$		
T(RA)	$(\alpha, \frac{1}{2}, \frac{1}{2})$		
S(ZU)	$(\alpha, 0, \frac{1}{2})$		
Λ(ΓZ)	$(0, 0, \alpha)$	E, σ_{v_x}	σ_h, C_{2y}
W(XR)	$(0, \frac{1}{2}, \alpha)$		
V(MA)	$(\frac{1}{2}, \frac{1}{2}, \alpha)$		
G(YU)	$(\frac{1}{2}, 0, \alpha)$		

Table 11.3 Character table for antiferromagnetic NiF_2.

Γ(0, 0, 0):

C_{2h}	Γ_1^+	Γ_2^+	Γ_1^-	Γ_2^-	Γ_3^+	Γ_4^+	Γ_3^-	Γ_4^-
$\{E\|0\}$	1	1	1	1	1	1	1	1
$\{\bar{E}\|0\}$	1	1	1	1	−1	−1	−1	−1
$\{C_{2x}\|\boldsymbol{\tau}\}$	1	−1	1	−1	i	−i	i	−i
$\{\bar{C}_{2x}\|\boldsymbol{\tau}\}$	1	−1	1	−1	−i	i	−i	i
$\{I\|0\}$	1	1	−1	−1	1	1	−1	−1
$\{\bar{I}\|0\}$	1	1	−1	−1	−1	−1	1	1
$\{\sigma_{v_x}\|\boldsymbol{\tau}\}$	1	−1	−1	1	i	−i	−i	i
$\{\bar{\sigma}_{v_x}\|\boldsymbol{\tau}\}$	1	−1	−1	1	−i	i	i	−i
Coreps	a	a	a	a	a	a	a	a

Table 11.4 Character table for antiferromagnetic NiF_2. $X(0, \frac{1}{2}, 0)$, $Y(\frac{1}{2}, 0, 0)$, $Z(0, 0, \frac{1}{2})$, $A(\frac{1}{2}, \frac{1}{2}, \frac{1}{2})$:

$X(n = 2)$ $Z(n = 3)$	$Y(n = 1)$ $A(n = 1, 2, \text{ or } 3)$	X_1 Z_1 Y_2 A_2	X_2 Z_2 Y_1 A_1
$\{E\|0\}$	$\{E\|0\}$	2	2
$\{E\|\boldsymbol{t}_n\}$	$\{E\|\boldsymbol{t}_n\}$	-2	-2
$\{\bar{E}\|0\}$	$\{\bar{E}\|\boldsymbol{t}_n\}$	2	-2
$\{\bar{E}\|\boldsymbol{t}_n\}$	$\{\bar{E}\|0\}$	-2	2
$\{I\|0\}$, $\{I\|\boldsymbol{t}_n\}$	$\{I\|0\}$, $\{I\|\boldsymbol{t}_n\}$	0	0
$\{\bar{I}\|0\}$, $\{\bar{I}\|\boldsymbol{t}_n\}$	$\{\bar{I}\|0\}$, $\{\bar{I}\|\boldsymbol{t}_n\}$	0	0
$\{C_{2x}\|\tau\}$, $\{C_{2x}\|\tau + \boldsymbol{t}_n\}$	$\{C_{2x}\|\tau\}$, $\{C_{2x}\|\tau + \boldsymbol{t}_n\}$	0	0
$\{\bar{C}_{2x}\|\tau\}$, $\{\bar{C}_{2x}\|\tau + \boldsymbol{t}_n\}$	$\{\bar{C}_{2x}\|\tau\}$, $\{\bar{C}_{2x}\|\tau + \boldsymbol{t}_n\}$	0	0
$\{\sigma_{v_x}\|\tau\}$, $\{\sigma_{v_x}\|\tau + \boldsymbol{t}_n\}$	$\{\sigma_{v_x}\|\tau\}$, $\{\sigma_{v_x}\|\tau + \boldsymbol{t}_n\}$	0	0
$\{\bar{\sigma}_{v_x}\|\tau\}$, $\{\bar{\sigma}_{v_x}\|\tau + \boldsymbol{t}_n\}$	$\{\bar{\sigma}_{v_x}\|\tau\}$, $\{\bar{\sigma}_{v_x}\|\tau + \boldsymbol{t}_n\}$	0	0
	Coreps X	a	a
	Coreps Y	a	a
	Coreps Z	a	a
	Coreps A	a	a

Table 11.5 Character table for antiferromagnetic NiF_2. $U(\frac{1}{2}, 0, \frac{1}{2})$, $M(\frac{1}{2}, \frac{1}{2}, 0)$: direct product of group given below with $(\{E|0\} + \{E|\boldsymbol{t}_n\})$.

$U(n = 1 \text{ or } 3)$ $M(n = 1 \text{ or } 2)$	U_1^+ M_1^+	U_2^+ M_2^+	U_1^- M_1^-	U_2^- M_2^-	U_3^+ M_3^+	U_4^+ M_4^+	U_3^- M_3^-	U_4^- M_4^-
$\{E\|0\}$	1	1	1	1	1	1	1	1
$\{C_{2x}\|\tau\}$	i	$-$i	i	$-$i	1	-1	1	-1
$\{\bar{E}\|\boldsymbol{t}_n\}$	-1	-1	-1	-1	1	1	1	1
$\{\bar{C}_{2x}\|\tau + \boldsymbol{t}_n\}$	$-$i	i	$-$i	i	1	-1	1	-1
$\{I\|0\}$	1	1	-1	-1	1	1	-1	-1
$\{\sigma_{v_x}\|\tau\}$	i	$-$i	$-$i	i	1	-1	-1	1
$\{\bar{I}\|\boldsymbol{t}_n\}$	-1	-1	1	1	1	1	-1	-1
$\{\bar{\sigma}_{v_x}\|\tau + \boldsymbol{t}_n\}$	$-$i	i	i	$-$i	1	-1	-1	1
Coreps U	a	a	a	a	a	a	a	a
Coreps M	c ——	—— c	c ——	—— c	c ——	—— c	c ——	—— c

Table 11.6 Character table for antiferromagnetic NiF_2.

$R(0, \frac{1}{2}, \frac{1}{2})$:

Direct product of $(\{E|0\} + \{E|\boldsymbol{t}_n\})$ with group isomorphic with $\Gamma(n = 2 \text{ or } 3)$, replace labels Γ_1^+ by R_1^+, Γ_2^+ by R_2^+, etc.

Coreps: all belong to case (c) and they come from pairs (R_1^+, R_2^+), (R_1^-, R_2^-), (R_3^+, R_4^+), (R_3^-, R_4^-).

Table 11.7 Character table for antiferromagnetic NiF_2.

Δ		Δ_1	Δ_2	Δ_3	Δ_4
B		B_1	B_2	B_3	B_4
Λ		Λ_1	Λ_2	Λ_3	Λ_4
W		W_1	W_2	W_3	W_4
	Σ	Σ_1	Σ_2	Σ_3	Σ_4
	T	T_1	T_2	T_3	T_4
$\{E\|0\}$	$\{E\|0\}$	1	1	1	1
$\{\bar{E}\|0\}$	$\{\bar{E}\|0\}$	1	1	-1	-1
$\{\sigma_{v_x}\|\tau\}$	$\{C_{2x}\|\tau\}$	ω	$-\omega$	$i\omega$	$-i\omega$
$\{\bar{\sigma}_{v_x}\|\tau\}$	$\{\bar{C}_{2x}\|\tau\}$	ω	$-\omega$	$-i\omega$	$i\omega$
	Coreps Δ	a	a	a	a
	Coreps B	c ——	c	c ——	c
	Coreps Λ	a	a	a	a
	Coreps W	c ——	c	c ——	c
	Coreps Σ	a	a	a	a
	Coreps T	c ——	c	c ——	c

$\omega = \exp(-i\boldsymbol{k}\cdot\boldsymbol{\tau})$.

Table 11.8 Character table for antiferromagnetic NiF_2.

D		D_1	D_2	D_3	D_4
H		H_1	H_2	H_3	H_4
V		V_1	V_2	V_3	V_4
G		G_1	G_2	G_3	G_4
	C	C_1	C_2	C_3	C_4
	S	S_1	S_2	S_3	S_4
$\{E\|0\}$	$\{E\|0\}$	1	1	1	1
$\{\bar{E}\|0\}$	$\{\bar{E}\|0\}$	1	1	-1	-1
$\{\sigma_{v_x}\|\tau\}$	$\{C_{2x}\|\tau\}$	$i\omega$	$-i\omega$	ω	$-\omega$
$\{\bar{\sigma}_{v_x}\|\tau\}$	$\{\bar{C}_{2x}\|\tau\}$	$i\omega$	$-i\omega$	$-\omega$	ω
	Coreps D	c ——	c	c ——	c
	Coreps H	a	a	a	a
	Coreps V	c ——	c	c ——	c
	Coreps G	a	a	a	a
	Coreps C	c ——	c	c ——	c
	Coreps S	a	a	a	a

$\omega = \exp(-i\boldsymbol{k}\cdot\boldsymbol{\tau})$.

Y and U (see figure 11.4) and several additional lines of symmetry in the orthorhombic Brillouin zone.

The points and lines of symmetry and the appropriate little groups are identified in table 11.2.

The unitary subgroup $P2_1/b$ of the magnetic space group of antiferromagnetic NiF_2 is one of the ordinary (Fedorov) space groups and its

Table 11.9 Compatibility tables for paramagnetic NiF_2.

Γ_1^+	Γ_2^+	Γ_3^+	Γ_4^+	Γ_5^+	Γ_1^-	Γ_2^-	Γ_3^-	Γ_4^-	Γ_5^-	Γ_6^+	Γ_7^+	Γ_6^-	Γ_7^-
Λ_1	Λ_2	Λ_3	Λ_4	Λ_5	Λ_2	Λ_1	Λ_4	Λ_3	Λ_5	Λ_6	Λ_7	Λ_6	Λ_7
Σ_1	Σ_2	Σ_2	Σ_1	$\Sigma_3 + \Sigma_4$	Σ_3	Σ_4	Σ_4	Σ_3	$\Sigma_1 + \Sigma_2$	Σ_5	Σ_5	Σ_5	Σ_5
Δ_1	Δ_2	Δ_1	Δ_2	$\Delta_3 + \Delta_4$	Δ_3	Δ_4	Δ_3	Δ_4	$\Delta_1 + \Delta_2$	Δ_5	Δ_5	Δ_5	Δ_5

M_1^+	M_2^+	M_3^+	M_4^+	M_5^+	M_1^-	M_2^-	M_3^-	M_4^-	M_5^-	M_6^+	M_7^+	M_6^-	M_7^-
V_1	V_2	V_3	V_4	V_5	V_3	V_4	V_1	V_2	V_5	V_6	V_7	V_6	V_7
Y_1	Y_2	Y_2	Y_1	$Y_3 + Y_4$	Y_3	Y_4	Y_4	Y_3	$Y_1 + Y_2$	Y_5	Y_5	Y_5	Y_5
Σ_1	Σ_1	Σ_2	Σ_2	$\Sigma_3 + \Sigma_4$	Σ_3	Σ_3	Σ_4	Σ_4	$\Sigma_1 + \Sigma_2$	Σ_5	Σ_5	Σ_5	Σ_5

Z_1	Z_2	Z_3	Z_4	Z_5
$S_1 + S_4$	$S_2 + S_3$	$S_1 + S_3$	$S_2 + S_4$	$2S_5$
$\Lambda_1 + \Lambda_4$	$\Lambda_2 + \Lambda_3$	Λ_5	Λ_5	$\Lambda_6 + \Lambda_7$
U_1	U_1	U_1	U_1	$U_2 + U_3 + U_4 + U_5$

A_1	A_2	A_3	A_4	A_5
$S_1 + S_4$	$S_2 + S_3$	$S_1 + S_3$	$S_2 + S_4$	$2S_5$
$V_1 + V_2$	$V_3 + V_4$	V_5	V_5	$V_6 + V_7$
T_1	T_1	T_1	T_1	$T_2 + T_3 + T_4 + T_5$

R_1^+	R_1^-	R_2^+	R_3^+	R_4^+	R_5^+	R_2^-	R_3^-	R_4^-	R_5^-
U_1	U_1	U_2	U_3	U_4	U_5	U_3	U_2	U_5	U_4
W_1	W_1	W_2	W_3	W_4	W_5	W_4	W_5	W_2	W_3
T_1	T_1	T_2	T_3	T_4	T_5	T_5	T_4	T_3	T_2

X_1	X_2	X_3	X_4
W_1	W_1	$W_2 + W_4$	$W_3 + W_5$
$Y_1 + Y_2$	$Y_3 + Y_4$	Y_5	Y_5
$\Delta_1 + \Delta_2$	$\Delta_3 + \Delta_4$	Δ_5	Δ_5

Table 11.10 Compatibility tables for antiferromagnetic NiF_2.

Γ_1^+	Γ_2^+	Γ_1^-	Γ_2^-	Γ_3^+	Γ_4^+	Γ_3^-	Γ_4^-
Σ_1	Σ_2	Σ_1	Σ_2	Σ_3	Σ_4	Σ_3	Σ_4
Δ_1	Δ_2	Δ_2	Δ_1	Δ_3	Δ_4	Δ_4	Δ_3
Λ_1	Λ_2	Λ_2	Λ_1	Λ_3	Λ_4	Λ_4	Λ_3

X_1	X_2
Δ_1, Δ_2	Δ_3, Δ_4
$(W_1 + W_2)$	$(W_3 + W_4)$
$(C_1 + C_2)$	$(C_3 + C_4)$

Y_1	Y_2
Σ_1, Σ_2	Σ_3, Σ_4
$(D_1 + D_2)$	$(D_3 + D_4)$
G_1, G_2	G_3, G_4

Z_1	Z_2
Λ_1, Λ_2	Λ_3, Λ_4
S_1, S_2	S_3, S_4
$(B_1 + B_2)$	$(B_3 + B_4)$

A_1	A_2
$(T_1 + T_2)$	$(T_3 + T_4)$
$(V_1 + V_2)$	$(V_3 + V_4)$
H_1, H_2	H_3, H_4

U_1^+	U_2^+	U_1^-	U_2^-	U_3^+	U_4^+	U_3^-	U_4^-
S_2	S_1	S_2	S_1	S_4	S_3	S_4	S_3
H_2	H_1	H_2	H_1	H_4	H_3	H_4	H_3
G_2	G_1	G_2	G_1	G_4	G_3	G_4	G_3

R_1^+	R_2^+	R_1^-	R_2^-	R_3^+	R_4^+	R_3^-	R_4^-
B_2	B_1	B_1	B_2	B_3	B_4	B_4	B_3
W_2	W_1	W_1	W_2	W_3	W_4	W_4	W_3
T_2	T_1	T_2	T_1	T_4	T_3	T_4	T_3

M_1^+	M_2^+	M_1^-	M_2^-	M_3^+	M_4^+	M_3^-	M_4^-
D_2	D_1	D_1	D_2	D_4	D_3	D_3	D_4
C_2	C_1	C_2	C_1	C_4	C_3	C_4	C_3
V_2	V_1	V_1	V_2	V_4	V_3	V_3	V_4

irreducible representations can therefore be found from the tables given by Faddeyev [49] or Kovalev [36]. Alternatively, since it is a rather intricate task to extract the relevant information from these tables for any given space group, it is probably just as easy to derive the space-group representations of $P2_1/b$ *ab initio*. In tables 11.3–11.8 we give the character tables of the magnetic little groups of the points and lines of symmetry of antiferromagnetic NiF_2. It is possible to write the magnetic little group M^k of the magnetic space group M as

$$M^k = H^k + RG_I^k \tag{11.1}$$

where H^k consists of those unitary elements of M that send a $\boldsymbol{k}$ into $+\boldsymbol{k} + \boldsymbol{K}_0$, where $\boldsymbol{K}_0$ is a reciprocal lattice vector, and RG_I^k consists of those antiunitary elements of M that send $\boldsymbol{k}$ into $-\boldsymbol{k} + \boldsymbol{K}_0$ (see Bradley and Cracknell [10]). In tables 11.3–11.8 only the characters of the unitary elements of the magnetic little groups are given. In the last row of each table we indicate whether the coreps of the magnetic little groups belong to case (a), case (b) or case (c).

The effect of magnetic ordering can be seen from the compatibility tables 11.9 and 11.10 between the unitary representations of H_k and H'_k for corresponding points in the non-magnetic and magnetic Brillouin zones. The compatibility tables for H_k ($P4_2/mnm$) have been taken from [12].

11.3 Symmetry properties of NiO

The transition element oxides MnO, FeO, CoO and NiO are known to be antiferromagnetic substances [50] and their spin arrangements have been studied by neutron diffraction experiments (see [51–53]). NiO has the NaCl structure (Fm3m, O_h^5) whereas below T_N the crystal becomes slightly distorted from the cubic structure to a rhombohedral one [54]. This deformation is presumably due to the magnetostriction which accompanies the magnetic ordering, and the amount of the distortion is found to increase with decreasing temperature. The distortion consists of contraction of the original cubic unit cell along any one of the four $\langle 111 \rangle$ axes (figure 11.5).

The antiferromagnetic ordering of the magnetic moments in NiO is shown in figure 11.6. Only the magnetic ions are shown. The spins of all the ions represented by full circles are parallel to one another and antiparallel to the spins of all the ions represented by open circles. This antiferromagnetic structure is compatible with the neutron diffraction experiments of Roth [51, 52]. The spins in NiO are arranged in ferromagnetic sheets parallel to the (111) plane (see [51–53, 55–57]. Roth [51, 52] showed from his neutron diffraction data that the direction of sublattice magnetization of the ferromagnetic (111) sheets was actually in the plane of the ferromagnetic sheet and he assumed that it was parallel or antiparallel to $[1\bar{1}0]$.

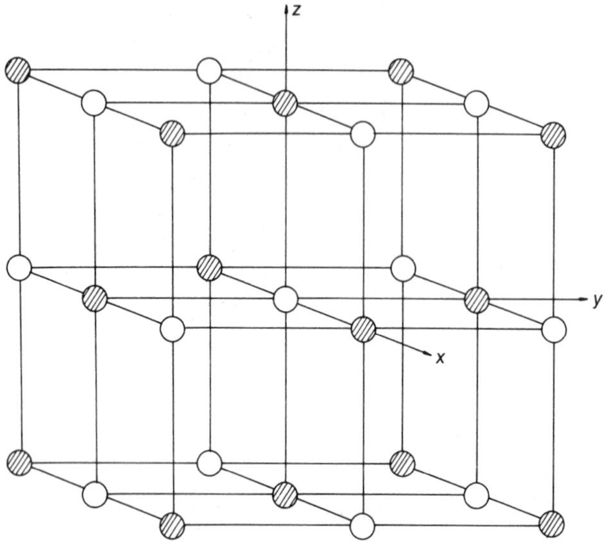

Figure 11.5 The FCC chemical unit cell of NiO.

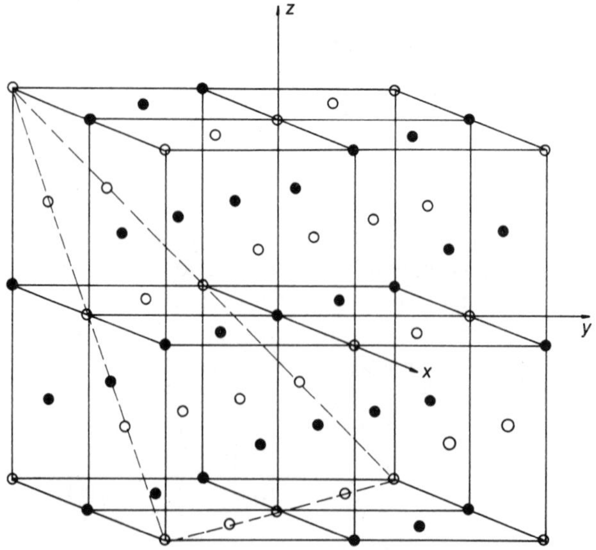

Figure 11.6 The two sublattices of antiferromagnetic NiO. The full and open circles represent spin-up and spin-down Ni^{2+} ions respectively. O^{2-} ions are omitted.

Later, from the observed behaviour of S domains in applied magnetic fields, Yamada [56] showed that the direction of sublattice magnetization was parallel or antiparallel to $[11\bar{2}]$ and not to $[1\bar{1}0]$. This means that within each ferromagnetic (111) sheet the spins are aligned in the direction of the intersection of the sheet with the vertical plane through the z axis and [110], that is along $\pm(\boldsymbol{t}_2 + \boldsymbol{t}_3)$ in figure 11.7. With this picture the symmetry group of the ordered phase is C_c2/c. As for NiF_2, we can now use the principal results of Chapter 10 to classify the eigenstates of NiO both above and below T_N by finding the irreducible corepresentations of the group of the wavevector for each of the points and lines of symmetry of the Brillouin zone of the appropriate magnetic space group.

11.3.1 Paramagnetic NiO ($T > T_N$)

Above T_N, NiO has the NaCl structure. The non-magnetic unit cell therefore has the face-centred cubic Bravais lattice (figure 11.5). In terms of unit vectors $\hat{\boldsymbol{i}}, \hat{\boldsymbol{j}}$ and $\hat{\boldsymbol{k}}$ along the x, y and z axes the primitive translations can be defined as

$$\begin{aligned} \boldsymbol{t}_1 &= \tfrac{1}{2}a(\hat{\boldsymbol{i}} + \hat{\boldsymbol{j}}) \\ \boldsymbol{t}_2 &= \tfrac{1}{2}a(-\hat{\boldsymbol{j}} + \hat{\boldsymbol{k}}) \\ \boldsymbol{t}_3 &= \tfrac{1}{2}a(-\hat{\boldsymbol{i}} + \hat{\boldsymbol{k}}) \end{aligned} \qquad (11.2)$$

where a is the length of a cube edge. The Brillouin zone for this FCC Bravais lattice is shown in figure 11.8. We have summarized in table 11.11 the

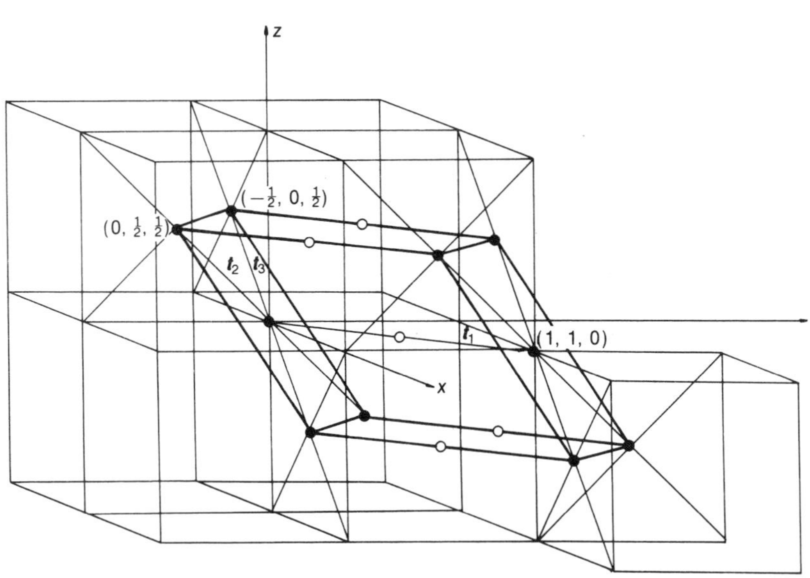

Figure 11.7 The unit cell of the magnetic space group of antiferromagnetic NiO.

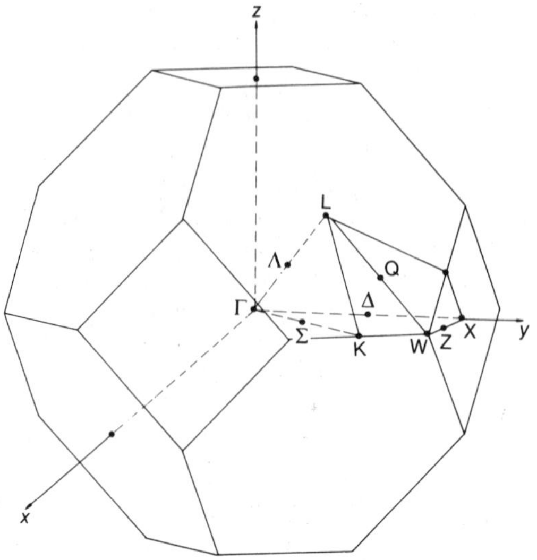

Figure 11.8 The paramagnetic Brillouin zone of NiO with the points and lines of symmetry indicated.

important points and lines of symmetry in the Brillouin zone of the FCC Bravais lattice.

The space group of this lattice is $Fm3m(O_h^5)$. Since this is a symmorphic space group we need consider only the product of $m3m(O_h)$ with $\{E|\boldsymbol{T}_{n_1n_2n_3}\}$ of the translation group, where

$$\boldsymbol{T}_{n_1n_2n_3} = n_1\boldsymbol{t}_1 + n_2\boldsymbol{t}_2 + n_3\boldsymbol{t}_3 \tag{11.3}$$

and n_1, n_2, n_3 are integers.

The paramagnetic space group has in addition to the above operations their products with the time reversal operator R; this is just the grey space group Fm3′m in the notation of Shubnikov and Belov [11].

The little cogroup, $G^{\boldsymbol{k}}$, for any point or line of symmetry in Fm3m, is one of the 32 point groups. Therefore, the magnetic little cogroup, $M^{\boldsymbol{k}}$, for any point or line of symmetry in Fm3′m is simply related to the little cogroup, $G^{\boldsymbol{k}}$, for that point or line in Fm3m. There are two possibilities: if $\boldsymbol{k}$ is a point of symmetry for which $\boldsymbol{k}$ is equivalent to $-\boldsymbol{k}$ (i.e. $\boldsymbol{k}$ and $-\boldsymbol{k}$ are separated by a reciprocal lattice vector) then $M^{\boldsymbol{k}}$ in Fm3′m is simply the grey point group derived from $\boldsymbol{k}$ for that point of symmetry in Fm3m. If $\boldsymbol{k}$ is a point or line of symmetry for which $\boldsymbol{k}$ is not equivalent to $-\boldsymbol{k}$, then $M^{\boldsymbol{k}}$ is one of the 58 black and white groups and can be written as

$$M^{\boldsymbol{k}} = G^{\boldsymbol{k}} + RIG^{\boldsymbol{k}} \tag{11.4}$$

where $G^{\boldsymbol{k}}$ is the little cogroup for that point or line of symmetry in Fm3m and I is the inversion operation. In either case the irreducible corepresentations

Table 11.11 Points and lines of symmetry in the Brillouin zone of FCC NiO (above T_N).

Point or line of symmetry	k	Group of k
Γ	$\frac{2\pi}{a}(0, 0, 0)$	m3m (O_h)
Λ	$\frac{2\pi}{a}(\alpha, \alpha, \alpha)$	3m (C_{3v})
L	$\frac{2\pi}{a}(\frac{1}{2}, \frac{1}{2}, \frac{1}{2})$	$\bar{3}$m (D_{3d})
Δ	$\frac{2\pi}{a}(0, \beta, 0)$	4mm (C_{4v})
X	$\frac{2\pi}{a}(0, 1, 0)$	4/mmm (D_{4h})
Σ	$\frac{2\pi}{a}(\alpha, \alpha, 0)$	2mm (C_{2v})
K	$\frac{2\pi}{a}(\frac{3}{4}, \frac{3}{4}, 0)$	2mm (C_{2v})
W	$\frac{2\pi}{a}(\frac{1}{2}, 1, 0)$	$\bar{4}$2m (D_{2d})
S	$\frac{2\pi}{a}(\alpha, 1, \alpha)$	2mm (C_{2v})
U	$\frac{2\pi}{a}(\frac{1}{2}, 1, \frac{1}{2})$	2mm (C_{2v})
Z	$\frac{2\pi}{a}(\alpha, 1, 0)$	2mm (C_{2v})
Q	$\frac{2\pi}{a}(\frac{1}{2}, \frac{1}{2}+\alpha, \frac{1}{2}-\alpha)$	2 (C_2)

of M^k can then easily be written down, since the irreducible corepresentations of the grey point groups and of the black and white point groups have been given by various authors (see [59, 60, 12]).

11.3.2 Antiferromagnetic NiO ($T < T_N$)

The antiferromagnetic structure of NiO below T_N is shown in figure 11.6. This structure has two magnetic sublattices and the magnetic unit cell is twice as large as the chemical unit cell. As mentioned earlier, below T_N the crystal becomes slightly distorted from the cubic structure to a rhombohedral one, the distortion being temperature sensitive. At room temperature the rhombohedral angle α is about $90° 4.2'$ and the unit cell is a slightly distorted cube. The variation of the lattice constant a and the rhombohedral angle α with temperature has been given by Slack [58]. His results are summarized in table 11.12 for three different temperatures.

To preserve the trigonal symmetry the direction of sublattice magnetization would have to be along [111]. Since the direction of sublattice magnetization is $[11\bar{2}]$ (see [56, 57]), the symmetry group of the ordered phase is C_c2/c and further distortions, from trigonal to monoclinic symmetry, may occur.

The unit cell corresponding to the black and white magnetic space group C_c2/c is shown in figure 11.7. The primitive translations are defined as

$$\begin{aligned} \boldsymbol{t}_1 &= a(\hat{\boldsymbol{i}} + \hat{\boldsymbol{j}}) \\ \boldsymbol{t}_2 &= \tfrac{1}{2}a(-\hat{\boldsymbol{j}} + \hat{\boldsymbol{k}}) \\ \boldsymbol{t}_3 &= \tfrac{1}{2}a(-\hat{\boldsymbol{i}} + \hat{\boldsymbol{k}}) \end{aligned} \tag{11.5}$$

where we neglect the very small distortions.

The Brillouin zone corresponding to this magnetic structure is shown in figure 11.9. The reciprocal lattice vectors are

$$\begin{aligned} \boldsymbol{g}_1 &= \frac{2\pi}{a}(\tfrac{1}{2}\hat{\boldsymbol{i}} + \tfrac{1}{2}\hat{\boldsymbol{j}} + \tfrac{1}{2}\hat{\boldsymbol{k}}) \\ \boldsymbol{g}_2 &= \frac{2\pi}{a}(-\hat{\boldsymbol{i}} - \hat{\boldsymbol{j}} + \hat{\boldsymbol{k}}) \\ \boldsymbol{g}_3 &= \frac{2\pi}{a}(-\hat{\boldsymbol{i}} + \hat{\boldsymbol{j}} + \hat{\boldsymbol{k}}). \end{aligned} \tag{11.6}$$

Since we have ignored the small distortions from the cubic dimensions the geometrical shape of the antiferromagnetic Brillouin zone in figure 11.9 does have a three-fold axis along $\boldsymbol{g}_1$; indeed, if the direction of sublattice magnetization were along the [111] direction then the crystal would be

Table 11.12 Values of lattice constant a_0 and rhombohedral angle α at three different temperatures (see [58]).

T (K)	a_0(Å)	α
0	4.1725	$90°6'$
297	4.1770	$90°4.2'$
573	4.1930	$90°$

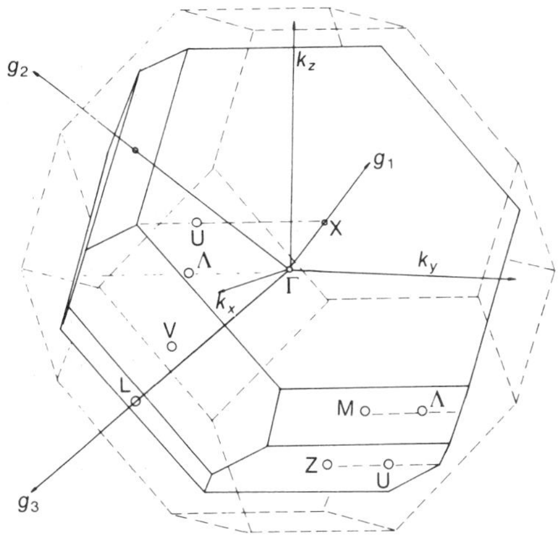

Figure 11.9 The antiferromagnetic Brillouin zone of NiO; the paramagnetic Brillouin zone is indicated in broken lines.

described by a trigonal (rhombohedral) magnetic space group. However, the spins have been assumed to be in some direction in the vertical plane through $\boldsymbol{g}_1$ so that the existence of a three-fold axis along $\boldsymbol{g}_1$ is incompatible with the symmetry of the spin system. In the notation of Dimmock and Wheeler [13] the magnetic structure has the following point-group symmetry operations, E, C_{2b}, I, σ_{d_h}, which make up the point group 2/m (C_{2h}). The magnetic space group of this structure is based on the black and white monoclinic Bravais lattice C_c[11] and is C_c2/c, which contains the elements $M = \{E|0\}$, $\{C_{2b}|\boldsymbol{\tau}\}$, $\{I|0\}$, $\{\sigma_{d_h}|\boldsymbol{\tau}\}R\{E|\boldsymbol{\tau}\}$, $R\{C_{2b}|0\}$, $R\{I|\boldsymbol{\tau}\}$, $R\{\sigma_{d_h}|0\}$, together with the products of these elements with $\{E|T_{n_1n_2n_3}\}$ and $\{\bar{E}|T_{n_1n_2n_3}\}$. Here $\boldsymbol{\tau} = \frac{1}{2}a(\boldsymbol{i} + \hat{\boldsymbol{j}}) = \frac{1}{2}\boldsymbol{t}_1$ is a non-primitive translation which carries a black (white) Bravais lattice point to a position formerly occupied by a white (black) Bravais lattice point and $\boldsymbol{t}_1$, $\boldsymbol{t}_2$ and $\boldsymbol{t}_3$ are the primitive translation vectors of the antiferromagnetic structure shown in figure 11.7. In table 11.13 the coordinates of the special points and lines of symmetry are given and the elements of G^k are identified. These are the unitary elements of M^k, and the antiunitary elements of M^k are given by multiplying G^k by $R\{I|\boldsymbol{\tau}\}$.

In tables 11.14–11.19 we have given the character table of the unitary subgroup G^k of the wavevector for each of the points and lines of symmetry for the magnetic space group C_c2/c shown in figure 11.9. M^k can be identified for each of the points of symmetry from table 11.13 and the last row of each of tables 11.14–11.19 indicates, as before, whether the irreducible coreps of M^k belong to case (a), case (b) or case (c).

It is interesting to study in a little more detail the symmetry properties of the magnetic Brillouin zone of NiO. This is because, for NiO, the magnetic

Table 11.13 Symmetry points in the magnetic Brillouin zone of antiferromagnetic NiO.

Point or line of symmetry	$\boldsymbol{k}$	$\boldsymbol{k}$	$\bar{G}^k$
Γ	$\frac{2\pi}{a}(0, 0, 0)$	0	$E, C_{2b}, I, \sigma_{d_h}$
M	$\frac{2\pi}{a}(\frac{1}{2}, \frac{1}{2}, -\frac{1}{2})$	$\boldsymbol{g}_1 - \frac{1}{2}\boldsymbol{g}_2 - \frac{1}{2}\boldsymbol{g}_3$	$E, C_{2b}, I, \sigma_{d_h}$
X	$\frac{2\pi}{a}(\frac{1}{4}, \frac{1}{4}, \frac{1}{4})$	$\frac{1}{2}\boldsymbol{g}_1$	$E, C_{2b}, I, \sigma_{d_h}$
Z	$\frac{2\pi}{a}(\frac{1}{4}, \frac{1}{4}, -\frac{3}{4})$	$\frac{1}{2}\boldsymbol{g}_1 - \frac{1}{2}\boldsymbol{g}_2 - \frac{1}{2}\boldsymbol{g}_3$	$E, C_{2b}, I, \sigma_{d_h}$
L	$\frac{2\pi}{a}(\frac{1}{2}, -\frac{1}{2}, -\frac{1}{2})$	$\frac{1}{2}\boldsymbol{g}_2$	E, I
V	$\frac{2\pi}{a}(\frac{3}{4}, -\frac{1}{4}, -\frac{1}{4})$	$\frac{1}{2}\boldsymbol{g}_1 - \frac{1}{2}\boldsymbol{g}_3$	E, I
Λ	$\frac{2\pi}{a}(\alpha, -\alpha, 0)$	$\alpha(\boldsymbol{g}_2 - \boldsymbol{g}_3)$	E, C_{2b}
U	$\frac{2\pi}{a}(\alpha, -\alpha, \frac{1}{4})$	$\frac{1}{2}\boldsymbol{g}_1 + \alpha(\boldsymbol{g}_2 - \boldsymbol{g}_3)$	E, C_{2b}
B	$\frac{2\pi}{a}(\frac{1}{2}\alpha, \frac{1}{2}\alpha, \frac{1}{2}\alpha + 2\beta)$	$\boldsymbol{g}_1 + \beta(\boldsymbol{g}_2 + \boldsymbol{g}_3)$	E, σ_{d_h}

and non-magnetic sublattices are dissimilar. In figure 11.9 we have shown the antiferromagnetic Brillouin zone inside the paramagnetic FCC Brillouin zone; $\boldsymbol{g}_2$ and $\boldsymbol{g}_3$ of (11.6) are the same as would be obtained for the cubic paramagnetic phase, while $\boldsymbol{g}_1$ is half of the corresponding $\boldsymbol{g}_1$ for the cubic phase. The volume of the antiferromagnetic Brillouin zone is thus half of the volume of the Brillouin zone of the cubic phase.

Therefore, in general, the zone boundaries will be closer to Γ than in the paramagnetic phase. In terms of the reduced wavevector ζ the phonon Brillouin zone boundaries in the [001], [110] and [111] directions for $\boldsymbol{k} = (2\pi/a)\ (0, 0, \zeta)$, $\boldsymbol{k} = (2\pi/a)\ (\zeta, \zeta, 0)$ and $\boldsymbol{k} = (2\pi/a)\ (\zeta, \zeta, \zeta)$ occur at $\zeta = 1.0$, 0.75 and 0.5 respectively. For the magnetic Brillouin zone shown in figure 11.9 the boundaries are at $\zeta = 0.75$ in the [001] direction, at $\zeta = 0.375$ in the [110] direction, and at $\zeta = 0.25$ in the [111] direction. Because [111] is along the direction of $\boldsymbol{g}_1$ the magnon dispersion curves along [111] from

Table 11.14 Character table for antiferromagnetic NiO. Γ and M.

Γ:

Γ	Γ_1^+	Γ_2^+	Γ_1^-	Γ_2^-	Γ_3^+	Γ_4^+	Γ_3^-	Γ_4^-
$\{E\|0\}$	1	1	1	1	1	1	1	1
$\{\bar{E}\|0\}$	1	1	1	1	−1	−1	−1	−1
$\{C_{2b}\|\tau\}$	1	−1	1	−1	i	−i	i	−i
$\{\bar{C}_{2b}\|\tau\}$	1	−1	1	−1	−i	i	−i	i
$\{I\|0\}$	1	1	−1	−1	1	1	−1	−1
$\{\bar{I}\|0\}$	1	1	−1	−1	−1	−1	1	1
$\{\sigma_{d_h}\|\tau\}$	1	−1	−1	1	i	−i	−i	i
$\{\bar{\sigma}_{d_h}\|\tau\}$	1	−1	−1	1	−i	i	i	−i
Coreps	a	a	a	a	c ——	c	c ——	c

$M(\boldsymbol{g}_1 - \frac{1}{2}\boldsymbol{g}_2 - \frac{1}{2}\boldsymbol{g}_3)$:

Direct product of $(\{E|0\} - \{E|\boldsymbol{t}_n\})$ with the group M^k for $(n = 2$ or $3)$.

Coreps: same as Γ.

Table 11.15 Character table for antiferromagnetic NiO: Z and X.

Z(n = 1, 2 or 3) X(n = 1)	Z_1 X_1	Z_2 X_2
$\{E\|0\}$	2	2
$\{\bar{E}\|0\}$	2	−2
$\{E\|\boldsymbol{t}_n\}$	−2	−2
$\{\bar{E}\|\boldsymbol{t}_n\}$	−2	2
$\{C_{2b}\|\tau\}$, $\{C_{2b}\|\tau + \boldsymbol{t}_n\}$	0	0
$\{\bar{C}_{2b}\|\tau\}$, $\{\bar{C}_{2b}\|\tau + \boldsymbol{t}_n\}$	0	0
$\{I\|0\}$	0	0
$\{\bar{I}\|0\}$	0	0
$\{I\|\boldsymbol{t}_n\}$	0	0
$\{\bar{I}\|\boldsymbol{t}_n\}$	0	0
$\{\sigma_{d_h}\|\tau\}$, $\{\sigma_{d_h}\|\tau + \boldsymbol{t}_n\}$	0	0
$\{\bar{\sigma}_{d_h}\|\tau\}$, $\{\bar{\sigma}_{d_h}\|\tau + \boldsymbol{t}_n\}$	0	0
Coreps X	a	a
Coreps Z	a	a

$\zeta = 0$ to $\zeta = 0.5$ should be symmetrical about $\zeta = 0.25$ and therefore flat at $\zeta = 0.25$. The behaviour of the magnon branches along [001] for $\zeta > 0.75$ and along [110] for $\zeta > 0.375$ is more complicated because these are not along the direction of a reciprocal lattice vector and there is no reflection

Table 11.16 Character table for antiferromagnetic NiO: L and V.

$L(n=2)$ $V(n=1 \text{ or } 3)$	L_1^+ V_1^+	L_1^- V_1^-	L_2^+ V_2^+	L_2^- V_2^-
$\{E\|0\}$	1	1	1	1
$\{\bar{E}\|0\}$	1	1	−1	−1
$\{E\|\boldsymbol{t}_n\}$	−1	−1	−1	−1
$\{\bar{E}\|\boldsymbol{t}_n\}$	−1	−1	1	1
$\{I\|0\}$	1	−1	1	−1
$\{\bar{I}\|0\}$	1	−1	−1	1
$\{I\|\boldsymbol{t}_n\}$	−1	1	−1	1
$\{\bar{I}\|\boldsymbol{t}_n\}$	−1	1	1	−1
Coreps L	a	a	a	a
Coreps V	c ——	c	c ——	c

Table 11.17 Character table for antiferromagnetic NiO: Λ.

Λ	Λ_1	Λ_2	Λ_3	Λ_4
$\{E\|0\}$	1	1	1	1
$\{\bar{E}\|0\}$	1	1	−1	−1
$\{C_{2b}\|\boldsymbol{\tau}\}$	1	−1	i	−i
$\{\bar{C}_{2b}\|\boldsymbol{\tau}\}$	1	−1	−i	i
Coreps Λ	a	a	c ——	c

Table 11.18 Character table for antiferromagnetic NiO: U.

U	U_1	U_2	U_3	U_4
$\{E\|0\}$	1	1	1	1
$\{\bar{E}\|0\}$	1	1	−1	−1
$\{C_{2b}\|-\boldsymbol{\tau}\}$	i	−i	−1	1
$\{\bar{C}_{2b}\|-\boldsymbol{\tau}\}$	i	−i	1	−1
Coreps U	a	a	c ——	c

plane of symmetry normal to either of these directions. The magnon branches in these directions are therefore not symmetrical about $\zeta = 0.75$ for the [001] direction, nor about $\zeta = 0.375$ for the [110] direction, and neither are the branches flat at those points. The magnon branches as calculated directly by Morgan and Joshua (see [24, 25]) do exhibit just the symmetry properties

Table 11.19 Character table for antiferromagnetic NiO: B.

B	B_1	B_2	B_3	B_4
$\{E\|0\}$	1	1	1	1
$\{\bar{E}\|0\}$	1	1	-1	-1
$\{\sigma_{d_h}\|\tau\}$	ξ	$-\xi$	$i\xi$	$-i\xi$
$\{\bar{\sigma}_{d_h}\|\tau\}$	ξ	$-\xi$	$-i\xi$	$i\xi$
Coreps B	a	a	c ——	c

$\xi = \exp(-i\boldsymbol{k}\cdot\boldsymbol{\tau})$.

that we described above for the [111], [110] and [001] directions. The six directions which were [110], [011], [101], [$1\bar{1}0$], [$10\bar{1}$] and [$0\bar{1}1$] in the cubic phase and which cut the Brillouin zone boundaries at a distance of $\frac{3}{4}\sqrt{2}(2\pi/a)$ from Γ in the cubic phase are now different and cut the Brillouin zone boundaries at different distances from Γ. If we ignore the distortions, then [110], [011] and [101] cut the Brillouin zone boundaries at $\frac{3}{8}\sqrt{2}(2\pi/a)$ from Γ while [$1\bar{1}0$], [$10\bar{1}$] and [$0\bar{1}1$] cut the Brillouin zone boundary at $\frac{11}{16}\sqrt{2}(2\pi/a)$ from Γ. There will thus be different dispersion curves for the magnon frequency against $\boldsymbol{k}$ in these directions. A multidomain structure can exist below the Néel temperature even in a crystal which is, crystallographically speaking, a single crystal, because as the crystal is cooled below the Néel temperature there is a choice of four equivalent directions, [111], [$\bar{1}11$], [$1\bar{1}1$] and [$11\bar{1}$], for the normal to the ferromagnetic sheets. In different parts of the crystal a different one of these directions might be chosen so that a crystal is obtained that magnetically has a domain structure while being crystallographically a single crystal.

11.4 Crystal field splittings in NiO and NiF_2

In this section we shall briefly study the different energy level splittings of the Ni^{2+} ion when placed in various crystal field surroundings which lower the symmetry of the originally free ion. The general theory for crystal field splittings has been given by various authors (see [5, 8, 72–75]). Cracknell, in his review paper [76], has given a more advanced account using the theory of magnetic groups which is particularly relevant for our study.

We now study the splittings of the Ni^{2+} ions when subjected to the following symmetry fields which occur in NiO:

(a) free atom;
(b) cubic field (O_h);
(c) cubic field plus spin–orbit coupling;

(d) rhombohedral field (D_3);
(e) orthorhombic field ($\underline{2}/\underline{m}$).

The ground state term of the Ni^{2+} ion in NiO is ^3F, which gives rise to an orbital degeneracy $(2L + 1) = 7$ and a spin degeneracy of $(2S + 1) = 3$, since $S = 1$ for the Ni^{2+} ion in NiO. Hence there is a total degeneracy of 21. Using the tables of Koster *et al* [35] we can see that for the O_h field, $\Gamma(21)$ would split as follows:

$$\Gamma(21) = \Gamma_2(3) + \Gamma_4(9) + \Gamma_5(9).$$

The number in parentheses gives the degeneracy of the level.

The separation of the lowest $\Gamma_2(3)$ level and the next $\Gamma_5(9)$ level is of the order of 8000 cm^{-1} [77] and hence we need consider only the lowest level when studying the magnons in NiO.

Including spin–orbit coupling we find that the spin function belongs to D_1, and hence upon taking the direct product $D_1 \otimes \Gamma_2$, the required splittings are obtained. Since $D_1 \equiv \Gamma_4$ of O_h we get

$$\Gamma_4 \otimes \Gamma_2 = \Gamma_5(3).$$

In the antiferromagnetic state there is a rhombohedral distortion (D_3), and so by comparison of the character table of D_3 with the Γ_5 representation of O_h we find that

$$\Gamma_5(3) = \Gamma_1(1) + \Gamma_3(2).$$

In the magnetic (orthorhombic) state for which the point group is $\underline{2}/\underline{m}$ we find, using the method given by Cracknell [76], that the $\Gamma_3(2)$ level further splits into $2D_1(1)$ levels. Figure 11.10 shows the crystal field splittings of the F term of the Ni^{2+} ion in NiO.

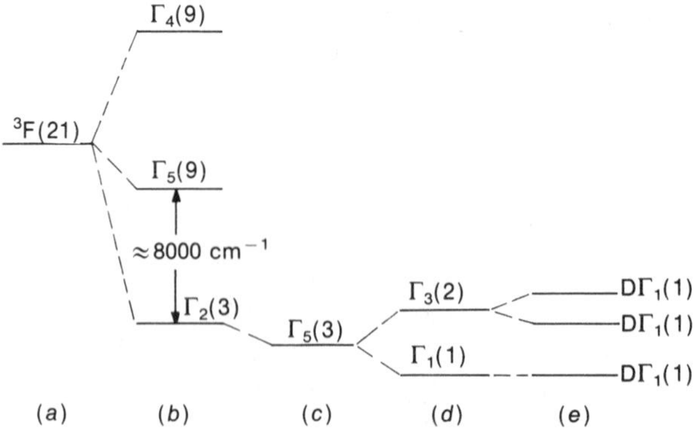

Figure 11.10 The splitting of the ^3F term of the Ni^{2+} ion by various crystal fields and by spin–orbit coupling: (*a*) free atom; (*b*) cubic field (O); (*c*) cubic field plus spin–orbit coupling; (*d*) rhombohedral field (D_3); (*e*) magnetic point group $\underline{2}/\underline{m}$.

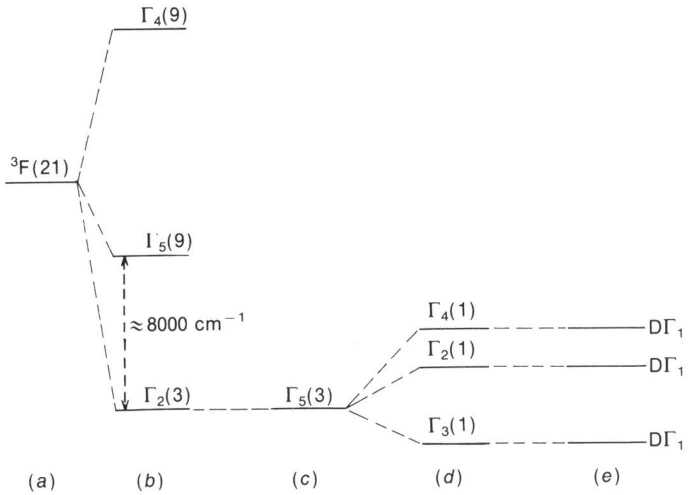

Figure 11.11 Crystal field splittings of the Ni^{2+} ion in NiF_2: (*a*) free atom; (*b*) cubic field; (*c*) cubic field plus spin–orbit coupling; (*d*) orthorhombic (D_{2h}); (*e*) magnetic point group $\underline{2}/\underline{m}$.

In a similar manner we can obtain the crystal field splittings of the Ni^{2+} ion in NiF_2. These are shown in figure 11.11.

11.5 Symmetry properties of spin waves in NiF_2 and NiO

The low-lying elementary excitations of a spin system are called magnons. These are wave-like excitations in that the excitation is shared by all the magnetic ions in the crystal. The amplitude of the excitation is the same for ions at equivalent positions in all the unit cells. The usual treatment of magnons in terms of pure spins placed on lattice sites (see [61, 15]) holds good for NiF_2 and NiO.

A magnon which has a wavevector $\boldsymbol{k}$ must belong to one of the irreducible corepresentations of the group of the wavevector $\boldsymbol{k}$. The assignment of a magnon to the appropriate irreducible corepresentation can be determined by studying the transformation properties of the magnon creation operators. If i and j label the two sublattices for an antiferromagnet then the positive and negative spin components could be taken as S_i^z and S_j^z. The transition operators for representative ions on the two sublattices could be written as a_i^+ and a_j^+. Since for the normal magnons the transitions would involve a unit change of the z component of spin we may write

$$a_i^+ = S_i^-/(2S)^{1/2} \qquad a_j^+ = S_j^+/(2S)^{1/2} \tag{11.7}$$

where

$$S_j^+ = S_j^x + iS_j^y \qquad S_i^- = S_i^x - iS_i^y \tag{11.8}$$

are step-up and step-down spin operators and

$$S = S^z + n \tag{11.9}$$

S being the spin of a magnetic ion and n the spin deviation being defined by $n = a^+a$. The creation and annihilation operators that we have defined satisfy the usual commutation relation for bosons, namely

$$[a_i, a_j^+] = \delta_{ij}. \tag{11.10}$$

The spin wave operators are defined by the Fourier components of a_i^+ and a_j^+ as follows:

$$a_k^+ = \sqrt{N} \sum_i \exp(i\boldsymbol{k}\cdot\boldsymbol{r}_i)a_i^+ \tag{11.11}$$

$$b_k^+ = \sqrt{N} \sum_j \exp(i\boldsymbol{k}\cdot\boldsymbol{r}_j)a_j^+. \tag{11.12}$$

Here N is the number of magnetic ions on each sublattice and $\boldsymbol{r}_i$ and $\boldsymbol{r}_j$ are the positions of the ion on sublattices i and j. The true magnon creation operators for the entire crystal are formed by constructing those linear combinations of the a_k^+ and b_k^+ which diagonalize the Hamiltonian representing the interaction between magnetic ions. These operators which diagonalize the Hamiltonian are of the form (see [61, 15])

$$\begin{aligned} \alpha_{\downarrow k}^+ &= U_k a_k^+ - V_k b_{-k} \\ \alpha_{\uparrow k}^+ &= -V_k a_{-k} + U_k b_k^+ \end{aligned} \tag{11.13}$$

where $\alpha_{\downarrow k}^+$ and $\alpha_{\uparrow k}^+$ are creation operators for the two types of magnon of wavevector $\boldsymbol{k}$ corresponding to $S^z = -1$ and $S^z = +1$. U_k and V_k are solutions of the simultaneous equations

$$U_k^2 - V_k^2 = 1 \tag{11.14}$$

$$V_k/U_k = -[E_0 + V_1(\boldsymbol{k}) - E(\boldsymbol{k})]/V_2(\boldsymbol{k}). \tag{11.15}$$

Here E_0 is the molecular field energy associated with a unit change of spin component and $E(\boldsymbol{k})$ is the energy eigenvalue obtained by diagonalizing the exchange interaction Hamiltonian. $V_1(\boldsymbol{k})$ and $V_2(\boldsymbol{k})$ are given by

$$\begin{aligned} \mathrm{V}_1(\boldsymbol{k}) &= \sum_{i'} V_{ii'} \exp[i\boldsymbol{k}\cdot(\boldsymbol{r}_i - \boldsymbol{r}_{i'})] \\ &= \sum_{j'} V_{jj'} \exp[i\boldsymbol{k}\cdot(\boldsymbol{r}_j - \boldsymbol{r}_{j'})] \end{aligned} \tag{11.16}$$

and

$$V_2(\boldsymbol{k}) = \sum_j V_{ij} \exp[\mathrm{i}\boldsymbol{k}\cdot(\boldsymbol{r}_i - \boldsymbol{r}_j)]. \tag{11.17}$$

The V_{ij} etc measure the strengths of the exchange interactions between the ions indicated by the subscripts. By combining equations (11.7), (11.11), (11.12) and (11.13) the magnon creation operators can be written as

$$\alpha_{\downarrow k}^{+} = \frac{1}{\sqrt{2SN}}\left(U_k \sum_{i'} \exp(\mathrm{i}\boldsymbol{k}\cdot\boldsymbol{r}_i)\boldsymbol{S}_i^{-} - V_k \sum_j \exp(\mathrm{i}\boldsymbol{k}\cdot\boldsymbol{r}_j)\boldsymbol{S}_j^{-}\right) \tag{11.18}$$

and

$$\alpha_{\uparrow k}^{+} = \frac{1}{\sqrt{2SN}}\left(U_k \sum_{j} \exp(\mathrm{i}\boldsymbol{k}\cdot\boldsymbol{r}_j)\boldsymbol{S}_j^{+} - V_k \sum_i \exp(\mathrm{i}\boldsymbol{k}\cdot\boldsymbol{r}_i)\boldsymbol{S}_i^{+}\right). \tag{11.19}$$

The assignment of a magnon to the appropriate irreducible corepresentation (see §§11.2, 11.3) can now be found by studying the transformation properties of (11.18) and (11.19). In the next two sections we assign the magnons of NiF_2 and NiO to the appropriate points and lines of symmetry over the entire Brillouin zone.

11.5.1 Symmetries of magnons in NiF_2

The symmetries of magnons in NiF_2 are determined by assigning the magnon creation operators (11.18), (11.19) to the irreducible corepresentations of M^k. If the direction of magnetization of the sublattice i in the crystal is ξ then the operators $\boldsymbol{S}_i^+$ and $\boldsymbol{S}_i^-$ which appear in equations (11.18) and (11.19) are $\boldsymbol{S}_\zeta + \mathrm{i}\boldsymbol{S}_\eta$ and $\boldsymbol{S}_\zeta - \mathrm{i}\boldsymbol{S}_\eta$, where ξ, ζ and η form a right-handed orthogonal set of axes (Dimmock and Wheeler [12]). In NiF_2 we then set up axes ξ, ζ and η on each sublattice so that η is along the $+z$ direction for sublattice i (at the corners of the unit cell) and along the $-z$ direction for sublattice j; these axes are shown in figure 11.12 where ζ is in the x–y plane. We have already written the magnetic little group as

$$\mathrm{M}^k = \mathrm{H}^k + R\mathrm{G}_I^k.$$

The magnons in antiferromagnetic NiF_2 can then be assigned to the coreps of M^k by using equations (11.18) and (11.19). The results are given in table 11.20 where the magnons belonging to the coreps of M^k are derived from the representations of H^k listed in the table. Where two representations of H^k stick together forming a corep of M^k that belongs to case (c) these representations of H^k have been written in parentheses, for example $(R_1^+ + R_2^+)$. The magnon symmetries for the various points and lines of symmetry in the Brillouin zone are illustrated in figure 11.13.

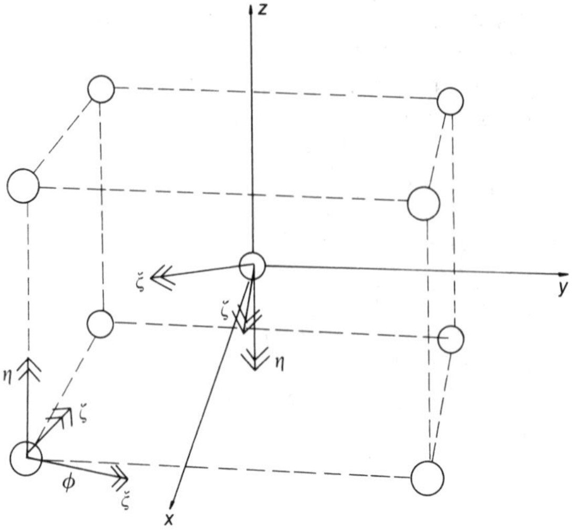

Figure 11.12 The axes ξ, ζ, η for NiF_2 sublattice magnetization.

Table 11.20 Symmetries of magnons in NiF_2.

Point or line	*k*	Magnons	Point or line	*k*	Magnons
Γ	(0, 0, 0)	Γ_1^+, Γ_2^+	X	$(0, \frac{1}{2}, 0)$	X_1
Y	$(\frac{1}{2}, 0, 0)$	Y_1	Z	$(0, 0, \frac{1}{2})$	Z_1
U	$(\frac{1}{2}, 0, \frac{1}{2})$	U_1^+, U_2^+	R	$(0, \frac{1}{2}, \frac{1}{2})$	$(R_1^+ + R_2^+)$
M	$(\frac{1}{2}, \frac{1}{2}\ 0)$	$(M_1^+ + M_2^+)$	A	$(\frac{1}{2}, \frac{1}{2}, \frac{1}{2})$	A_1
Δ	$(0, \alpha, 0)$	(Δ_1, Δ_2)	D	$(\frac{1}{2}, \alpha, 0)$	$(D_1 + D_2)$
H	$(\frac{1}{2}, \alpha, \frac{1}{2})$	H_1, H_2	B	$(0, \alpha, \frac{1}{2})$	$(B_1 + B_2)$
Σ	$(\alpha, 0, 0)$	Σ_1, Σ_2	C	$(\alpha, \frac{1}{2}, 0)$	$(C_1 + C_2)$
T	$(\alpha, \frac{1}{2}, \frac{1}{2})$	$(T_1 + T_2)$	S	$(\alpha, 0, \frac{1}{2})$	S_1, S_2
Λ	$(0, 0, \alpha)$	Λ_1, Λ_2	W	$(0, \frac{1}{2}, \alpha)$	$(W_1 + W_2)$
V	$(\frac{1}{2}, \frac{1}{2}, \alpha)$	$(V_1 + V_2)$	G	$(\frac{1}{2}, 0, \alpha)$	G_1, G_2

It will be noticed that the magnon branches are two-fold degenerate at many of the points of symmetry and along several of the lines of symmetry. Furthermore, the magnon branches are degenerate along all the sides and at all the corners of the rectangle ARXM; however, at general points on the surface of the Brillouin zone within this rectangle the magnon energies are non-degenerate. There is thus no general double degeneracy over any surface in this theory. Compatibility tables between the points and lines of symmetry are given in table 11.10.

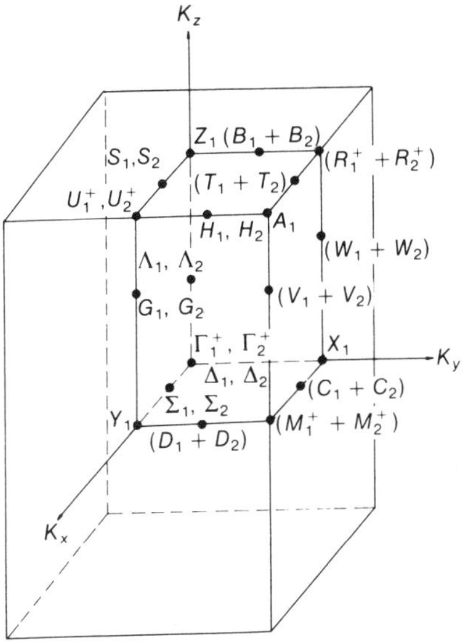

Figure 11.13 Magnon symmetries in antiferromagnetic NiF_2.

It is interesting to note that the true electronically excited states of a magnetic crystal are really the excitons, so that by considering the exciton theory [15] one avoids having to assume that the orbital angular momentum of the Ni^{2+} ions is quenched. It is well known that the orbital angular momentum of the Ni^{2+} ions is not completely quenched. Whether or not one actually uses the word magnon to describe one of these excitons depends on whether the orbital angular momentum is negligible or not. It is therefore useful to relate exciton theory to the single-ion molecular field treatment. The way in which the excited states of single magnetic ions, compatible with the ionic site symmetry, combine to produce excitons that are compatible with the (magnetic) space-group symmetry of the crystal has been described in detail by Loudon [15].

The site group of a single Ni^{2+} ion consists of the elements $\{E|0\}$, $\{I|0\}$, $R\{C_2|0\}$, $R\{\sigma_h|0\}$, that is of the magnetic point group 2′/m′. The single-valued and double-valued representations of the unitary subgroup ($\bar{1}$, (C_i)) of this group are given by Koster *et al* [35] and are reproduced in table 11.21. In the last column of this table we identify case (a), (b) or (c) of the coreps of 2′/m′ (see [59, 60]).

Since Ni^{2+} has an even number of electrons the wavefunctions must belong to single-valued corepresentations of 2′/m′, that is to the coreps derived from Γ_1 representations of the unitary subgroup. The single-ion energy level scheme for each sublattice therefore consists of Γ_1 levels and the site symmetry of

Table 11.21 Character table for the site group of Ni^{2+} ion.

	E	$\bar{E}$	I	$\bar{I}$	2′/m′
Γ_1^+	1	1	1	1	a
Γ_1^-	1	1	−1	−1	a
Γ_2^+	1	−1	1	−1	a
Γ_2^-	1	−1	−1	1	a

the transition operator a^+ which destroys an electron in the ground state (Γ_g) and creates an electron in an excited state (Γ_e) is given by the inner Kronecker product

$$\Gamma_t = \Gamma_e \otimes \Gamma_g^* = \Gamma_1 \otimes \Gamma_1 = \Gamma_1^+. \tag{11.20}$$

We now combine the single-ion transitions (see figure 11.14) on all the sites of the crystal. The space-group representation to which the exciton derived from the single-ion transition with transition operator Γ_t has character (R) equal to zero if R is not in the site group and is twice the character of R in Γ_t if R is in the site group. The space-group representation generated from $\Gamma_t = \Gamma_1^+$ therefore has the characters given in table 11.22. This representation is reducible and yields the space-group representations $\Gamma_1^+ + \Gamma_2^+$ at $\boldsymbol{k} = 0$. In other words there is a Davydov splitting of the magnon frequencies at $\boldsymbol{k} = 0$. Whether or not one observes these Davydov splittings is governed by the space-group selection rules for first-order radiative processes.

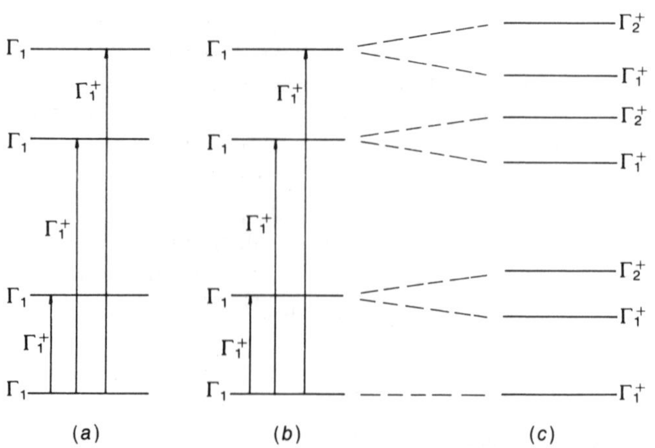

Figure 11.14 Exciton states for NiF_2 at $\boldsymbol{k} = 0$. Similar results hold for NiO (see [24]).

Table 11.22 The $k = 0$ space-group representation generated from Γ_1^+ of the site group.

	$\{E\|0\}$	$\{E\|0\}$	$\{C_{2x}\|\tau\}$	$\{\bar{C}_{2x}\|\tau\}$	$\{I\|0\}$	$\{\bar{I}\|0\}$	$\{\sigma_{v_x}\|\tau\}$	$\{\bar{\sigma}_{v_x}\|\tau\}$
Γ_1^+ site group	2	2	0	0	2	2	0	0

11.5.2 Symmetries of magnons in NiO

The symmetries of magnons in NiO are determined in exactly the same manner as for NiF_2 (§11.5.1). We once again study the transformation properties of $\alpha_{\downarrow k}^+$ and $\alpha_{\uparrow k}^+$ under the members of M^k for NiO. S^+ and S^- in $\alpha_{\downarrow k}^+$ and $\alpha_{\uparrow k}^+$ are defined, as before, by

$$S^+ = S_\zeta + iS_\eta \qquad S^- = S_\zeta - iS_\eta$$

where ξ, ζ and η form a right-handed orthogonal set of axes and ξ is along the direction of sublattice magnetization. We choose η along the two-fold rotation axes C_{2b} and this fixes the direction of ζ. The effect of the rotational parts of the space-group operations of NiO on S^+ and S^- can readily be seen to be

$$\begin{aligned} ES^+ &= S^+ \qquad & ES^- &= S^- \\ C_{2b}S^+ &= -S^- \qquad & C_{2b}S^- &= -S^+ \\ IS^+ &= S^+ \qquad & IS^- &= S^- \\ \sigma_{d_h}S^+ &= -S^- \qquad & \sigma_{d_h}S^- &= -S^+. \end{aligned} \tag{11.21}$$

The results of the symmetry assignments for magnons in NiO are given in table 11.23.

Table 11.23 Magnon symmetries in NiO.

Point or line	Magnon
Γ	Γ_1^+, Γ_2^+
M	(M_1^+, M_2^+)
X	X_1
Z	Z_1
L	L_1^+, L_1^+
V	V_1^+, V_1^+
Λ	Λ_1, Λ_2
U	U_1, U_2
B	B_1, B_2

12

SPACE-GROUP SELECTION RULES FOR MAGNETIC CRYSTALS

12.1 Introduction

The theory of selection rules for phonons in crystals belonging to one of the Fedorov groups can be extensively found in the literature (see [62–71]). Much less, however, has been done for crystals belonging to one of the Shubnikov or magnetic groups. Because of the fact that Shubnikov groups are antiunitary the ordinary representation theory on which the theoretical studies of selection rules are based are no longer applicable and have to be replaced by the theory of corepresentation outlined in Chapter 10 (see also [17, 30, 48]).

The usual method of determining the selection rules for phonons in non-magnetic crystals is the reduction of the Kronecker products of the space-group representations. This procedure is readily extended to the case of magnetic crystals in that one now reduces the Kronecker products of coreps of non-unitary groups. The appropriate method of reducing the Kronecker products of coreps of non-unitary groups has been dealt with in great detail by Cracknell [48] who has also applied the method to the case of antiferromagnetic MnF_2. We therefore summarize below the relevant portions of Cracknell [48].

If Δ^i, Δ^j and Δ^k are representations belonging to the halving unitary group G of the magnetic space group $M = G + RG$, then

$$\Delta^i \otimes \Delta^j = \sum_k c_{ij,k} \Delta^k \tag{12.1}$$

where $c_{ij,k}$ are the Clebsch–Gordan coefficients. If these coefficients are known then Bradley and Davies [30] have shown that it is a relatively simple matter to obtain the Clebsch–Gordan coefficients in the reduction of the inner Kronecker product of the coreps of M derived from Δ^i and Δ^j. If $D\Delta^i$, $D\Delta^j$ and $D\Delta^k$ denote the coreps corresponding to the representations Δ^i, Δ^j and Δ^k, then

$$D\Delta^i \otimes D\Delta^j = \sum_k d_{ij,k} D\Delta^k \tag{12.2}$$

and the coefficients $d_{ij,k}$ can be readily expressed in terms of $c_{ij,k}$. The actual relationship between $d_{ij,k}$ and $c_{ij,k}$ depends on the properties of $D\Delta^i$, $D\Delta^j$ and $D\Delta^k$, that is whether they belong to case (a) or case (b) or case (c). For any of these cases the expressions for $d_{ij,k}$ for various possible combinations that can arise were worked out by Bradley and Davies [30] and as such we shall not repeat their work here. In the next two sections we use the method of Cracknell [48] and Bradley and Davies [30] to reduce the inner Kronecker products of the space group of NiF_2. In §12.4 we reduce and tabulate the symmetrized cube and the antisymmetrized square of the irreducible representation of the point groups. These are the products which are relevant to Landau's theory of second-order phase transitions. Finally, in §12.5 we make use of Bradley's theory [71] to determine the selection rules for two-magnon absorption in NiF_2 and NiO.

12.2 Kronecker products for the unitary representation $P2_1/b$ of NiF_2

Following the general theory given by Bradley [71] and Cracknell [48] we require to determine the coefficients $C_{pq,r}^{km,k}$ in

$$\Gamma_p^k \otimes \Gamma_q^m = \sum_{h,r} C_{pq,r}^{km,h} \Gamma_r^h \qquad (12.3)$$

Table 12.1 Table giving the sets of values k, m and h which satisfy the condition $k + m \approx h$.

k	m	h	k	m	h	k	m	h	k	m	h
Γ	Γ	Γ	X	Γ	X	A	Γ	A	R	Γ	R
	Y	Y		Y	M		Y	R		Y	A
	X	X		X	Γ		X	U		X	Z
	Z	Z		Z	R		Z	M		Z	X
	A	A		A	U		A	Γ		A	Y
	U	U		U	A		U	X		U	M
	R	R		R	Z		R	Y		R	Γ
	M	M		M	Y		M	Z		M	U
Y	Γ	Y	Z	Γ	Z	U	Γ	U	M	Γ	M
	Y	Γ		Y	U		Y	Z		Y	X
	X	M		X	R		X	A		X	Y
	Z	U		Z	Γ		Z	Y		Z	A
	A	R		A	M		A	X		A	Z
	U	Z		U	Y		U	Γ		U	R
	R	A		R	X		R	M		R	U
	M	X		M	A		M	R		M	Γ

where k, m and h are wavevectors of three points in the Brillouin zone and the condition given by Bradley in his equation (46) implies that

$$k + m \approx h. \tag{12.4}$$

The set of values k, m and h which satisfy this condition are given in table 12.1 and are the same as those given by Cracknell for MnF_2.

12.3 Kronecker products for the coreps of the antiferromagnetic space group of NiF_2

We illustrate the method of Cracknell [48] and Bradley and Davies [30] by working out the Kronecker products of the coreps of the antiferromagnetic space group of NiF_2. As typical examples we work out the products for $\{DX \otimes DY\}$ and $\{DY \otimes DZ\}$. The results are summarized in tables 12.2 and 12.3.

Table 12.2 Kronecker products for X and Y.

$$DX_1 \otimes DY_1 = DM_1^+ + DM_1^-$$
$$DX_1 \otimes DY_2 = DM_3^+ + DM_3^-$$
$$DX_2 \otimes DY_1 = DM_3^+ + DM_3^-$$
$$DX_2 \otimes DY_2 = DM_1^+ + DM_1^-$$

where

$$DM_1^+ = \{M_1^+ + M_2^+\}$$
$$DM_1^- = \{M_1^- + M_2^-\}$$
$$DM_3^+ + \{M_3^+ + M_4^+\}$$
$$DM_3^- = \{M_1^- + M_4^-\}$$

(See table 11.5 for M)

Table 12.3 Kronecker products for Y and Z.

$$DY_1 \otimes DZ_1 = DU_1^+ + DU_2^+ + DU_1^- + DU_2^-$$
$$DY_1 \otimes DZ_2 = DU_3^+ + DU_4^+ + DU_3^- + DU_4^-$$
$$DY_2 \otimes DZ_1 = DU_3^+ + DU_4^+ + DU_3^- + DU_4^-$$
$$DY_2 \otimes DZ_2 = DU_1^+ + DU_2^+ + DU_1^- + DU_2^-$$

12.4 The reduction of the symmetrized cube and the antisymmetrized square of the irreducible representations of point groups

The symmetrized square of the degenerate point-group irreducible representations were tabulated by Jahn and Teller [84]. Certain other products of the point-group irreducible representations, namely the symmetrized cube and the antisymmetrized square, are of interest in the Landau theory of second-order phase transitions (see [78–81]) and it therefore seems useful to evaluate and tabulate these products in the notation of Koster *et al* [35]. The character of an element R in $[\Gamma]^3$, the symmetrized cube of the representation Γ, is given by Lyubarskii [79] as

$$[\chi]^3(R) = \tfrac{1}{3}\chi(R^3) + \tfrac{1}{2}\chi(R^2)\chi(R) + \tfrac{1}{6}\{\chi(R)\}^3 \tag{12.5}$$

and in $\{\Gamma\}^2$, the antisymmetrized square, by

$$\{\chi\}^2(R) = \tfrac{1}{2}\{\chi(R)\}^2 - \tfrac{1}{2}\chi(R^2). \tag{12.6}$$

For a non-degenerate representation of a point group, equations (12.5) and (12.6) can be simplified, since in this case $\chi(R)$ is just equal to $D_{11}(R)$, the sole element of the one-dimensional matrix representative of R; also $D_{11}(R)^2 = \{D_{11}(R)\}^2$ and $D_{11}(R^3) = \{D_{11}(R)\}^3$ from the definition of a representation.

Therefore

$$\begin{aligned}[\chi]^3(R) &= \tfrac{1}{3}D_{11}(R^3) + \tfrac{1}{2}D_{11}(R^2)D_{11}(R) + \tfrac{1}{6}\{D_{11}(R)\}^3 \\ &= \tfrac{1}{3}\{D_{11}(R)\}^3 + \tfrac{1}{2}\{D_{11}(R)\}^3 + \tfrac{1}{6}\{D_{11}(R)\}^3 \\ &= \{\chi(R)\}^3. \end{aligned} \tag{12.7}$$

Similarly, one can show that

$$\{\chi\}^2(R) = 0 \tag{12.8}$$

for a non-degenerate representation. The expression on the right-hand side of (12.7) is simply the character of R in the triple inner Kronecker product $\Gamma \otimes \Gamma \otimes \Gamma$. We therefore conclude that if Γ is a non-degenerate representation then $\{\Gamma\}^2 = 0$, while $[\Gamma]^3$ is identical with $\Gamma \otimes \Gamma \otimes \Gamma$ and so can easily be found by a repeated use of the multiplication tables in Koster *et al* [35]. There is therefore no need for us to tabulate $\{\Gamma\}^2$ and $[\Gamma]^3$ for the non-degenerate representations of the point groups. If Γ is real as well as being non-degenerate, then $\Gamma \otimes \Gamma \otimes \Gamma$ is simply equal to Γ; the way in which complex representations fit into the Landau theory requires special consideration anyway (see [78]). Barker and Loudon [82] assume the result of (12.7), while not stating it explicitly, and use $\Gamma \otimes \Gamma \otimes \Gamma$ rather than $[\Gamma]^3$ in discussing the Landau theory.

If Γ is a degenerate representation, the above simple results do not hold and the determination of $\{\Gamma\}^2$ and $[\Gamma]^3$ is not so straightforward. For the degenerate point-group representations $\{\Gamma\}^2$ and $[\Gamma]^3$ have been evaluated using (12.5) and (12.6) and reduced; they are given in the third and fifth columns of table 12.4, in the notation of Koster *et al* [35]. The symmetrized cubes of the degenerate single-valued representations of the groups O and T_d have been given previously by Birman [83]. In the fourth column of table 12.4, $\Gamma \otimes \Gamma \otimes \Gamma$ is given, which can again be determined from Koster *et al* [35]. For example, for Γ_4 of O or T_d, table 82 of Koster *et al* gives

$$\begin{aligned}\Gamma_4 \otimes \Gamma_4 \otimes \Gamma_4 &= \Gamma_4 \otimes (\Gamma_1 + \Gamma_3 + \Gamma_4 + \Gamma_5)\\ &= \Gamma_1 + \Gamma_2 + 2\Gamma_2 + 4\Gamma_4 + 3\Gamma_5.\end{aligned} \tag{12.9}$$

Groups that are direct products of a point group G with the group C_1 are not included in table 12.4 since their product representations can be found

Table 12.4 $[\Gamma]^3$ and $\{\Gamma\}^2$ for crystallographic point groups.

Group	Rep.	$\{\Gamma\}^2$	$\Gamma\otimes\Gamma\otimes\Gamma$	$[\Gamma]^3$
(6) D_2				
(7) C_{2v}	Γ_5	Γ_1	$4\Gamma_5$	$2\Gamma_5$
(12) D_4				
(13) C_{4v}				
(14) D_{2d}	Γ_5	Γ_2	$4\Gamma_5$	$2\Gamma_5$
	Γ_6	Γ_1	$3\Gamma_6 + \Gamma_7$	$\Gamma_6 + \Gamma_7$
	Γ_7	Γ_1	$\Gamma_6 + 3\Gamma_7$	$\Gamma_6 + \Gamma_7$
(18) D_3				
(19) C_{3v}	Γ_3	Γ_2	$\Gamma_1 + \Gamma_2 + 3\Gamma_3$	$\Gamma_1 + \Gamma_2 + \Gamma_3$
	Γ_4	Γ_1	$3\Gamma_4 + \Gamma_5 + \Gamma_6$	$\Gamma_4 + \Gamma_5 + \Gamma_6$
(24) D_6				
(25) C_{6v}				
(26) D_{3h}	Γ_5	Γ_2	$\Gamma_3 + \Gamma_4 + 3\Gamma_5$	$\Gamma_3 + \Gamma_4 + \Gamma_5$
	Γ_6	Γ_2	$\Gamma_1 + \Gamma_2 + 3\Gamma_6$	$\Gamma_1 + \Gamma_2 + \Gamma_6$
	Γ_7	Γ_1	$3\Gamma_7 + \Gamma_9$	$\Gamma_7 + \Gamma_9$
	Γ_8	Γ_1	$3\Gamma_8 + \Gamma_9$	$\Gamma_8 + \Gamma_9$
	Γ_9	Γ_1	$4\Gamma_9$	$2\Gamma_9$
(28) T	Γ_4	Γ_4	$2\Gamma_1 + 2\Gamma_2 + 2\Gamma_3 + 7\Gamma_4$	$\Gamma_1 + 3\Gamma_4$
	Γ_5	Γ_1	$2\Gamma_5 + \Gamma_6 + \Gamma_7$	$\Gamma_6 + \Gamma_7$
	Γ_6	Γ_3	$2\Gamma_5 + \Gamma_6 + \Gamma_7$	$\Gamma_6 + \Gamma_7$
	Γ_7	Γ_2	$2\Gamma_5 + \Gamma_6 + \Gamma_7$	$\Gamma_6 + \Gamma_7$
(30) O				
(31) T_d	Γ_3	Γ_2	$\Gamma_1 + \Gamma_2 + 3\Gamma_3$	$\Gamma_1 + \Gamma_2 + \Gamma_3$
	Γ_4	Γ_4	$\Gamma_1 + \Gamma_2 + 2\Gamma_3 + 4\Gamma_4 + 3\Gamma_5$	$\Gamma_2 + 2\Gamma_4 + \Gamma_5$
	Γ_5	Γ_4	$\Gamma_1 + \Gamma_2 + 2\Gamma_3 + 3\Gamma_4 + 4\Gamma_5$	$\Gamma_1 + \Gamma_4 + 2\Gamma_5$
	Γ_6	Γ_1	$2\Gamma_6 + \Gamma_8$	Γ_8
	Γ_7	Γ_1	$2\Gamma_7 + \Gamma_8$	Γ_8
	Γ_8	$\Gamma_1 + \Gamma_3 + \Gamma_5$	$5\Gamma_6 + 5\Gamma_7 + 11\Gamma_8$	$\Gamma_6 + \Gamma_7 + 4\Gamma_8$

from those of G by adding plus and minus signs obeying the usual rule. It would be a very arduous task to evaluate these products for all the representations of all the 230 space groups and we make no attempt to do this.

12.5 Selection rules for two-magnon absorption in NiF_2 and NiO

Of considerable interest to experimentalists are the selection rules for the creation of two magnons with equal and opposite $\boldsymbol{k}$ vectors by the absorption of a photon. The selection rule for two-magnon electric dipole absorption for a general wavevector $\boldsymbol{k}$ has been discussed by Loudon [15]. The momentum of the incoming photon is so small that the two magnons that are created have equal and opposite $\boldsymbol{k}$ vectors. For each of the special points of symmetry in the magnetic Brillouin zone $\boldsymbol{k}$ and $-\boldsymbol{k}$ are equivalent so that, to all intents and purposes, the two magnons are created at the same special point.

Since there are two types of magnon of wavevector $\boldsymbol{k}$ and two of wavevector $-\boldsymbol{k}$ it is therefore possible to construct four different zero-wavevector two-magnon states [15] as follows:

$$\begin{aligned}|0,-\rangle &= |\uparrow\boldsymbol{k}, \downarrow-\boldsymbol{k}\rangle - |\downarrow\boldsymbol{k}, \uparrow-\boldsymbol{k}\rangle \\ |0,+\rangle &= |\uparrow\boldsymbol{k}, \downarrow-\boldsymbol{k}\rangle + |\downarrow\boldsymbol{k}, \uparrow-\boldsymbol{k}\rangle \\ |2,+\rangle &= |\uparrow\boldsymbol{k}, \uparrow-\boldsymbol{k}\rangle \\ |-2,+\rangle &= |\downarrow\boldsymbol{k}, \downarrow-\boldsymbol{k}\rangle.\end{aligned} \qquad (12.10)$$

The numbers in the two-magnon kets represent the S^z quantum numbers of the states. In determining the parity of these states we assume that each magnetic ion is a centre of inversion symmetry I. The inversion properties of the magnons are

$$\{I|0\}|\downarrow\boldsymbol{k}\rangle = |\downarrow-\boldsymbol{k}\rangle \qquad \{I|0\}|\uparrow\boldsymbol{k}\rangle = |\uparrow-\boldsymbol{k}\rangle. \qquad (12.11)$$

Application of these transformation rules to the two-magnon states (12.10) shows that $|0,-\rangle$ has odd parity while the other three states have even parity. Thus the $|0,-\rangle$ state is inactive in Raman scattering. In table 12.5

Table 12.5 Symmetry characters of the zero-wavevector two-magnon states for all the symmetry points in the Brillouin zone of antiferromagnetic NiF_2.

	Γ	X	Y	M	U	Z	R	A
$\|2+\rangle + \|-2+\rangle$	Γ_1^+	Γ_1^+	Γ_1^+	Γ_1^+	Γ_1^+	Γ_1^+	Γ_1^+	Γ_1^+
$\|2+\rangle - \|-2+\rangle$	Γ_2^+	Γ_2^+	Γ_2^+	Γ_2^+	Γ_2^+	Γ_2^+	Γ_2^+	Γ_2^+
$\|0+\rangle$	Γ_1^+	0	0	Γ_2^+	Γ_2^+	0	Γ_1^+	0
$\|0-\rangle$	0	Γ_1^-	Γ_2^-	0	0	Γ_1^-	0	Γ_2^-

Table 12.6 Selection rules for two-magnon absorption in NiF_2.

Point	Electric dipole $[\Gamma_i]^2$	Yes or No	Magnetic dipole $\Gamma_i \otimes \Gamma_j$	Yes or No
X	$[X_1]^2 = \Gamma_1^+ + \Gamma_2^+ + \Gamma_1^-$	Yes	$X_1 \otimes X_1 = \Gamma_1^+ + \Gamma_2^+ + \Gamma_1^- + \Gamma_2^-$	Yes
Y	$[Y_1]^2 = \Gamma_1^+ + \Gamma_2^+ + \Gamma_2^-$	Yes	$Y_1 \otimes Y_1 = \Gamma_1^+ + \Gamma_2^+ + \Gamma_1^- + \Gamma_2^-$	Yes
Z	$[Z_1]^2 = \Gamma_1^+ + \Gamma_2^+ + \Gamma_1^-$	Yes	$Z_1 \otimes Z_1 = \Gamma_1^+ + \Gamma_2^+ + \Gamma_1^- + \Gamma_2^-$	Yes
A	$[A_1]^2 = \Gamma_1^+ + \Gamma_2^+ + \Gamma_2^-$	Yes	$A_1 \otimes A_1 = \Gamma_1^+ + \Gamma_2^+ + \Gamma_1^- + \Gamma_2^-$	Yes
U	–	No	$U_1^+ \otimes U_1^+ = \Gamma_2^+$	Yes
			$U_2^+ \otimes U_2^+ = \Gamma_2^+$	Yes
M	–	No	$M_1^+ \otimes M_1^+ = \Gamma_2^+$	Yes
			$M_1^+ \otimes M_2^+ = \Gamma_1^+$	Yes
			$M_2^+ \otimes M_2^+ = \Gamma_1^+$	Yes
R	–	No	$R_1^+ \otimes R_1^+ = \Gamma_1^+$	Yes
			$R_1^+ \otimes R_2^+ = \Gamma_2^+$	Yes
			$R_2^+ \otimes R_2^+ = \Gamma_1^+$	Yes

Table 12.7 Selection rules for two-magnon absorption in NiO.

Point	Electric dipole $[\Gamma_i]^2$	Yes or No	Magnetic dipole $\Gamma_i \otimes \Gamma_j$	Yes or No
M	–	No	$M_1^+ \otimes M_1^+ = \Gamma_1^+$	Yes
			$M_1^+ \otimes M_2^+ = \Gamma_2^+$	Yes
			$M_2^+ \otimes M_2^+ = \Gamma_1^+$	Yes
X	$[X_1]^2 = \Gamma_1^+ + \Gamma_2^+ + \Gamma_1^-$	Yes	$X_1 \otimes X_1 = \Gamma_1^+ + \Gamma_2^+ + \Gamma_1^- + \Gamma_2^-$	Yes
Z	$[Z_1]^2 = \Gamma_1^+ + \Gamma_2^+ + \Gamma_1^-$	Yes	$Z_1 \otimes Z_1 = \Gamma_1^+ + \Gamma_2^+ + \Gamma_1^- + \Gamma_2^-$	Yes
L	–	No	$L_1^+ \otimes L_1^+ = \Gamma_1^+ + \Gamma_2^+ + M_1^- + M_2^-$	Yes
V	–	No	$V_1^+ \otimes V_1^+ = \Gamma_1^+ + \Gamma_2^+ + M_1^- + M_2^-$	Yes
			$V_1^+ \otimes V_1^- = \Gamma_1^- + \Gamma_2^- + M_1^+ + M_2^+$	No
			$V_1^- \otimes V_1^- = \Gamma_1^+ + \Gamma_2^+ + M_1^- + M_2^-$	Yes

we give the symmetries of the four zero-wavevector states in (12.10) for all the symmetry points in the Brillouin zone of antiferromagnetic NiF_2.

The selection rules connecting different excitations at different points in the Brillouin zone are easily determined (see [62]). The components of the electric dipole operator transform like ξ, η and ζ, that is like Γ_2^-, Γ_1^- and Γ_2^- respectively, while the components of the magnetic dipole operator transform like L_ξ, L_η, L_ζ, that is like Γ_2^+, Γ_1^+ and Γ_2^+ respectively. For each of the points X, Y, Z, A, U, M, R in NiF_2 and M, X, Z, L, V in NiO, we give in tables 12.6 and 12.7 the ordinary Kronecker product $R_i \otimes R_j$ for each of the representations given in tables 11.20 and 11.23. We also give the symmetrized square $[R_i]^2$ for the degenerate representations and note that for non-degenerate representations

$$[R_i]^2 = R_i \otimes R_i. \tag{12.12}$$

In evaluating $R_i \otimes R_j$ we have made use of the theory of Bradley [71]. It is an interesting feature that the Kronecker products of two representations at either L or V for NiO contain representations at points other than Γ, for example

$$V_1^+ \otimes V_1^+ = \Gamma_1^+ + \Gamma_2^+ + M_1^- + M_2^-. \tag{12.13}$$

However, these representations of M are not relevant for two-magnon absorption because the incoming photon which is absorbed to create the two magnons has momentum very nearly equal to zero, so that it belongs only to representations at Γ and not at any other special points in the Brillouin zone. The selection rules for two-magnon absorption of unpolarized radiation are indicated in tables 12.6 and 12.7 for NiF_2 and NiO respectively.

13

CONCLUSION

The chief usefulness of Part II consists essentially in the detailed study of the symmetry properties of two magnetic crystals, NiF_2 and NiO. This study has been the outcome of the results of recent neutron diffraction experiments which have now clearly revealed that the symmetry properties of crystals possessing a non-vanishing time-averaged distribution of current density, as is the case in strongly magnetic crystals, must be described in terms of Shubnikov or magnetic groups. Since these groups are antiunitary we discussed the formalism called corepresentation theory which is used when determining the symmetry properties and the energy eigenstates of a magnetic crystal.

To lead us up to corepresentation theory we had to review briefly magnetic point groups and space groups in order to work out the coreps of the magnetic space groups for all the points and lines of symmetry in the appropriate Brillouin zones of antiferromagnetic NiF_2 and NiO. We have shown from the deduction of coreps that there are three specific effects of magnetic ordering on the crystal eigenstates:

(1) there is a lifting of some eigenfunction degeneracies due to the reduction in crystal symmetry brought about through the magnetic ordering of the lattice (i.e. the introduction of a non-vanishing time-averaged magnetic moment density) and possible magnetostriction;
(2) new Brillouin zone surfaces are introduced owing to change in translational symmetry; and
(3) The symmetry in $\boldsymbol{k}$ space may be reduced.

Furthermore, when the magnetic and non-magnetic lattices are different, additional features are introduced in the dispersion curves.

We have also discussed the use of magnetic space groups in determining the selection rules for magnetic crystals. Using the works of Cracknell [48] and Bradley and Davies [30] and by means of examples (using NiF_2) the method of reduction of Kronecker products of corepresentations of non-unitary groups has been shown. The symmetrized cube and the antisymmetrized square which are of importance in Landau's theory of second-order phase transitions have also been reduced and tabulated. Finally, by using an earlier theory of Bradley [71] the selection rules for two-magnon electric dipole and magnetic dipole absorption for NiF_2 and NiO have been determined.

APPENDIX A

THE EULER ANGLES AND THE ROTATION–REPRESENTATION MATRICES

The rotational operations of symmetry of a sphere are described specifically in terms of three angles α, β, γ called the Euler angles. If we take the three Cartesian coordinate axes as three rotation axes passing through the centre

Table A.1 The Euler angles for the group O (Altmann S L and Cracknell A P 1957 *Proc. Camb. Philos. Soc.* **53** 349).

Element	α	β	γ
E	0	0	0
C_{2x}	π	π	0
C_{2y}	0	π	0
C_{2z}	π	0	0
C_{31}^+	0	$\pi/2$	$\pi/2$
C_{32}^+	π	$\pi/2$	$-\pi/2$
C_{33}^+	π	$\pi/2$	$\pi/2$
C_{34}^+	0	$\pi/2$	$-\pi/2$
C_{31}^-	$\pi/2$	$\pi/2$	π
C_{32}^-	$-\pi/2$	$\pi/2$	0
C_{33}^-	$\pi/2$	$\pi/2$	0
C_{34}^-	$-\pi/2$	$\pi/2$	π
C_{4x}^+	$\pi/2$	$\pi/2$	$-\pi/2$
C_{4y}^+	0	$-\pi/2$	0
C_{4z}^+	$-\pi/2$	0	0
C_{4x}^-	$\pi/2$	$-\pi/2$	$-\pi/2$
C_{4y}^-	0	$\pi/2$	0
C_{4z}^-	$\pi/2$	0	0
C_{2a}	$\pi/2$	π	π
C_{2b}	$\pi/2$	π	0
C_{2c}	0	$\pi/2$	π
C_{2d}	$\pi/2$	$\pi/2$	$\pi/2$
C_{2e}	π	$\pi/2$	0
C_{2f}	$-\pi/2$	$\pi/2$	$-\pi/2$

of the sphere, then the Euler angles could be defined as follows (see Tinkham [2]):

(i) Considering the z axis as the rotation axis, rotate the other two Cartesian axes (i.e. the x and y axes) through an angle α.

(ii) Next, considering the x axis in its rotated position as the rotation axis, rotate the other two Cartesian axes (i.e. the y and z axes) through an angle β.

Table A.2 Matrix elements for the symmetry elements of group O.

$$E = \begin{pmatrix} 0 & 0 & 0 \\ 0 & 1 & 0 \\ 0 & 0 & 1 \end{pmatrix} \quad C_{31}^{-} = \begin{pmatrix} 0 & -1 & 0 \\ 0 & 0 & 1 \\ -1 & 0 & 0 \end{pmatrix} \quad C_{4y}^{-} = \begin{pmatrix} 0 & 0 & -1 \\ 0 & 1 & 0 \\ 1 & 0 & 0 \end{pmatrix}$$

$$C_{2x} = \begin{pmatrix} 1 & 0 & 0 \\ 0 & -1 & 0 \\ 0 & 0 & -1 \end{pmatrix} \quad C_{32}^{-} = \begin{pmatrix} 0 & -1 & 0 \\ 0 & 0 & -1 \\ 1 & 0 & 0 \end{pmatrix} \quad C_{4z}^{-} = \begin{pmatrix} 0 & 1 & 0 \\ -1 & 0 & 0 \\ 0 & 0 & 1 \end{pmatrix}$$

$$C_{2y} = \begin{pmatrix} -1 & 0 & 0 \\ 0 & 1 & 0 \\ 0 & 0 & -1 \end{pmatrix} \quad C_{33}^{-} = \begin{pmatrix} 0 & 1 & 0 \\ 0 & 0 & 1 \\ 1 & 0 & 0 \end{pmatrix} \quad C_{2a} = \begin{pmatrix} 0 & 1 & 0 \\ 1 & 0 & 0 \\ 0 & 0 & -1 \end{pmatrix}$$

$$C_{2z} = \begin{pmatrix} -1 & 0 & 0 \\ 0 & -1 & 0 \\ 0 & 0 & 1 \end{pmatrix} \quad C_{34}^{-} = \begin{pmatrix} 0 & 1 & 0 \\ 0 & 0 & -1 \\ -1 & 0 & 0 \end{pmatrix} \quad C_{2b} = \begin{pmatrix} 0 & -1 & 0 \\ -1 & 0 & 0 \\ 0 & 0 & -1 \end{pmatrix}$$

$$C_{31}^{+} = \begin{pmatrix} 0 & 0 & -1 \\ -1 & 0 & 0 \\ 0 & 1 & 0 \end{pmatrix} \quad C_{4x}^{+} = \begin{pmatrix} 1 & 0 & 0 \\ 0 & 0 & 1 \\ 0 & -1 & 0 \end{pmatrix} \quad C_{2c} = \begin{pmatrix} 0 & 0 & -1 \\ 0 & -1 & 0 \\ -1 & 0 & 0 \end{pmatrix}$$

$$C_{32}^{+} = \begin{pmatrix} 0 & 0 & 1 \\ -1 & 0 & 0 \\ 0 & -1 & 0 \end{pmatrix} \quad C_{4y}^{+} = \begin{pmatrix} 0 & 0 & 1 \\ 0 & 1 & 0 \\ -1 & 0 & 0 \end{pmatrix} \quad C_{2d} = \begin{pmatrix} -1 & 0 & 0 \\ 0 & 0 & 1 \\ 0 & 1 & 0 \end{pmatrix}$$

$$C_{33}^{+} = \begin{pmatrix} 0 & 0 & 1 \\ 1 & 0 & 0 \\ 0 & 1 & 0 \end{pmatrix} \quad C_{4z}^{+} = \begin{pmatrix} 0 & -1 & 0 \\ 1 & 0 & 0 \\ 0 & 0 & 1 \end{pmatrix} \quad C_{2e} = \begin{pmatrix} 0 & 0 & 1 \\ 0 & -1 & 0 \\ 1 & 0 & 0 \end{pmatrix}$$

$$C_{34}^{+} = \begin{pmatrix} 0 & 0 & -1 \\ 1 & 0 & 0 \\ 0 & -0 & 0 \end{pmatrix} \quad C_{4x}^{-} = \begin{pmatrix} 1 & 0 & 0 \\ 0 & 0 & -1 \\ 0 & 1 & 0 \end{pmatrix} \quad C_{2f} = \begin{pmatrix} -1 & 0 & 0 \\ 0 & 0 & -1 \\ 0 & -1 & 0 \end{pmatrix}$$

Table A.3 Matrix elements of the 2D spin representation of group O.

$E = \begin{pmatrix} 1 & 0 \\ 0 & 1 \end{pmatrix}$	$C_{31}^{-} = \frac{1}{2}\begin{pmatrix} 1-i & -1+i \\ 1+i & 1+i \end{pmatrix}$	$C_{4y}^{-} = \frac{1}{\sqrt{2}}\begin{pmatrix} 1 & -1 \\ 1 & 1 \end{pmatrix}$
$C_{2x} = \begin{pmatrix} 0 & -i \\ -i & 0 \end{pmatrix}$	$C_{32}^{-} = \frac{1}{2}\begin{pmatrix} 1-i & 1-i \\ -1-i & 1+i \end{pmatrix}$	$C_{4z}^{-} = \frac{1}{\sqrt{2}}\begin{pmatrix} 1-i & 0 \\ 0 & 1+i \end{pmatrix}$
$C_{2y} = \begin{pmatrix} 0 & -1 \\ 1 & 0 \end{pmatrix}$	$C_{33}^{-} = \frac{1}{2}\begin{pmatrix} -1-i & -1-i \\ 1-i & -1+i \end{pmatrix}$	$C_{2a} = \frac{1}{\sqrt{2}}\begin{pmatrix} 0 & -1+i \\ 1+i & 0 \end{pmatrix}$
$C_{2z} = \begin{pmatrix} -i & 0 \\ 0 & i \end{pmatrix}$	$C_{34}^{-} = \frac{1}{2}\begin{pmatrix} 1+i & -1-i \\ 1-i & 1-i \end{pmatrix}$	$C_{2b} = \frac{1}{\sqrt{2}}\begin{pmatrix} 0 & -1-i \\ 1-i & 0 \end{pmatrix}$
$C_{31}^{+} = \frac{1}{2}\begin{pmatrix} -1-i & -1+i \\ 1+i & -1+i \end{pmatrix}$	$C_{4x}^{+} = \frac{1}{\sqrt{2}}\begin{pmatrix} 1 & -i \\ -i & 1 \end{pmatrix}$	$C_{2c} = \frac{1}{\sqrt{2}}\begin{pmatrix} -i & i \\ i & i \end{pmatrix}$
$C_{32}^{+} = \frac{1}{2}\begin{pmatrix} 1+i & -1+i \\ 1+i & 1-i \end{pmatrix}$	$C_{4y}^{+} = \frac{1}{\sqrt{2}}\begin{pmatrix} 1 & 1 \\ -1 & 1 \end{pmatrix}$	$C_{2d} = \frac{1}{\sqrt{2}}\begin{pmatrix} -i & -1 \\ 1 & i \end{pmatrix}$
$C_{33}^{+} = \frac{1}{2}\begin{pmatrix} 1-i & -1-i \\ 1-i & 1+i \end{pmatrix}$	$C_{4z}^{+} = \frac{1}{\sqrt{2}}\begin{pmatrix} 1+i & 0 \\ 0 & 1-i \end{pmatrix}$	$C_{2e} = \frac{1}{\sqrt{2}}\begin{pmatrix} -i & -i \\ -i & i \end{pmatrix}$
$C_{34}^{+} = \frac{1}{2}\begin{pmatrix} 1-i & 1+i \\ -1+i & 1+i \end{pmatrix}$	$C_{4x}^{-} = \frac{1}{\sqrt{2}}\begin{pmatrix} 1 & i \\ i & 1 \end{pmatrix}$	$C_{2f} = \frac{1}{\sqrt{2}}\begin{pmatrix} i & -1 \\ 1 & -i \end{pmatrix}$

(iii) Finally, considering the y axis in its rotated position as the rotation axis, rotate the other two Cartesian axes (i.e. the z and x axes) through an angle γ.

By such operations and using appropriate values for α, β and γ one can accomplish any desired rotational symmetry operation. As an illustration we have given in table A.1 the Euler angles for all the rotational symmetry operations of the cubic group O. The matrix representation of a rotation operation in terms of α, β, γ is given by:

$$R(\alpha, \beta, \gamma) = \begin{pmatrix} C_\alpha C_\beta C_\gamma - S_\alpha S_\gamma & C_\alpha C_\beta S_\gamma + S_\alpha C_\gamma & -C_\alpha S_\beta \\ -S_\alpha C_\beta C_\gamma - C_\alpha S_\gamma & -S_\alpha C_\beta S_\gamma + C_\alpha C_\gamma & S_\alpha S_\beta \\ S_\beta C_\gamma & S_\beta S_\gamma & C_\beta \end{pmatrix}. \quad (A.1)$$

Here C_α stands for $\cos\alpha$ etc and S_α for $\sin\alpha$ etc. Using equation (A.1) the matrix elements of the group O are given in table A.2.

Similarly the matrix representations of the two-dimensional spin representation are given by:

$$D^{1/2}(\alpha, \beta, \gamma) = \begin{pmatrix} \exp[\frac{1}{2}\mathrm{i}(\alpha+\gamma)]\cos\beta/2 & \exp[\frac{1}{2}\mathrm{i}(\alpha-\gamma)]\sin\beta/2 \\ -\exp[-\frac{1}{2}\mathrm{i}(\alpha-\gamma)]\sin\beta/2 & \exp[-\frac{1}{2}\mathrm{i}(\alpha+\gamma)]\cos\beta/2 \end{pmatrix}. \quad (A.2)$$

Using equation (A.2) the matrix elements of the two-dimensional spin representation of O are given in table A.3. For more details readers are referred to Tinkham [2].

APPENDIX B

MULTIPLICATION TABLE FOR THE DOUBLE GROUPS ^dO, ^dO_h AND ^dT_d

B.1 Double group ^dO

For convenience we shall label all the symmetry operations by the numbers 1–24. The double-group multiplication table is then as shown in table B.1.

B.2 Double group ^dO_h

The multiplication table given in table B.1 and the table of matrix elements for O given in Appendix A are appropriate for the double group ^dO_h (and also ^dT_d). For ^dO_h we now have the following operations:

1–24 as in the group O and 1′ to 24′ (each equal to an unprimed operation together with inversion), plus 48 corresponding barred operations. Note that the matrix elements for the primed and unprimed operations are the same but going from unbarred to barred operations changes the sign of the matrix elements.

The rule for the ^dO_h multiplication is as follows:

(a) The product of a primed and unprimed operation is primed.
(b) The product of two primed operations is unprimed.
(c) If one of the factors is barred, the product changes from unbarred to barred and vice versa.
(d) If both the factors are barred the product is as given in the table for ^dO.

B.3 Double group ^dT_d

Replace C_{2a}, C_{2b}, C_{2c}, C_{2d}, C_{2e}, C_{2f} in table B.1 by σ_{d_a}, σ_{d_b}, σ_{d_c}, σ_{d_d}, σ_{d_e}, σ_{d_f} respectively and also change C_{4x}^+, C_{4y}^+, C_{4z}^+, C_{4x}^-, C_{4y}^- and C_{4z}^- by S_{4x}^+, S_{4y}^+, S_{4z}^+, S_{4x}^-, S_{4y}^- and S_{4z}^- respectively. The rest of the table in table B.1 remains the same.

Table B.1

		1	2	3	4	5	6	7	8	9	10	11	12	13	14	15	16	17	18	19	20	21	22	23	24
	^dO	E	C_{2x}	C_{2y}	C_{2z}	C_{33}^+	C_{31}^+	C_{32}^+	C_{34}^+	C_{33}^-	C_{31}^-	C_{32}^-	C_{34}^-	C_{4x}^-	C_{4x}^+	C_{4y}^+	C_{4y}^-	C_{4z}^+	C_{4z}^-	C_{2f}	C_{2d}	C_{2c}	C_{2e}	C_{2a}	C_{2b}
1	E	1	2	3	4	5	6	7	8	9	10	11	12	13	14	15	16	17	18	19	20	21	22	23	24
2	C_{2x}	2	$\bar{1}$	$\bar{4}$	3	7	8	$\bar{5}$	$\bar{6}$	$\bar{12}$	$\bar{11}$	10	9	$\bar{14}$	13	$\bar{21}$	$\bar{22}$	$\bar{24}$	23	20	$\bar{19}$	15	16	$\bar{18}$	17
3	C_{2y}	3	4	$\bar{1}$	$\bar{2}$	8	$\bar{7}$	6	$\bar{5}$	$\bar{10}$	9	$\bar{12}$	11	$\bar{20}$	19	$\bar{16}$	15	$\bar{23}$	$\bar{24}$	$\bar{14}$	13	22	$\bar{21}$	17	18
4	C_{2z}	4	$\bar{3}$	2	$\bar{1}$	6	$\bar{5}$	$\bar{8}$	7	$\bar{11}$	12	9	$\bar{10}$	$\bar{19}$	$\bar{20}$	$\bar{22}$	21	$\bar{18}$	17	13	14	$\bar{16}$	15	24	$\bar{23}$
5	C_{33}^+	5	8	6	7	$\bar{9}$	12	10	11	1	$\bar{4}$	$\bar{2}$	$\bar{3}$	18	24	14	20	16	22	$\bar{23}$	$\bar{17}$	$\bar{19}$	$\bar{13}$	$\bar{21}$	$\bar{15}$
6	C_{31}^+	6	7	$\bar{5}$	$\bar{8}$	11	$\bar{10}$	12	9	4	1	3	$\bar{2}$	17	$\bar{23}$	$\bar{20}$	14	21	15	$\bar{24}$	18	$\bar{13}$	19	16	22
7	C_{32}^+	7	$\bar{6}$	8	$\bar{5}$	12	9	$\bar{11}$	10	2	$\bar{3}$	1	4	23	17	13	$\bar{19}$	$\bar{22}$	16	18	24	$\bar{20}$	14	$\bar{15}$	21
8	C_{34}^+	8	$\bar{5}$	$\bar{7}$	6	10	11	9	$\bar{12}$	3	2	$\bar{4}$	1	$\bar{24}$	18	19	13	15	$\bar{21}$	$\bar{17}$	23	14	20	$\bar{22}$	16
9	C_{33}^-	9	$\bar{11}$	$\bar{12}$	10	1	3	4	2	$\bar{5}$	7	8	6	$\bar{22}$	15	$\bar{24}$	17	$\bar{20}$	13	$\bar{21}$	16	$\bar{23}$	18	$\bar{19}$	14
10	C_{31}^-	10	$\bar{12}$	11	9	$\bar{3}$	1	2	$\bar{4}$	8	$\bar{6}$	5	7	$\bar{21}$	16	18	23	13	20	22	$\bar{15}$	17	24	$\bar{14}$	$\bar{19}$
11	C_{32}^-	11	9	$\bar{10}$	12	$\bar{4}$	$\bar{2}$	1	3	6	8	$\bar{7}$	5	15	22	$\bar{23}$	18	14	19	$\bar{16}$	$\bar{21}$	24	$\bar{17}$	13	20
12	C_{34}^-	12	10	9	$\bar{11}$	$\bar{2}$	4	$\bar{3}$	1	7	5	6	$\bar{8}$	16	21	17	24	$\bar{19}$	14	15	22	$\bar{18}$	$\bar{23}$	20	$\bar{13}$
13	C_{4x}^-	13	$\bar{14}$	19	$\bar{20}$	16	15	$\bar{22}$	$\bar{21}$	$\bar{24}$	23	18	17	2	1	8	7	9	10	$\bar{4}$	$\bar{3}$	6	5	$\bar{11}$	12
14	C_{4x}^+	14	13	$\bar{20}$	$\bar{19}$	22	21	16	15	17	18	$\bar{23}$	24	1	$\bar{2}$	6	5	12	11	3	$\bar{4}$	$\bar{8}$	$\bar{7}$	10	$\bar{9}$
15	C_{4y}^+	15	$\bar{22}$	$\bar{16}$	21	18	$\bar{23}$	17	$\bar{24}$	$\bar{20}$	13	19	14	9	4	3	1	6	8	$\bar{12}$	10	$\bar{2}$	$\bar{4}$	7	5
16	C_{4y}^-	16	$\bar{21}$	15	$\bar{22}$	24	17	23	18	13	20	14	$\bar{19}$	10	12	1	$\bar{3}$	7	5	11	$\bar{9}$	4	$\bar{2}$	$\bar{6}$	$\bar{8}$
17	C_{4z}^+	17	23	$\bar{24}$	$\bar{18}$	14	$\bar{20}$	$\bar{19}$	13	$\bar{22}$	16	15	21	7	6	9	12	4	1	8	5	$\bar{10}$	11	$\bar{3}$	$\bar{2}$
18	C_{4z}^-	18	$\bar{24}$	$\bar{23}$	17	20	14	13	19	15	$\bar{21}$	22	16	8	5	11	10	1	$\bar{4}$	$\bar{7}$	$\bar{6}$	12	$\bar{9}$	2	$\bar{3}$
19	C_{2f}	19	$\bar{20}$	$\bar{13}$	14	$\bar{21}$	22	15	$\bar{16}$	$\bar{23}$	$\bar{24}$	$\bar{17}$	18	3	$\bar{4}$	$\bar{7}$	8	11	$\bar{12}$	$\bar{1}$	2	5	$\bar{6}$	9	10
20	C_{2d}	20	19	14	13	$\bar{15}$	16	$\bar{21}$	22	18	$\bar{17}$	24	23	$\bar{4}$	$\bar{3}$	5	$\bar{6}$	10	$\bar{9}$	$\bar{2}$	$\bar{1}$	7	$\bar{8}$	$\bar{12}$	$\bar{11}$
21	C_{2c}	21	16	$\bar{22}$	$\bar{15}$	$\bar{23}$	$\bar{18}$	24	17	$\bar{19}$	14	$\bar{20}$	$\bar{13}$	12	$\bar{10}$	4	$\bar{2}$	$\bar{8}$	6	9	11	$\bar{1}$	3	5	$\bar{7}$
22	C_{2e}	22	15	21	16	$\bar{17}$	24	18	$\bar{23}$	14	19	$\bar{13}$	20	11	$\bar{9}$	$\bar{2}$	$\bar{4}$	5	$\bar{7}$	$\bar{10}$	$\bar{12}$	$\bar{3}$	$\bar{1}$	8	$\bar{6}$
23	C_{2a}	23	$\bar{17}$	18	$\bar{24}$	$\bar{19}$	13	$\bar{14}$	20	$\bar{21}$	$\bar{15}$	16	$\bar{22}$	$\bar{6}$	7	10	$\bar{11}$	2	$\bar{3}$	5	$\bar{8}$	9	12	$\bar{1}$	4
24	C_{2b}	24	18	17	23	$\bar{13}$	$\bar{19}$	20	14	16	22	21	$\bar{15}$	5	$\bar{8}$	12	$\bar{9}$	$\bar{3}$	$\bar{2}$	6	$\bar{7}$	$\bar{11}$	$\bar{10}$	$\bar{4}$	$\bar{1}$

This multiplication table and the table of matrix elements for O are also appropriate for the double groups ^dT_d and ^dO_h. For O_h we have the following operation: $R_1 \ldots R_{24}$ as in O and $R'_1 \ldots R'_{24}$ (each equal to the unprimed operation plus inversion) and of course the 48 barred operations associated with these. The matrix elements for primed and unprimed operation are the same. Going from an unbarred to a barred operation changes the sign of the matrix elements.

APPENDIX C

THE 90 MAGNETIC POINT GROUPS LISTED IN THE INTERNATIONAL AND SCHOENFLIES NOTATIONS

	Ordinary group G		Invariant unitary subgroup H		Magnetic group M	
No.	Schoenflies	International	Schoenflies	International	Schoenflies	International
1	C_1	1	C_1	1	$C_1(C_1)$	1
2	C_i	$\bar{1}$	C_i	$\bar{1}$	$C_i(C_i)$	$\bar{1}$
3	C_i	$\bar{1}$	C_1	1	$C_i(C_1)$	$\underline{\bar{1}}$
4	C_{1h}	m	C_{1h}	m	$C_{1h}(C_{1h})$	m
5	C_{1h}	m	C_1	1	$C_{1h}(C_1)$	$\underline{m}$
6	C_2	2	C_2	2	$C_2(C_2)$	2
7	C_2	2	C_1	1	$C_2(C_1)$	$\underline{2}$
8	C_{2h}	2/m	C_{2h}	2/m	$C_{2h}(C_{2h})$	2/m
9	C_{2h}	2/m	C_i	$\bar{1}$	$C_{2h}(C_i)$	$\underline{2}/\underline{m}$
10	C_{2h}	2/m	C_2	2	$C_{2h}(C_2)$	$2/\underline{m}$
11	C_{2h}	2/m	C_{1h}	m	$C_{2h}(C_{1h})$	$\underline{2}/m$
12	C_{2v}	2mm	C_{2v}	2mm	$C_{2v}(C_{2v})$	2mm
13	C_{2v}	2mm	C_{1h}	m	$C_{2v}(C_{1h})$	$\underline{2m}m$
14	C_{2v}	2mm	C_2	2	$C_{2v}(C_2)$	$2\underline{mm}$
15	D_2	222	D_2	222	$D_2(D_2)$	222
16	D_2	222	C_2	2	$D_2(C_2)$	$2\underline{22}$
17	D_{2h}	mmm	D_{2h}	mmm	$D_{2h}(D_{2h})$	mmm
18	D_{2h}	mmm	C_{2h}	2/m	$D_{2h}(C_{2h})$	$\underline{mm}m$
19	D_{2h}	mmm	C_{2v}	2mm	$D_{2h}(C_{2v})$	$mm\underline{m}$
20	D_{2h}	mmm	D_2	222	$D_{2h}(D_2)$	$\underline{mmm}$
21	C_4	4	C_4	4	$C_4(C_4)$	4
22	C_4	4	C_2	2	$C_4(C_2)$	$\underline{4}$

23	S_4	$\bar{4}$	S_4	$\bar{4}$	$S_4(S_4)$	$\bar{4}$
24	S_4	$\bar{4}$	C_2	2	$S_4(C_2)$	$\underline{\bar{4}}$
25	C_{4h}	4/m	C_{4h}	4/m	$C_{4h}(C_{4h})$	4/m
26	C_{4h}	4/m	C_{2h}	2/m	$C_{4h}(C_{2h})$	$\underline{4}/m$
27	C_{4h}	4/m	C_4	4	$C_{4h}(C_4)$	$4/\underline{m}$
28	C_{4h}	4/m	S_4	$\bar{4}$	$C_{4h}(S_4)$	$\underline{4}/\underline{m}$
29	D_{2d}	$\bar{4}2m$	D_{2d}	$\bar{4}2m$	$D_{2d}(D_{2d})$	$\bar{4}2m$
30	D_{2d}	$\bar{4}2m$	C_{2v}	2mm	$D_{2d}(C_{2v})$	$\underline{\bar{4}2}m$
31	D_{2d}	$\bar{4}2m$	D_2	222	$D_{2d}(D_2)$	$\underline{\bar{4}}2\underline{m}$
32	D_{2d}	$\bar{4}2m$	S_4	$\bar{4}$	$D_{2d}(S_4)$	$\bar{4}\underline{2m}$
33	C_{4v}	4mm	C_{4v}	4mm	$C_{4v}(C_{4v})$	4mm
34	C_{4v}	4mm	C_{2v}	2mm	$C_{4v}(C_{2v})$	$\underline{4}m\underline{m}$
35	C_{4v}	4mm	C_4	4	$C_{4v}(C_4)$	$4\underline{mm}$
36	D_4	422	D_4	422	$D_4(D_4)$	422
37	D_4	422	D_2	222	$D_4(D_2)$	$\underline{4}22$
38	D_4	422	C_4	4	$D_4(C_4)$	$4\underline{22}$
39	D_{4h}	4/mmm	D_{4h}	4/mmm	$D_{4h}(D_{4h})$	4/mmm
40	D_{4h}	4/mmm	D_{2h}	mmm	$D_{4h}(D_{2h})$	$\underline{4}/mm\underline{m}$
41	D_{4h}	4/mmm	C_{4h}	4/m	$D_{4h}(C_{4h})$	$4/m\underline{mm}$
42	D_{4h}	4/mmm	D_{2d}	$\bar{4}2m$	$D_{4h}(D_{2d})$	$4/\underline{mm}m$
43	D_{4h}	4/mmm	C_{4v}	4mm	$D_{4h}(C_{4v})$	$4/\underline{m}mm$
44	D_{4h}	4/mmm	D_4	422	$D_{4h}(D_4)$	$4/\underline{mmm}$
45	C_3	3	C_3	3	$C_3(C_3)$	3
46	S_6	$\bar{3}$	S_6	$\bar{3}$	$S_6(S_6)$	$\bar{3}$
47	S_6	$\bar{3}$	C_3	3	$S_6(C_3)$	$\underline{\bar{3}}$
48	C_{3v}	3m	C_{3v}	3m	$C_{3v}(C_{3v})$	3m
49	C_{3v}	3m	C_3	3	$C_{3v}(C_3)$	$3\underline{m}$
50	D_3	32	D_3	32	$D_3(D_3)$	32
51	D_3	32	C_3	3	$D_3(C_3)$	$3\underline{2}$
52	D_{3d}	$\bar{3}m$	D_{3d}	$\bar{3}m$	$D_{3d}(D_{3d})$	$\bar{3}m$
53	D_{3d}	$\bar{3}m$	S_6	$\bar{3}$	$D_{3d}(S_6)$	$\bar{3}\underline{m}$
54	D_{3d}	$\bar{3}m$	C_{3v}	3m	$D_{3d}(C_{3v})$	$\underline{\bar{3}}m$
55	D_{3d}	$\bar{3}m$	D_3	32	$D_{3d}(D_3)$	$\underline{\bar{3}m}$
56	C_{3h}	$\bar{6}$	C_{3h}	$\bar{6}$	$C_{3h}(C_{3h})$	$\bar{6}$
57	C_{3h}	$\bar{6}$	C_3	3	$C_{3h}(C_3)$	$\underline{\bar{6}}$
58	C_6	6	C_6	6	$C_6(C_6)$	6
59	C_6	6	C_3	3	$C_6(C_3)$	$\underline{6}$

	Ordinary group G		Invariant unitary subgroup H		Magnetic group M	
No.	Schoenflies	International	Schoenflies	International	Schoenflies	International
60	C_{6h}	6/m	C_{6h}	6/m	$C_{6h}(C_{6h})$	6/m
61	C_{6h}	6/m	S_6	$\bar{3}$	$C_{6h}(S_6)$	$\underline{6}/\underline{m}$
62	C_{6h}	6/m	C_{3h}	$\bar{6}$	$C_{6h}(C_{3h})$	$\underline{6}/m$
63	C_{6h}	6/m	C_6	6	$C_{6h}(C_6)$	$6/\underline{m}$
64	D_{3h}	$\bar{6}m2$	D_{3h}	$\bar{6}m2$	$D_{3h}(D_{3h})$	$\bar{6}m2$
65	D_{3h}	$\bar{6}m2$	C_{3v}	3m	$D_{3h}(C_{3v})$	$\underline{\bar{6}}m\underline{2}$
66	D_{3h}	$\bar{6}m2$	D_3	32	$D_{3h}(D_3)$	$\underline{\bar{6}}\underline{m}2$
67	D_{3h}	$\bar{6}m2$	C_{3h}	$\bar{6}$	$D_{3h}(C_{3h})$	$\bar{6}\underline{m}\underline{2}$
68	C_{6v}	6mm	C_{6v}	6mm	$C_{6v}(C_{6v})$	6mm
69	C_{6v}	6mm	C_{3v}	3m	$C_{6v}(C_{3v})$	$\underline{6}m\underline{m}$
70	C_{6v}	6mm	C_6	6	$C_{6v}(C_6)$	$6\underline{m}\underline{m}$
71	D_6	62	D_6	62	$D_6(D_6)$	62
72	D_6	62	D_3	32	$D_6(D_3)$	$\underline{6}2$
73	D_6	62	C_6	6	$D_6(C_6)$	$6\underline{2}$
74	D_{6h}	6/mmm	D_{6h}	6/mmm	$D_{6h}(D_{6h})$	6/mmm
75	D_{6h}	6/mmm	D_{3d}	$\bar{3}m$	$D_{6h}(D_{3d})$	$\underline{6}/\underline{m}\underline{m}m$
76	D_{6h}	6/mmm	C_{6h}	6/m	$D_{6h}(C_{6h})$	$6/m\underline{m}\underline{m}$
77	D_{6h}	6/mmm	D_{3h}	$\bar{6}m2$	$D_{6h}(D_{3h})$	$\underline{6}/m\underline{m}m$
78	D_{6h}	6/mmm	C_{6v}	6mm	$D_{6h}(C_{6v})$	$6/\underline{m}mm$
79	D_{6h}	6/mmm	D_6	62	$D_{6h}(D_6)$	$6/\underline{m}\underline{m}\underline{m}$
80	T	23	T	23	T(T)	23
81	T_h	m3	T_h	m3	$T_h(T_h)$	m3
82	T_h	m3	T	23	$T_h(T)$	$\underline{m}3$
83	T_d	$\bar{4}3m$	T_d	$\bar{4}3m$	$T_d(T_d)$	$\bar{4}3m$
84	T_d	$\bar{4}3m$	T	23	$T_d(T)$	$\underline{\bar{4}}3\underline{m}$
85	O	43	O	43	O(O)	43
86	O	43	T	23	O(T)	$\underline{4}3$
87	O_h	m3m	O_h	m3m	$O_h(O_h)$	m3m
88	O_h	m3m	T_h	m3	$O_h(T_h)$	$m3\underline{m}$
89	O_h	m3m	T_d	$\bar{4}3m$	$O_h(T_d)$	$\underline{m}3m$
90	O_h	m3m	O	43	$O_h(O)$	$\underline{m}3\underline{m}$

APPENDIX D

CONSTRUCTION OF MAGNETIC POINT GROUPS BY REPRESENTATION ANALYSIS

All magnetic point groups belonging to type III are listed in the following tables (D.1–D.32) in the Schoenflies and International notations. In the International notation one takes the symbol for M and underlines those elements to which R has been adjoined. We also list the ordinary point groups (type I) not containing R at all, in which case H is of index 1. We do not list the type II groups which are obtained by adjoining R to all elements of type I. The magnetic order is indicated by F (ferromagnetic) and AF (antiferromagnetic).

Table D.1 Group C_1; 1.

Rep.	Character of elements	Unitary subgroup H		Magnetic group		Mag. order	No.
	E	Elements	Group	Schoenflies	Int.		
Γ_1	1	E	C_1	$C_1(C_1)$	1	F	1

Table D.2 Group C_i; $\bar{1}$.

Rep.	Character of elements		Unitary subgroup H		Magnetic group		Mag. order	No.
	E	I	Elements	Group	Schoenflies	Int.		
Γ_1^+	1	1	E, I	C_i	$C_i(C_i)$	$\bar{1}$	F	2
Γ_2^+	1	−1	E	C_1	$C_i(C_1)$	$\underline{\bar{1}}$	AF	3

Table D.3 Group C_2; 2.

Rep.	Character of elements		Unitary subgroup H		Magnetic group		Mag. order	No.
	E	C_2	Elements	Group	Schoenflies	Int.		
Γ_1	1	1	E, C_2	C_2	$C_2(C_2)$	2	F	4
Γ_2	1	−1	E	C_1	$C_2(C_1)$	$\underline{2}$	F	5

Table D.4 Group C_{1h}; M.

Rep.	Character of elements E	σ	Unitary subgroup H Elements	Group	Magnetic group Schoenflies	Int.	Mag. order	No.
Γ_1	1	1	E, σ	C_{1h}	$C_{1h}(C_{1h})$	M	F	6
Γ_2	1	−1	E	C_1	$C_{1h}(C_1)$	$\underline{M}$	F	7

Table D.5 Group C_{2h}; 2/M.

Rep.	Character of elements E	C_2	I	σ_h	Unitary subgroup H Elements	Group	Magnetic group Schoenflies	Int.	Mag. order	No.
Γ_1^+	1	1	1	1	E, C_2, I, σ_h	C_{2h}	$C_{2h}(C_{2h})$	2/M	F	8
Γ_2^+	1	−1	1	−1	E, I	C_i	$C_{2h}(C_i)$	$\underline{2}/\underline{M}$	F	9
Γ_1^-	1	1	−1	−1	E, C_2	C_2	$C_{2h}(C_2)$	$2/\underline{M}$	AF	10
Γ_2^-	1	−1	−1	1	E, σ_h	C_{1h}	$C_{2h}(C_{1h})$	$\underline{2}/M$	AF	11

Table D.6 Group D_2; 222.

Rep.	Character of elements E	C_2	C_2'	C_2''	Unitary subgroup H Elements	Group	Magnetic group Schoenflies	Int.	Mag. order	No.
Γ_1	1	1	1	1	E, C_2, C_2', C_2''	D_2	$D_2(D_2)$	222	AF	12
Γ_2	1	−1	1	−1	E, C_2'	}				
Γ_3	1	1	−1	−1	E, C_2	C_2	$D_2(C_2)$	$2\underline{22}$	F	13
Γ_4	1	−1	−1	1	E, C_2''	}				

Table D.7 Group C_{2v}; 2MM.

Rep.	Character of elements E	C_2	σ_v	σ_v'	Unitary subgroup H Elements	Group	Magnetic group Schoenflies	Int.	Mag. order	No.
Γ_1	1	1	1	1	$E, C_2, \sigma_v, \sigma_v'$	C_{2v}	$C_{2v}(C_{2v})$	2MM	AF	14
Γ_2	1	−1	1	−1	E, σ_v	C_{1h}	$C_{2v}(C_{1h})$	$\underline{2M}M$	F	15
Γ_3	1	−1	−1	1	E, σ_v'	}				
Γ_4	1	1	−1	−1	E, C_2	C_2	$C_{2v}(C_2)$	$2\underline{MM}$	F	16

Table D.8 Group D_{2h}; MMM.

Rep.	Character of elements E	C_2	C_2'	C_2''	I	σ_v	σ_v'	σ_v''	Unitary subgroup H Elements	Group	Magnetic group Schoenflies	Int.	Mag. order	No.
Γ_1^+	1	1	1	1	1	1	1	1	$E, C_2, C_2', C_2'', I, \sigma_v, \sigma_v', \sigma_v''$	D_{2h}	$D_{2h}(D_{2h})$	MMM	AF	17
Γ_2^+	1	−1	1	−1	1	−1	1	−1	E, C_2', I, σ_v'					
Γ_3^+	1	1	−1	−1	1	1	−1	−1	E, C_2, I, σ_v	C_{2h}	$D_{2h}(C_{2h})$	$\underline{MM}M$	F	18
Γ_4^+	1	−1	−1	1	1	−1	−1	1	E, C_2'', I, σ_v''					
Γ_1^-	1	1	1	1	−1	−1	−1	−1	E, C_2, C_2', C_2''	D_2	$D_{2h}(D_2)$	$\underline{MMM}$	AF	19
Γ_2^-	1	−1	1	−1	−1	1	−1	1	$E, C_2', \sigma_v, \sigma_v''$					
Γ_3^-	1	1	−1	−1	−1	−1	1	1	$E, C_2, \sigma_v', \sigma_v''$	C_{2v}	$D_{2h}(C_{2v})$	$MM\underline{M}$	AF	20
Γ_4^-	1	−1	−1	1	−1	1	1	−1	$E, C_2'', \sigma_v, \sigma_v'$					

Table D.9 Group C_4; 4.

Rep.	Character of elements E	C_4	C_2	C_4^{-1}	Unitary subgroup H Elements	Group	Magnetic group Schoenflies	Int.	Mag. order	No.
Γ_1	1	1	1	1	E, C_4, C_2, C_4^{-1}	C_4	$C_4(C_4)$	4	F	21
Γ_2	1	−1	1	−1	E, C_2	C_2	$C_4(C_2)$	$\underline{4}$	AF	22
Γ_3	1	i	−1	−i	–	–	–	–	–	
Γ_4	1	−i	−1	i	–	–	–	–	–	

Table D.10 Group S_4; $\bar{4}$.

Rep.	Character of elements E	S_4^{-1}	C_2	S_4	Unitary subgroup H Elements	Group	Magnetic group Schoenflies	Int.	Mag. order	No.
Γ_1	1	1	1	1	E, S_4^{-1}, C_2, S_4	S_4	$S_4(S_4)$	$\bar{4}$	F	23
Γ_2	1	−1	1	−1	E, C_2	C_2	$S_4(C_2)$	$\underline{\bar{4}}$	AF	24
Γ_3	1	i	−1	−i	–	–	–	–	–	
Γ_4	1	−i	−1	i	–	–	–	–	–	

Table D.11 Group C_{4h}; 4/M.

Rep.	Character of elements E	C_4	C_2	C_4^{-1}	I	S_4^{-1}	σ_h	S_4	Unitary subgroup H Elements	Group	Magnetic group Schoenflies	Int.	Mag. order	No.
Γ_1^+	1	1	1	1	1	1	1	1	$E, C_4, C_2, C_4^{-1}, I, S_4^{-1}, \sigma_h, S_4$	C_{4h}	$C_{4h}(C_{4h})$	4/M	F	25
Γ_2^+	1	−1	1	−1	1	−1	1	−1	E, C_2, I, σ_h	C_{2h}	$C_{4h}(C_{2h})$	$\underline{4}$/M	AF	26
Γ_3^+	1	i	−1	−i	1	i	−1	−i	–	–	–	–	–	
Γ_4^+	1	−i	−1	i	1	−i	−1	i	–	–	–	–	–	
Γ_1^-	1	1	1	1	−1	−1	−1	−1	E, C_4, C_2, C_4^{-1}	C_4	$C_{4h}(C_4)$	4/$\underline{M}$	AF	27
Γ_2^-	1	−1	1	−1	−1	1	−1	1	E, C_2, S_4^{-1}, S_4	S_4	$C_{4h}(S_4)$	$\underline{4/M}$	AF	28
Γ_3^-	1	i	−1	−i	−1	−i	1	i	–	–	–	–	–	
Γ_4^-	1	−i	−1	i	−1	i	1	−i	–	–	–	–	–	

Table D.12 Group D_4; 42.

Rep.	Character of elements E	$2C_4$	C_2	$2C_2'$	$2C_2''$	Unitary subgroup H Elements	Group	Magnetic group Schoenflies	Int.	Mag. order	No.
Γ_1	1	1	1	1	1	$E, 2C_4, C_2, 2C_2', 2C_2''$	D_4	$D_4(D_4)$	42	AF	29
Γ_2	1	1	1	−1	−1	$E, 2C_4, C_2$	C_4	$D_4(C_4)$	$4\underline{2}$	F	30
Γ_3	1	−1	1	1	−1	$E, C_2, 2C_2'$	D_2	$D_4(D_2)$	$\underline{4}2$	AF	31
Γ_4	1	−1	1	−1	1	$E, C_2, 2C_2''$	D_2				
Γ_5	2	0	−2	0	0	–	–	–	–	–	

Table D.13 Group C_{4v}; 4MM.

Rep.	Character of elements E	$2C_4$	C_2	$2\sigma_v$	$2\sigma_d$	Unitary subgroup H Elements	Group	Magnetic group Schoenflies	Int.	Mag. order	No.
Γ_1	1	1	1	1	1	$E, 2C_4, C_2, 2\sigma_v, 2\sigma_d$	C_{4v}	$C_{4v}(C_{4v})$	4MM	AF	32
Γ_2	1	1	1	−1	−1	$E, 2C_4, C_2$	C_4	$C_{4v}(C_4)$	$4\underline{MM}$	F	33
Γ_3	1	−1	1	−1	1	$E, C_2, 2\sigma_d$	C_{2v}	$C_{4v}(C_{2v})$	$\underline{4}M\underline{M}$	AF	34
Γ_4	1	−1	1	1	−1	$E, C_2, 2\sigma_v$					
Γ_5	2	0	−2	0	0	–	–	–	–	–	

Table D.14 Group D_{2d}; $\bar{4}2M$.

Rep.	Character of elements E	$2S_4$	C_2	$2C_2'$	$2\sigma_d$	Unitary subgroup H Elements	Group	Magnetic group Schoenflies	Int.	Mag. order	No.
Γ_1	1	1	1	1	1	$E, 2S_4, C_2, 2C_2', 2\sigma_d$	D_{2d}	$D_{2d}(D_{2d})$	$\bar{4}2M$	AF	35
Γ_2	1	1	1	−1	−1	$E, 2S_4, C_2$	S_4	$D_{2d}(S_4)$	$\bar{4}\underline{2M}$	F	36
Γ_3	1	−1	1	1	−1	$E, C_2, 2C_2'$	D_2	$D_{2d}(D_2)$	$\underline{\bar{4}2M}$	AF	37
Γ_4	1	−1	1	−1	1	$E, C_2, 2\sigma_d$	C_{2v}	$D_{2d}(C_{2v})$	$\underline{\bar{4}2}M$	AF	38
Γ_5	2	0	−2	0	0	–	–	–	–	–	

Table D.15 Group D_{4h}; 4/MMM.

Rep.	Character of elements E	$2C_4$	C_2	$2C'_2$	$2C''_2$	I	$2S_4$	σ_h	$2\sigma_v$	$2\sigma_d$	Unitary subgroup H Elements	Group	Magnetic group Schoenflies	Int.	Mag. order	No.
Γ_1^+	1	1	1	1	1	1	1	1	1	1	$E, 2C_4, C_2, 2C'_2, 2C''_2, I, 2S_4, \sigma_h, 2\sigma_v, 2\sigma_d$	D_{4h}	$D_{4h}(D_{4h})$	4/MMM	AF	39
Γ_2^+	1	1	1	−1	−1	1	1	1	−1	−1	$E, 2C_4, C_2, I, 2S_4, \sigma_h$	C_{4h}	$D_{4h}(C_{4h})$	$4/M\underline{MM}$	AF	40
Γ_3^+	1	−1	1	1	−1	1	−1	1	1	−1	$E, C_2, 2C'_2, I, \sigma_h, 2\sigma_v$	D_{2h}	$D_{4h}(D_{2h})$	$\underline{4}/MMM$	AF	41
Γ_4^+	1	−1	1	−1	1	1	−1	1	−1	1	$E, C_2, 2C''_2, I, \sigma_h, 2\sigma_d$					
Γ_5^+	2	0	−2	0	0	2	0	−2	0	0	–	–	–	–	–	
Γ_1^-	1	1	1	1	1	−1	−1	−1	−1	−1	$E, 2C_4, C_2, 2C'_2, 2C''_2$	D_4	$D_{4h}(D_4)$	$4/\underline{MMM}$	AF	42
Γ_2^-	1	1	1	−1	−1	−1	−1	−1	1	1	$E, 2C_4, C_2, 2\sigma_v, 2\sigma_d$	C_{4v}	$D_{4h}(C_{4v})$	$4/\underline{M}MM$	AF	43
Γ_3^-	1	−1	1	1	−1	−1	1	−1	−1	1	$E, C_2, 2C'_2, 2S_4, 2\sigma_d$	D_{2d}	$D_{4h}(D_{2d})$	$\underline{4}/\underline{MM}M$	AF	44
Γ_4^-	1	−1	1	−1	1	−1	1	−1	1	−1	$E, C_2, 2C''_2, 2S_4, 2\sigma_v$					
Γ_5^-	2	0	−2	0	0	−2	0	2	0	0	–	–	–	–	–	

Table D.16 Group C_3; 3.

Rep.	Character of elements E	C_3	C_3^{-1}	Unitary subgroup H Elements	Group	Magnetic group Schoenflies	Int.	Mag. order	No.
Γ_1	1	1	1	E, C_3, C_3^{-1}	C_3	$C_3(C_3)$	3	F	45
Γ_2	1	w^2	$-w$	–	–	–	–	–	
Γ_3	1	$-w$	w^2	–	–	–	–	–	

$w = \exp(\pi i/3)$.

Table D.17 Group C_{3i}; $\bar{3}$.

Rep.	Character of elements E	C_3	C_3^{-1}	I	S_6^{-1}	S_6	Unitary subgroup H Elements	Group	Magnetic group Schoenflies	Int.	Mag. order	No.
Γ_1^+	1	1	1	1	1	1	E, C_3, C_3^{-1}, I, S_6^{-1}, S_6	C_{3i}	$C_{3i}(C_{3i})$	$\bar{3}$	F	46
Γ_2^+	1	w^2	$-w$	1	w^2	$-w$	–	–	–	–	–	
Γ_3^+	1	$-w$	w^2	1	$-w$	w^2	–	–	–	–	–	
Γ_1^-	1	1	1	-1	-1	-1	E, C_3, C_3^{-1}	C_3	$C_{3i}(C_3)$	$\underline{\bar{3}}$	AF	47
Γ_2^-	1	w^2	$-w$	-1	$-w^2$	w	–	–	–	–	–	
Γ_3^-	1	$-w$	w^2	-1	w	$-w^2$	–	–	–	–	–	

$w = \exp(\pi i/3)$.

Table D.18 Group D_3; 32.

Rep.	Character of elements E	$2C_3$	$3C'_2$	Unitary subgroup H Elements	Group	Magnetic group Schoenflies	Int.	Mag. order	No.
Γ_1	1	1	1	E, $2C_3$, $3C'_2$	D_3	$D_3(D_3)$	32	AF	48
Γ_2	1	1	−1	E, $2C_3$	C_3	$D_3(C_3)$	$3\underline{2}$	F	49
Γ_3	2	−1	0	–	–	–	–	–	

Table D.19 Group C_{3v}; 3M.

Rep.	Character of elements E	$2C_3$	$3\sigma_v$	Unitary subgroup H Elements	Group	Magnetic group Schoenflies	Int.	Mag. order	No.
Γ_1	1	1	1	E, $2C_3$, $3\sigma_v$	C_{3v}	$C_{3v}(C_{3v})$	3M	AF	50
Γ_2	1	1	−1	E, $2C_3$	C_3	$C_{3v}(C_3)$	$3\underline{M}$	F	51
Γ_3	2	−1	0	–	–	–	–	–	

Table D.20 Group D_{3d}; $\bar{3}M$.

Rep.	Character of elements E	$2C_3$	$3C'_2$	I	$2S_6$	$3\sigma_d$	Unitary subgroup H Elements	Group	Magnetic group Schoenflies	Int.	Mag. order	No.
Γ_1^+	1	1	1	1	1	1	$E, 2C_3, 3C'_2, I, 2S_6, 3\sigma_d$	D_{3d}	$D_{3d}(D_{3d})$	$\bar{3}M$	AF	52
Γ_2^+	1	1	−1	1	1	−1	$E, 2C_3, I, 2S_6$	S_6	$D_{3d}(S_6)$	$\bar{3}\underline{M}$	F	53
Γ_3^+	2	−1	0	2	−1	0	–	–	–	–	–	
Γ_1^-	1	1	1	−1	−1	−1	$E, 2C_3, 3C'_2$	D_3	$D_{3d}(D_3)$	$\underline{\bar{3}M}$	AF	54
Γ_2^-	1	1	−1	−1	−1	1	$E, 2C_3, 3\sigma_d$	C_{3v}	$D_{3d}(C_{3v})$	$\underline{\bar{3}}M$	AF	55
Γ_3^-	2	−1	0	−2	1	0	–	–	–	–	–	

Table D.21 Group C_6; 6.

Rep.	Character of elements E	C_6	C_3	C_2	C_3^{-1}	C_6^{-1}	Unitary subgroup H Elements	Group	Magnetic group Schoenflies	Int.	Mag. order	No.
Γ_1	1	1	1	1	1	1	$E, C_6, C_3, C_2, C_3^{-1}, C_6^{-1}$	C_6	$C_6(C_6)$	6	F	56
Γ_2	1	$-w^2$	w^4	1	$-w^2$	w^4	–	–	–	–	–	
Γ_3	1	w^4	$-w^2$	1	w^4	$-w^2$	–	–	–	–	–	
Γ_4	1	−1	1	−1	1	−1	E, C_3, C_3^{-1}	C_3	$C_6(C_3)$	$\underline{6}$	AF	57
Γ_5	1	w^2	w^4	−1	$-w^2$	$-w^4$	–	–	–	–	–	
Γ_6	1	$-w^4$	$-w^2$	−1	w^4	w^2	–	–	–	–	–	

$w = \exp(\pi i/6)$.

Table D.22 Group C_{3h}; $\bar{6}$.

Rep.	Character of elements E	S_3^{-1}	C_3	σ_h	C_3^{-1}	S_3	Unitary subgroup H Elements	Group	Magnetic group Schoenflies	Int.	Mag. order	No.
Γ_1	1	1	1	1	1	1	$E, S_3^{-1}, C_3, \sigma_h, C_3^{-1}, S_3$	C_{3h}	$C_{3h}(C_{3h})$	$\bar{6}$	F	58
Γ_2	1	$-w^2$	w^4	1	$-w^2$	w^4	–	–	–	–	–	
Γ_3	1	w^4	$-w^2$	1	w^4	$-w^2$	–	–	–	–	–	
Γ_4	1	-1	1	-1	1	-1	E, C_3, C_3^{-1}	C_3	$C_{3h}(C_3)$	$\underline{\bar{6}}$	AF	59
Γ_5	1	w^2	w^4	-1	$-w^2$	$-w^4$	–	–	–	–	–	
Γ_6	1	$-w^4$	$-w^2$	-1	w^4	w^2	–	–	–	–	–	

$w = \exp(\pi i/6)$.

Table D.23 Group C_{6h}; 6/M.

Rep.	Character of elements E	C_6	C_3	C_2	C_3^{-1}	C_6^{-1}	I	S_3^{-1}	S_6^{-1}	σ_h	S_6	S_3	Unitary subgroup H Elements	Group	Magnetic group Schoenflies	Int.	Mag. order	No.
Γ_1^+	1	1	1	1	1	1	1	1	1	1	1	1	E, C_6, C_3, C_2 $C_3^{-1}, C_6^{-1}, I, S_3^{-1}$ $S_6^{-1}, \sigma_h, S_6, S_3$	C_{6h}	$C_{6h}(C_{6h})$	6/M	F	60
Γ_2^+	1	$-w^2$	w^4	1	$-w^2$	w^4	1	$-w^2$	w^4	1	$-w^2$	w^4	–	–	–	–	–	
Γ_3^+	1	w^4	$-w^2$	1	w^4	$-w^2$	1	w^4	$-w^2$	1	w^4	$-w^2$	–	–	–	–	–	
Γ_4^+	1	-1	1	-1	1	-1	1	-1	1	-1	1	-1	$E, C_3, C_3^{-1}, I, S_6^{-1}$, S_6	S_6	$C_{6h}(S_6)$	$\underline{6}/\underline{M}$	AF	61
Γ_5^+	1	w^2	w^4	-1	$-w^2$	$-w^4$	1	w^2	w^4	-1	$-w^2$	$-w^4$	–	–	–	–	–	
Γ_6^+	1	$-w^4$	$-w^2$	-1	w^4	w^2	1	$-w^4$	$-w^2$	-1	w^4	w^2	–	–	–	–	–	
Γ_1^-	1	1	1	1	1	1	-1	-1	-1	-1	-1	-1	E, C_6, C_3, C_2, C_3^{-1}, C_6^{-1}	C_6	$C_{6h}(C_6)$	$6/\underline{M}$	AF	62
Γ_2^-	1	$-w^2$	w^4	1	$-w^2$	w^4	-1	w^2	$-w^4$	-1	w^2	$-w^4$	–	–	–	–	–	
Γ_3^-	1	w^4	$-w^2$	1	w^4	$-w^2$	-1	$-w^4$	w^2	-1	$-w^4$	w^2	–	–	–	–	–	
Γ_4^-	1	-1	1	-1	1	-1	-1	1	-1	1	-1	1	$E, C_3, C_3^{-1}, S_3^{-1}$ σ_h, S_3	C_{3h}	$C_{6h}(C_{3h})$	$\underline{6}/M$	AF	63
Γ_5^-	1	w^2	w^4	-1	$-w^2$	$-w^4$	-1	$-w^2$	$-w^4$	1	w^2	w^4	–	–	–	–	–	
Γ_6^-	1	$-w^4$	$-w^2$	-1	w^4	w^2	-1	w^4	w^2	1	$-w^4$	$-w^2$	–	–	–	–	–	

$w = \exp(\pi i/6)$.

Table D.24 Group D_6; 62.

Rep.	Character of elements E	C_2	$2C_3$	$2C_6$	$3C'_2$	$3C''_2$	Unitary subgroup H Elements	Group	Magnetic group Schoenflies	Int.	Mag. order	No.
Γ_1	1	1	1	1	1	1	$E, C_2, 2C_3, 2C_6, 3C'_2, 3C''_2$	D_6	$D_6(D_6)$	62	AF	64
Γ_2	1	1	1	1	−1	−1	$E, C_2, 2C_3, 2C_6$	C_6	$D_6(C_6)$	$6\underline{2}$	F	65
Γ_3	1	−1	1	−1	1	−1	$E, 2C_3, 3C'_2$	D_3	$D_6(D_3)$	$\underline{6}2$	AF	66
Γ_4	1	−1	1	−1	−1	1	$E, 2C_3, 3C''_2$					
Γ_5	2	−2	−1	1	0	0	–	–	–	–	–	
Γ_6	2	2	−1	−1	0	0	–	–	–	–	–	

Table D.25 Group C_{6v}; 6MM.

Rep.	Character of elements E	C_2	$2C_3$	$2C_6$	$3\sigma_d$	$3\sigma_v$	Unitary subgroup H Elements	Group	Magnetic group Schoenflies	Int.	Mag. order	No.
Γ_1	1	1	1	1	1	1	$E, C_2, 2C_3, 2C_6, 3\sigma_d, 3\sigma_v$	C_{6v}	$C_{6v}(C_{6v})$	6MM	AF	67
Γ_2	1	1	1	1	−1	−1	$E, C_2, 2C_3, 2C_6$	C_6	$C_{6v}(C_6)$	$6\underline{MM}$	F	68
Γ_3	1	−1	1	−1	1	−1	$E, 2C_3, 3\sigma_d$	C_{3v}	$C_{6v}(C_{3v})$	$\underline{6}M\underline{M}$	AF	69
Γ_4	1	−1	1	−1	−1	1	$E, 2C_3, 3\sigma_v$					
Γ_5	2	−2	−1	1	0	0	–	–	–	–	–	
Γ_6	2	2	−1	−1	0	0	–	–	–	–	–	

Table D.26 Group D_{3h}; $\bar{6}M2$.

Rep.	Character of elements E	σ_h	$2C_3$	$2S_3$	$3C_2'$	$3\sigma_v$	Unitary subgroup H Elements	Group	Magnetic group Schoenflies	Int.	Mag. order	No.
Γ_1	1	1	1	1	1	1	$E, \sigma_h, 2C_3, 2S_3, 3C_2', 3\sigma_v$	D_{3h}	$D_{3h}(D_{3h})$	$\bar{6}M2$	AF	70
Γ_2	1	1	1	1	−1	−1	$E, \sigma_h, 2C_3, 2S_3$	C_{3h}	$D_{3h}(C_{3h})$	$\bar{6}\underline{M2}$	F	71
Γ_3	1	−1	1	−1	1	−1	$E, 2C_3, 3C_2'$	D_3	$D_{3h}(D_3)$	$\underline{\bar{6}M}2$	AF	72
Γ_4	1	−1	1	−1	−1	1	$E, 2C_3, 3\sigma_v$	C_{3v}	$D_{3h}(C_{3v})$	$\underline{\bar{6}}M\underline{2}$	AF	73
Γ_5	2	−2	−1	1	0	0	–	–	–	–	–	
Γ_6	2	2	−1	−1	0	0	–	–	–	–	–	

Table D.27 Group D_{6h}; 6/MMM.

Rep.	Character of elements E	C_2	$2C_3$	$2C_6$	$3C'_2$	$3C''_2$	I	σ_h	$2S_6$	$2S_3$	$3\sigma_d$	$3\sigma_v$	Unitary subgroup H Elements	Group	Magnetic group Schoenflies	Int.	Mag. order	No.
Γ_1^+	1	1	1	1	1	1	1	1	1	1	1	1	$E, C_2, 2C_3, 2C_6, 3C'_2, 3C''_2, I, \sigma_h, 2S_6, 2S_3, 3\sigma_d, 3\sigma_v$	D_{6h}	$D_{6h}(D_{6h})$	6/MMM	AF	74
Γ_2^+	1	1	1	1	−1	−1	1	1	1	1	−1	−1	$E, C_2, 2C_3, 2C_6, I, \sigma_h, 2S_6, 2S_3$	C_{6h}	$D_{6h}(C_{6h})$	$6/M\underline{MM}$	F	75
Γ_3^+	1	−1	1	−1	1	−1	1	−1	1	−1	1	−1	$E, 2C_3, 3C'_2, I, 2S_6, 3\sigma_d$	D_{3d} (for Γ_3^+ and Γ_4^+)	$D_{6h}(D_{3d})$	$\underline{6}/\underline{MM}M$	AF	76
Γ_4^+	1	−1	1	−1	−1	1	1	−1	1	−1	−1	1	$E, 2C_3, 3C''_2, I, 2S_6, 3\sigma_v$					
Γ_5^+	2	−2	−1	1	0	0	2	−2	−1	1	0	0	–	–	–	–	–	
Γ_6^+	2	2	−1	−1	0	0	2	2	−1	−1	0	0	–	–	–	–	–	
Γ_1^-	1	1	1	1	1	1	−1	−1	−1	−1	−1	−1	$E, C_2, 2C_3, 2C_6, 3C'_2, 3C''_2$	D_6	$D_{6h}(D_6)$	$6/\underline{MMM}$	AF	77
Γ_2^-	1	1	1	1	−1	−1	−1	−1	−1	−1	1	1	$E, C_2, 2C_3, 2C_6, 3\sigma_d, 3\sigma_v$	C_{6v}	$D_{6h}(C_{6v})$	$6/\underline{M}MM$	AF	78
Γ_3^-	1	−1	1	−1	1	−1	−1	1	−1	1	−1	1	$E, 2C_3, 3C'_2, \sigma_h, 2S_3, 3\sigma_v$	D_{3h} (for Γ_3^- and Γ_4^-)	$D_{6h}(D_{3h})$	$\underline{6}/M\underline{M}M$	AF	79
Γ_4^-	1	−1	1	−1	−1	1	−1	1	−1	1	1	−1	$E, 2C_3, 3C''_2, \sigma_h, 2S_3, 3\sigma_d$					
Γ_5^-	2	−2	−1	1	0	0	−2	2	1	−1	0	0	–	–	–	–	–	
Γ_6^-	2	2	−1	−1	0	0	−2	−2	1	1	0	0	–	–	–	–	–	

Table D.28 Group T; 23.

Rep.	Character of elements E	$3C_2$	$4C_3$	$4C_3^{-1}$	Unitary subgroup H Elements	Group	Magnetic group Schoenflies	Int.	Mag. order	No.
Γ_1	1	1	1	1	E, $3C_2$, $4C_3$, $4C_3^{-1}$	T	T(T)	23	AF	80
Γ_2	1	1	w	w^2	–	–	–	–	–	
Γ_3	1	1	w^2	w	–	–	–	–	–	
Γ_4	3	−1	0	0	–	–	–	–	–	

$w = \exp(2\pi i/3)$.

Table D.29 Group T_h; M3.

Rep.	Character of elements E	$3C_2$	$4C_3$	$4C_3^{-1}$	I	$3\sigma_h$	$4S_6^{-1}$	$4S_6$	Unitary subgroup H Elements	Group	Magnetic group Schoenflies	Int.	Mag. order	No.
Γ_1^+	1	1	1	1	1	1	1	1	$E, 3C_2, 4C_3, 4C_3^{-1}$, I, $3\sigma_h$, $4S_6^{-1}$, $4S_6$	T_h	$T_h(T_h)$	M3	AF	81
Γ_2^+	1	1	w	w^2	1	1	w	w^2	–	–	–	–	–	
Γ_3^+	1	1	w^2	w	1	1	w^2	w						
Γ_4^+	3	−1	0	0	3	−1	0	0	–	–	–	–	–	
Γ_1^-	1	1	1	1	−1	−1	−1	−1	E, $3C_2$, $4C_3$, $4C_3^{-1}$	T	$T_h(T)$	$\underline{\mathrm{M3}}$	AF	82
Γ_2^-	1	1	w	w^2	−1	−1	$-w$	$-w^2$	–	–	–	–	–	
Γ_3^-	1	1	w^2	w	−1	−1	$-w^2$	$-w$	–	–	–	–	–	
Γ_4^-	3	−1	0	0	−3	1	0	0	–	–	–	–	–	

$w = \exp(2\pi i/3)$.

Table D.30 Group O; 43.

Rep.	Character of elements E	$8C_3$	$3C_2$	$6C_4$	$6C_2'$	Unitary subgroup H Elements	Group	Magnetic group Schoenflies	Int.	Mag. order	No.
Γ_1	1	1	1	1	1	E, $8C_3$, $3C_2$, $6C_4$, $6C_2'$	O	O(O)	43	AF	83
Γ_2	1	1	1	−1	−1	E, $8C_3$, $3C_2$	T	O(T)	$\underline{4}3$	AF	84
Γ_3	2	−1	2	0	0	–	–	–	–	–	
Γ_4	3	0	−1	1	−1	–	–	–	–	–	
Γ_5	3	0	−1	−1	1	–	–	–	–	–	

Table D.31 Group T_d; $\bar{4}3M$.

Rep.	Character of elements E	$8C_3$	$3C_2$	$6S_4$	$6\sigma_d$	Unitary subgroup H Elements	Group	Magnetic group Schoenflies	Int.	Mag. order	No.
Γ_1	1	1	1	1	1	E, $8C_3$, $3C_2$, $6S_4$, $6\sigma_d$	T_d	$T_d(T_d)$	$\bar{4}3M$	AF	85
Γ_2	1	1	1	−1	−1	E, $8C_3$, $3C_2$	T	$T_d(T)$	$\underline{\bar{4}}3\underline{M}$	AF	86
Γ_3	2	−1	2	0	0	–	–	–	–	–	
Γ_4	3	0	−1	1	−1	–	–	–	–	–	
Γ_5	3	0	−1	−1	1	–	–	–	–	–	

Table D.32 Group O_h; M3M.

Rep.	Character of elements E	$8C_3$	$3C_2$	$6C_4$	$6C'_2$	I	$8S_6$	$3\sigma_h$	$6S_4$	$6\sigma_d$	Unitary subgroup H Elements	Group	Magnetic group Schoenflies	Int.	Mag. order	No.
Γ_1^+	1	1	1	1	1	1	1	1	1	1	E, $8C_3$, $3C_2$, $6C_4$, $6C'_2$, I, $8S_6$, $3\sigma_h$, $6S_4$, $6\sigma_d$	O_h	$O_h(O_h)$	M3M	AF	87
Γ_2^+	1	1	1	−1	−1	1	1	1	−1	−1	E, $8C_3$, $3C_2$, I, $8S_6$, $3\sigma_h$	T_h	$O_h(T_h)$	M3$\underline{\text{M}}$	AF	88
Γ_3^+	2	−1	2	0	0	2	−1	2	0	0	–	–	–	–	–	
Γ_4^+	3	0	−1	1	−1	3	0	−1	1	−1	–	–	–	–	–	
Γ_5^+	3	0	−1	−1	1	3	0	−1	−1	1	–	–	–	–	–	
Γ_1^-	1	1	1	1	1	−1	−1	−1	−1	−1	E, $8C_3$, $3C_2$, $6C_4$, $6C'_2$	O	$O_h(O)$	$\underline{\text{M}}$3$\underline{\text{M}}$	AF	89
Γ_2^-	1	1	1	−1	−1	−1	−1	−1	1	1	E, $8C_3$, $3C_2$, $6S_4$, $6\sigma_d$	T_d	$O_h(T_d)$	$\underline{\text{M}}$3M	AF	90
Γ_3^-	2	−1	2	0	0	−2	1	−2	0	0	–	–	–	–	–	
Γ_4^-	3	0	−1	1	−1	−3	0	1	−1	1	–	–	–	–	–	
Γ_5^-	3	0	−1	−1	1	−3	0	1	1	−1	–	–	–	–	–	

APPENDIX E

STEREOGRAPHIC PROJECTIONS OF THE 58 TYPE III MAGNETIC SYMMETRY GROUPS (Reproduced from Joshua [27])

$\bar{1}$ Triclinic

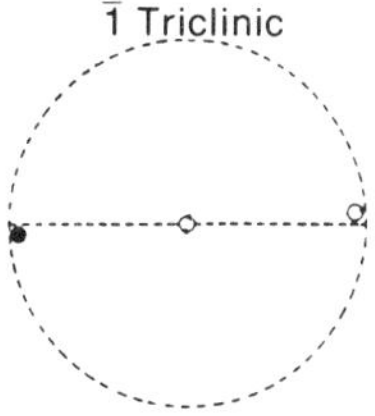

$\underline{2}$ Monoclinic

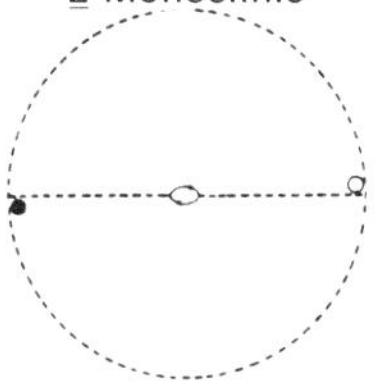

$\underline{m}$ Monoclinic

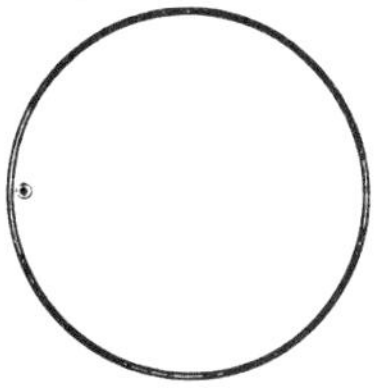

$2/\underline{m}$ Monoclinic

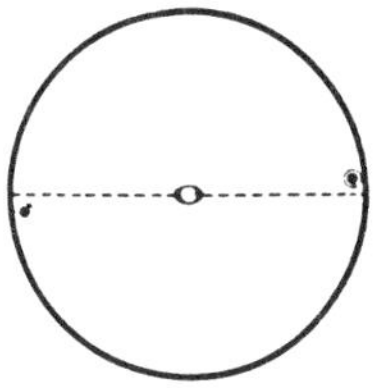

$\underline{2}/m$ Monoclinic

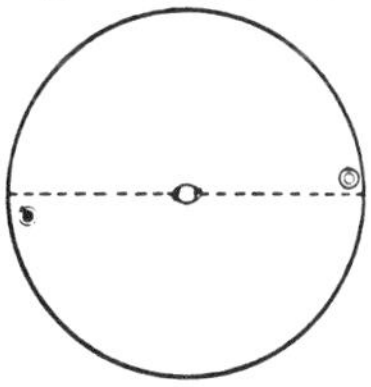

$\underline{2}/\underline{m}$ Monoclinic

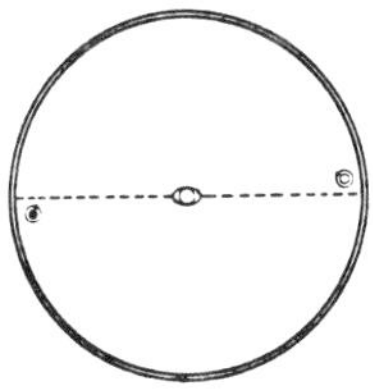

$2\underline{22}$ Orthorhombic

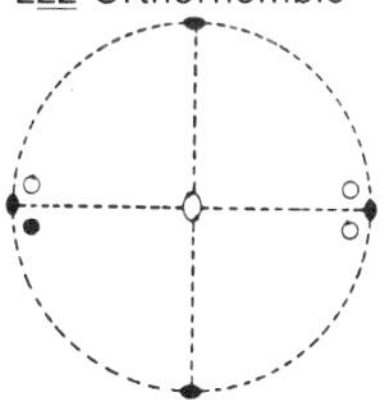

$2\underline{mm}$ Orthorhombic

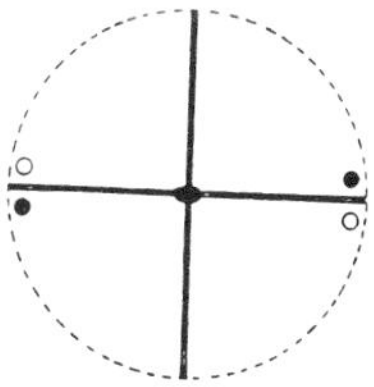

$\underline{2}m\underline{m}$ Orthorhombic

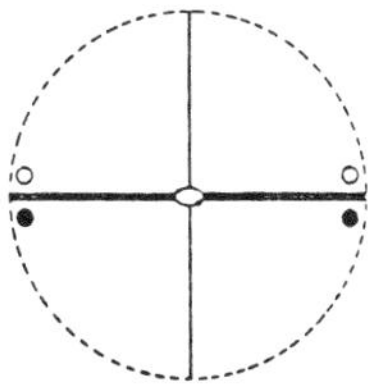

$\underline{mmm}$ Orthorhombic

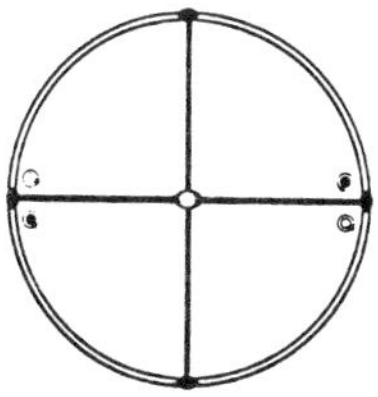

$mm\underline{m}$ Orthorhombic

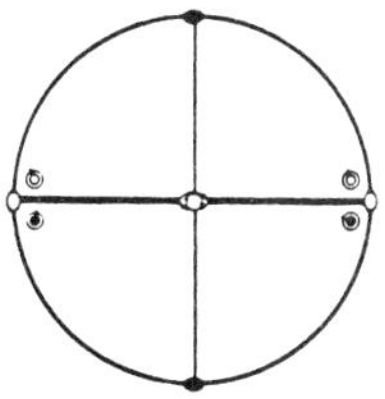

$\underline{mm}m$ Orthorhombic

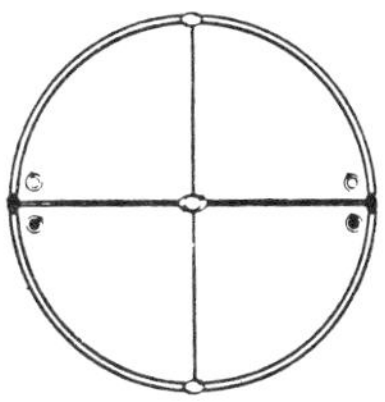

$\underline{4}$ Tetragonal

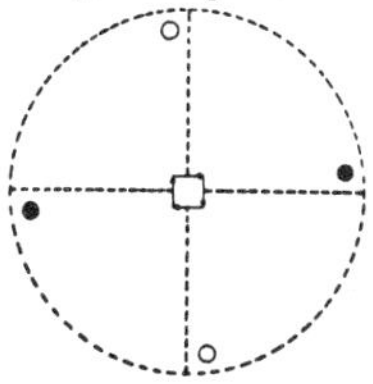

$\underline{\bar{4}}$ Tetragonal

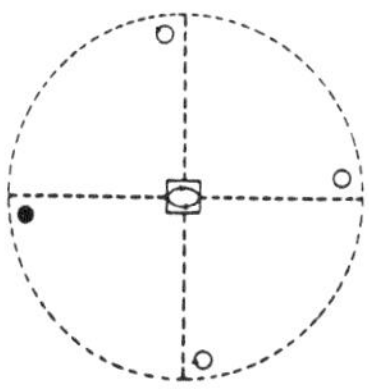

$4\underline{2}$ Tetragonal

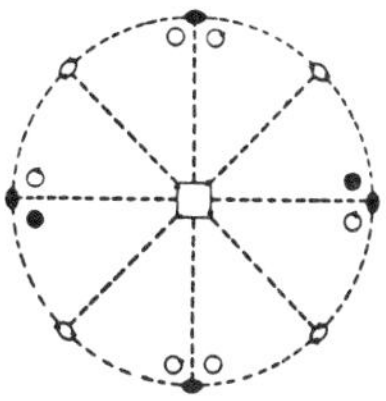

$\underline{4}2$ Tetragonal

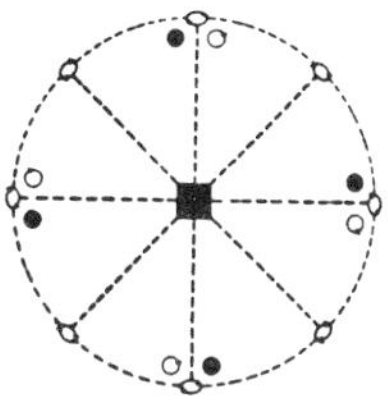

4/m Tetragonal

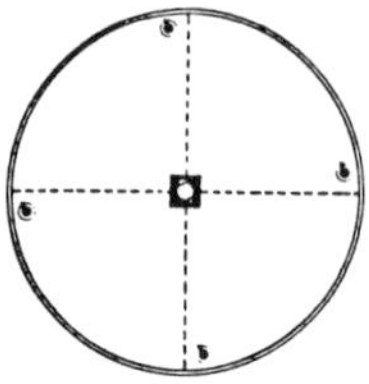

4/m Tetragonal

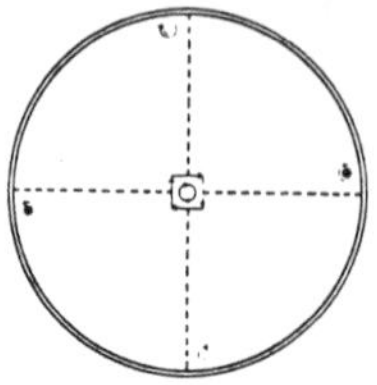

4/m Tetragonal

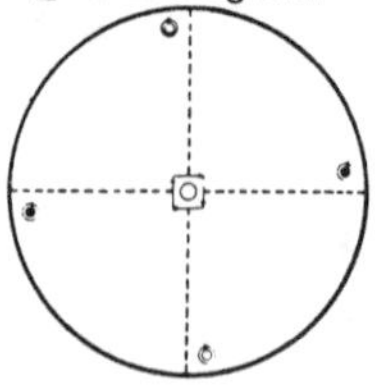

4mm Tetragonal

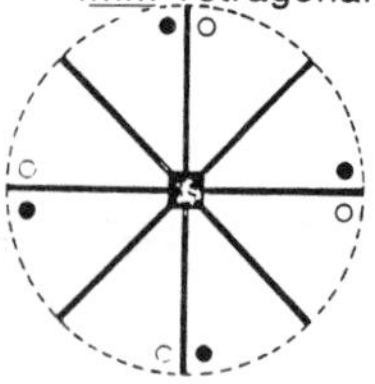

4mm Tetragonal

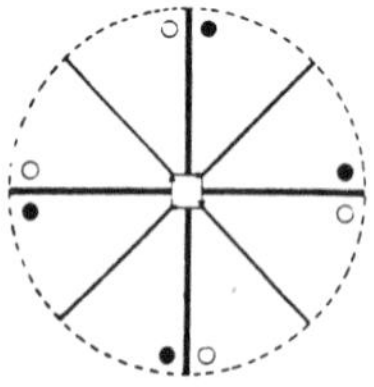

$\bar{4}2m$ Tetragonal

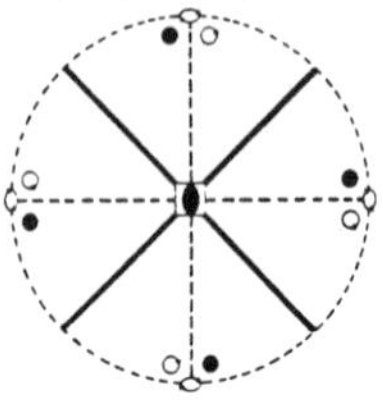

$\bar{4}2m$ Tetragonal

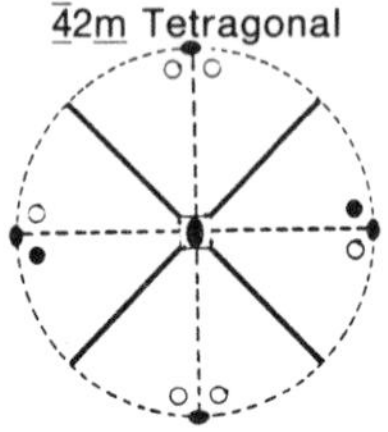

$\bar{4}2m$ Tetragonal

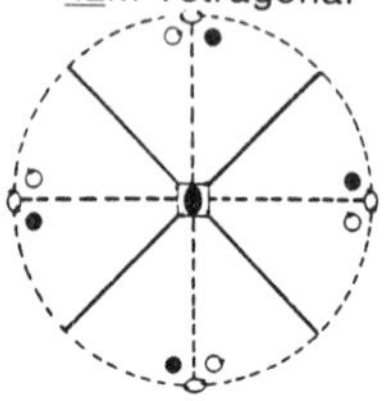

4/mmm Tetragonal

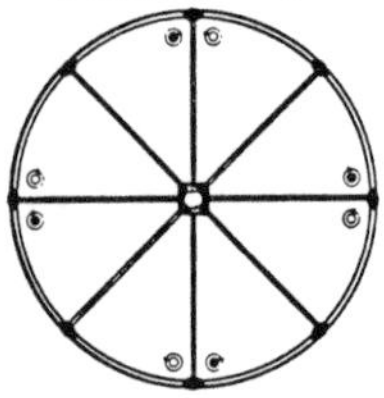

4/mmm Tetragonal

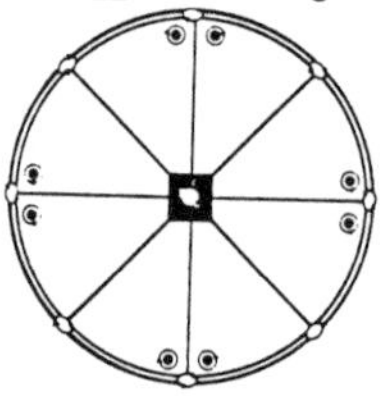

4/mmm Tetragonal

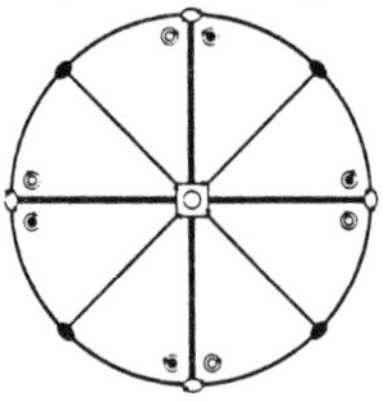

4/mmm Tetragonal

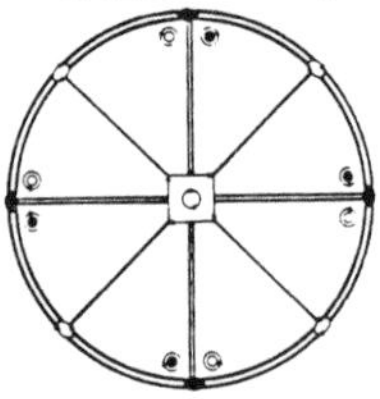

4/mmm Tetragonal

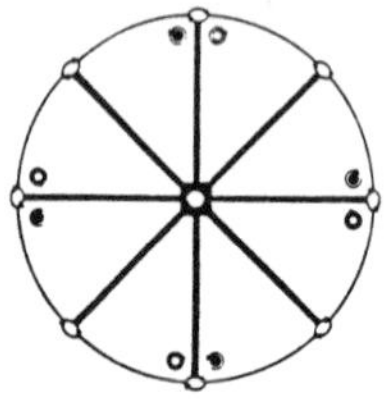

32 Rhombohedral

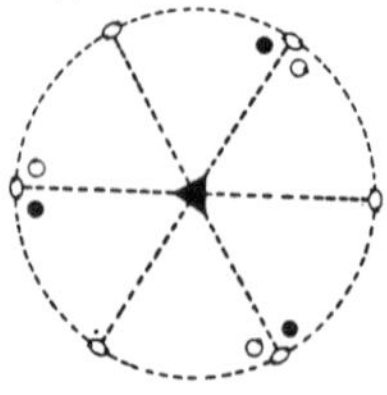

3m Rhombohedral

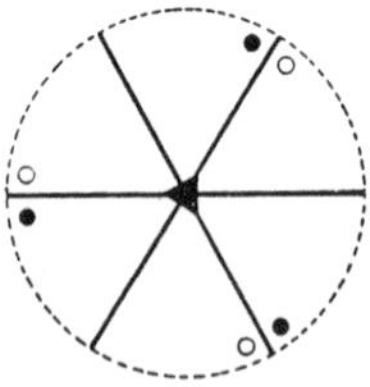

$\bar{6}$ Hexagonal

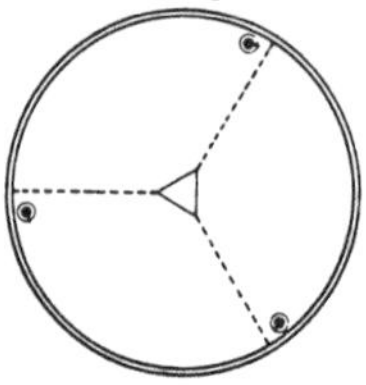

$\bar{6}m2$ Hexagonal

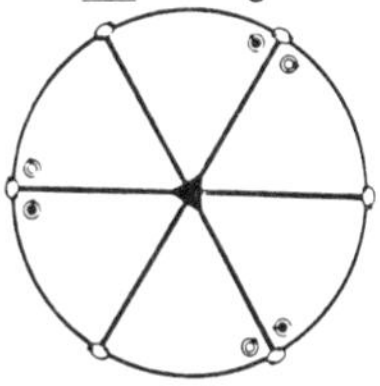

$\bar{6}m2$ Hexagonal

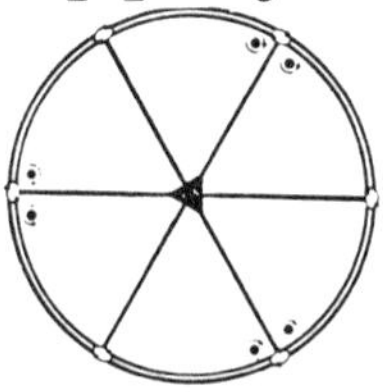

$\bar{6}m2$ Hexagonal

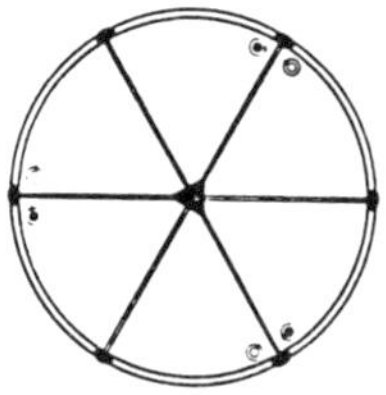

6 Hexagonal

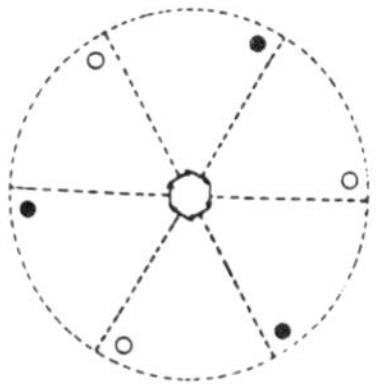

$\bar{3}$ Rhombohedral

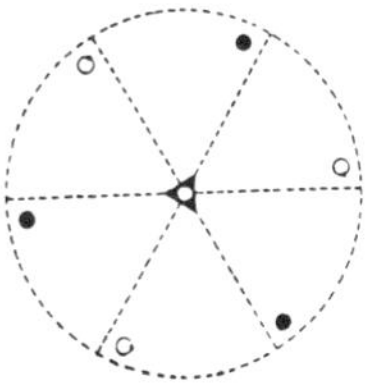

$\bar{3}m$ Rhombohedral

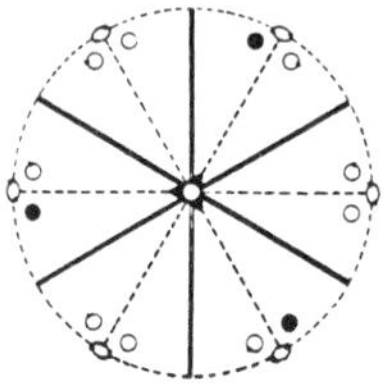

$\bar{3}m$ Rhombohedral

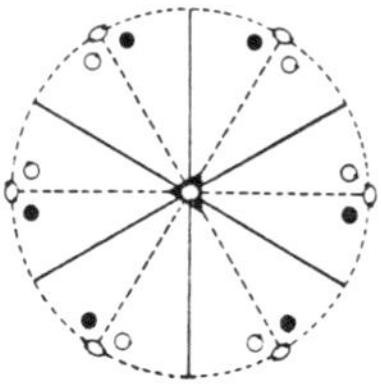

$\bar{3}m$ Rhombohedral

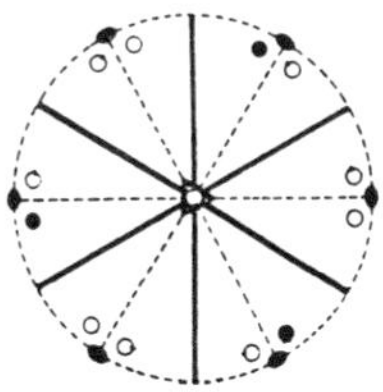

62 Hexagonal

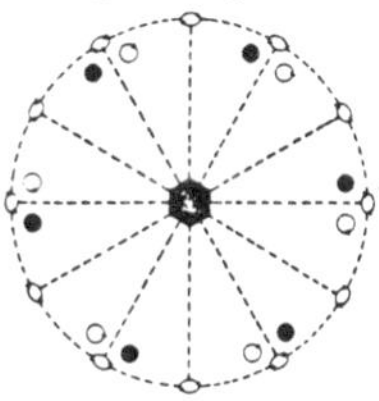

62 Hexagonal

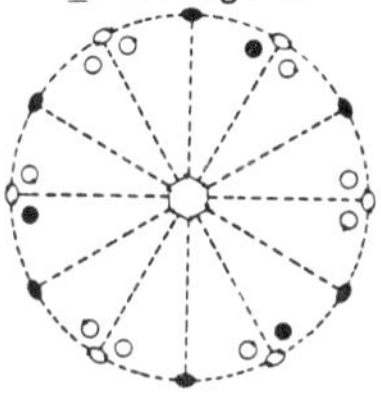

6/m Hexagonal

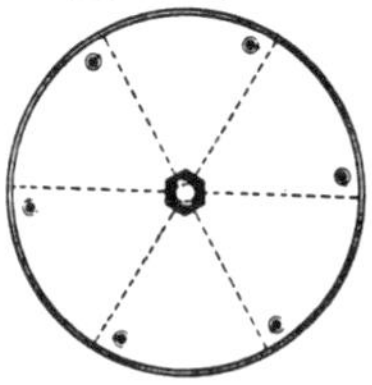

6/m Hexagonal

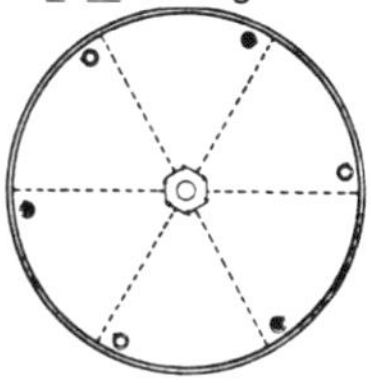

6/m Hexagonal

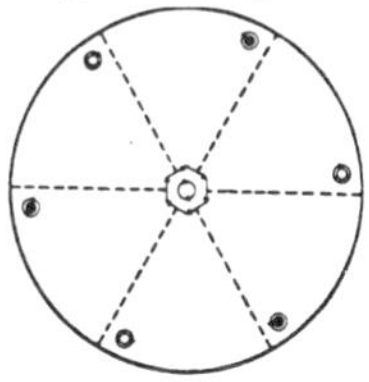

6mm Hexagonal

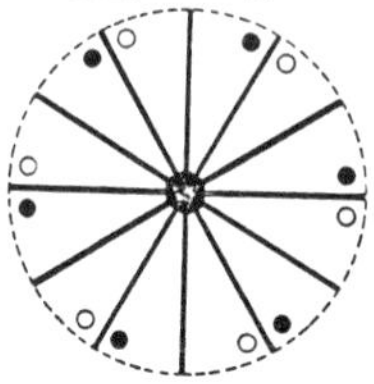

6/mm Hexagonal

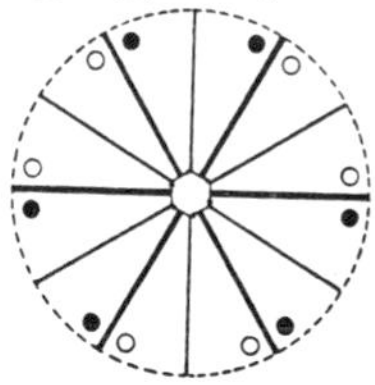

6/mmm Hexagonal

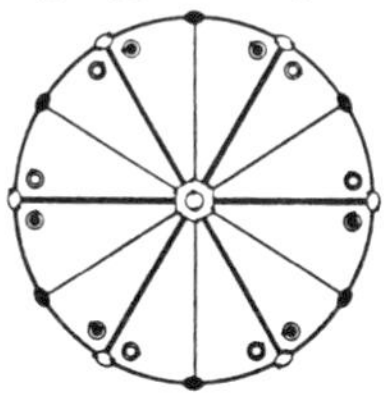

6/mmm Hexagonal

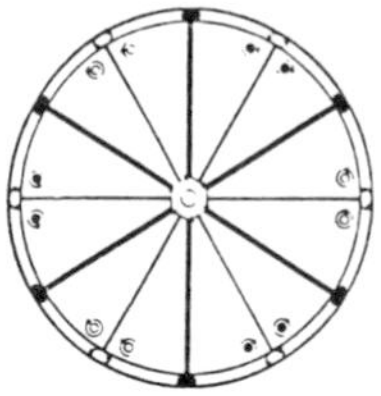

6/mmm Hexagonal

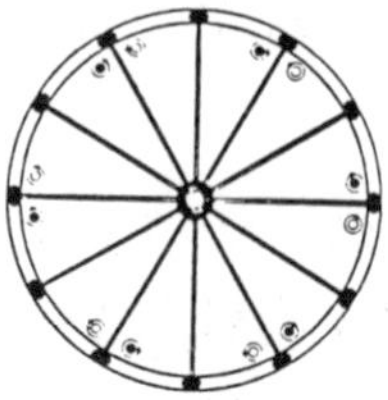

6/mmm Hexagonal

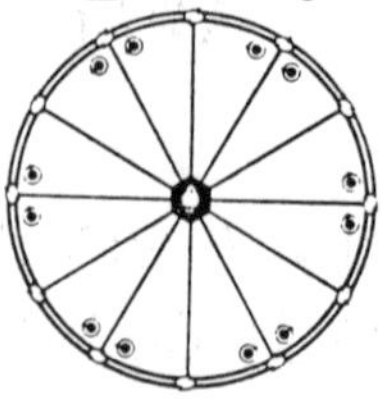

6/mmm Hexagonal

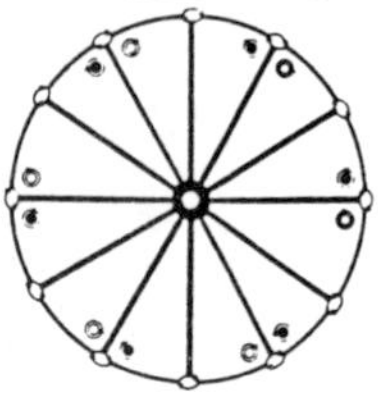

m3 Cubic

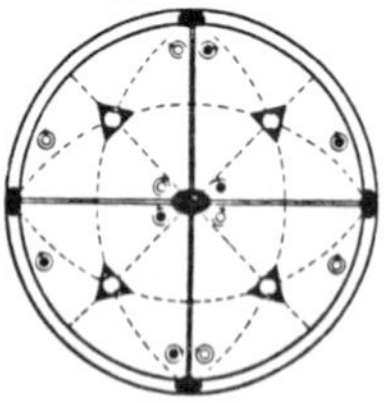

$\bar{4}3m$ Cubic

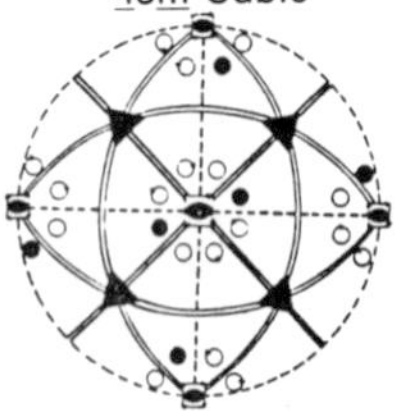

43 Cubic

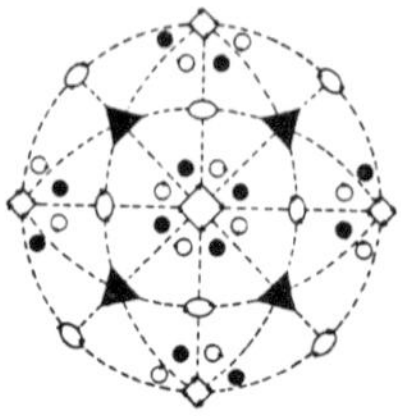

m3m Cubic

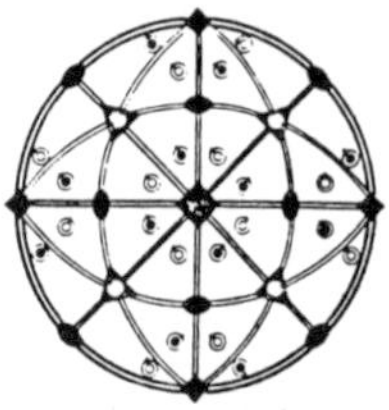

m3m Cubic

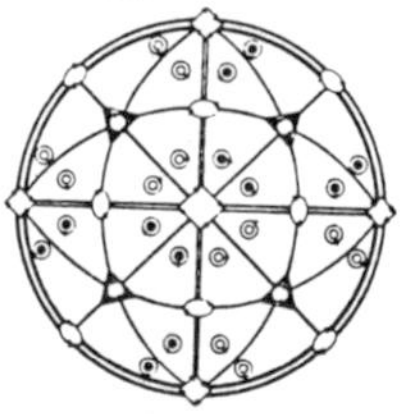

m3m Cubic

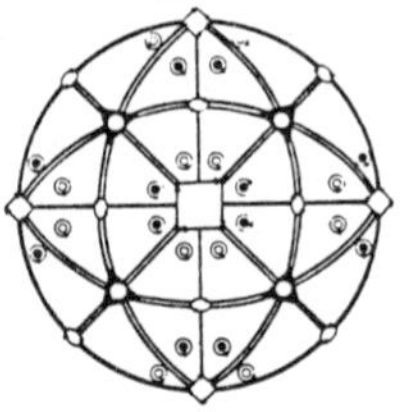

APPENDIX F

THE 14 'UNCOLOURED' AND 22 'COLOURED' BRAVAIS LATTICES (adapted from Belov *et al* [88])

Triclinic system

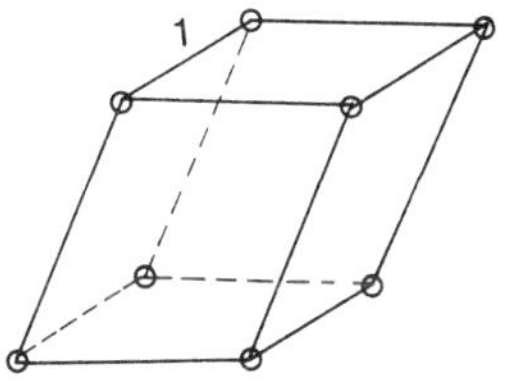

P

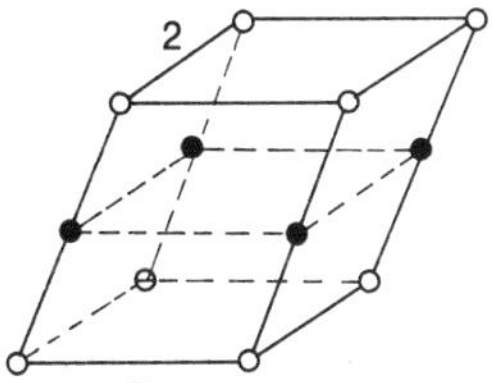

P_s

Monoclinic system

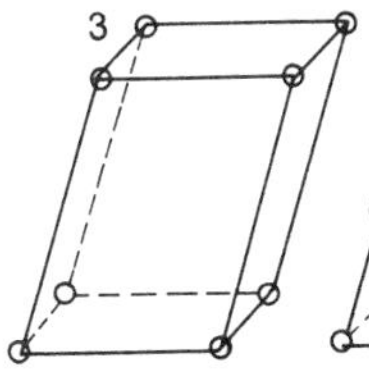

P

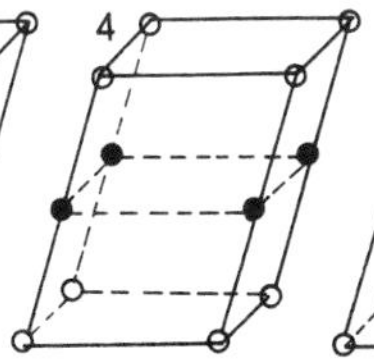

P_b

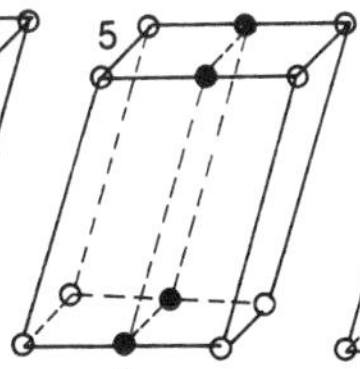

P_a

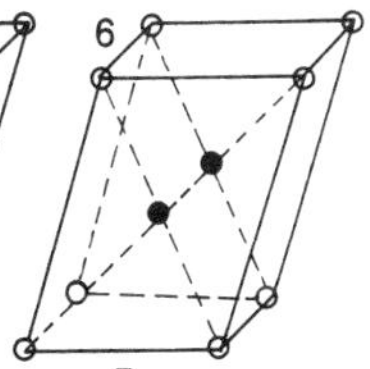

P_C

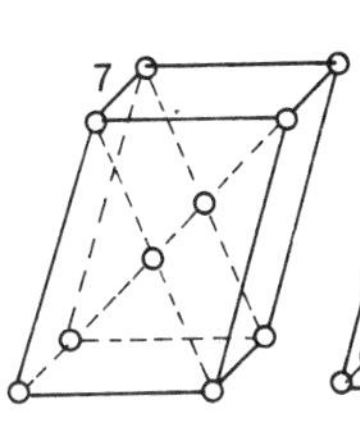

C

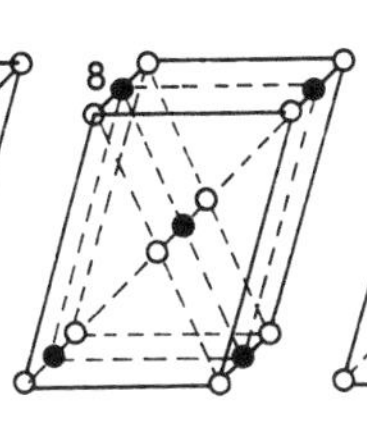

C_c

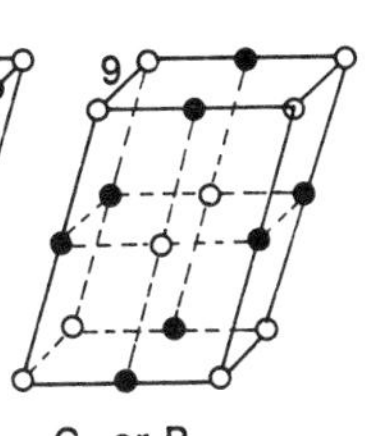

C_a or P_c

Orthorhombic system

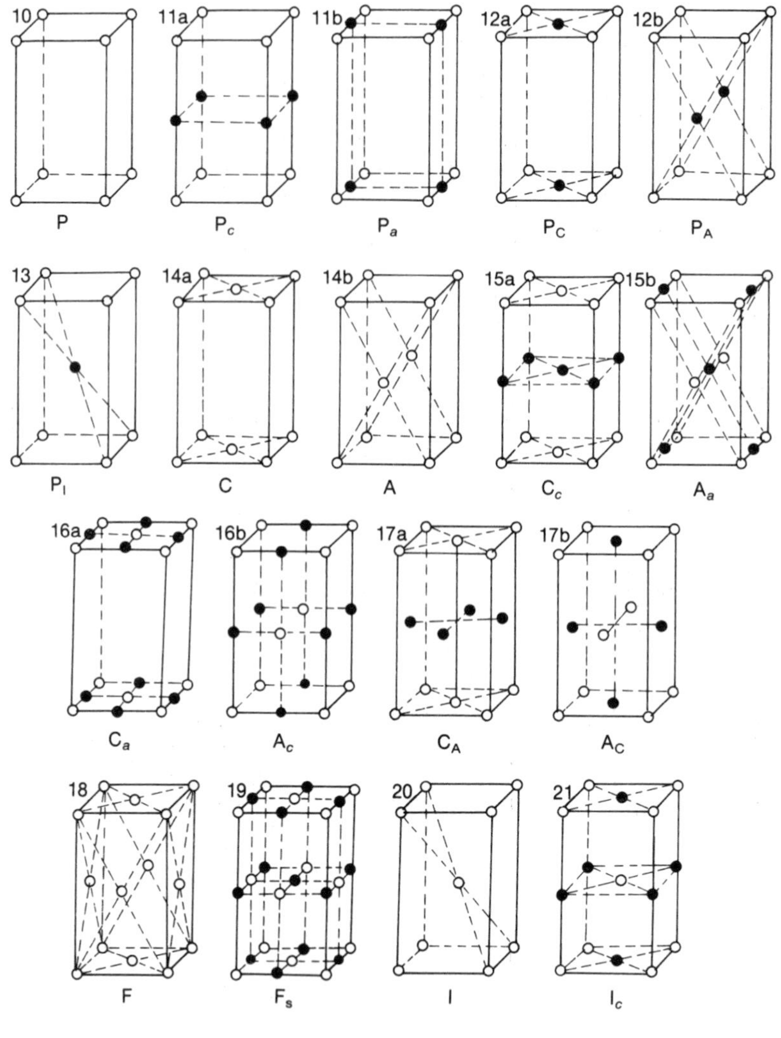

Tetragonal system

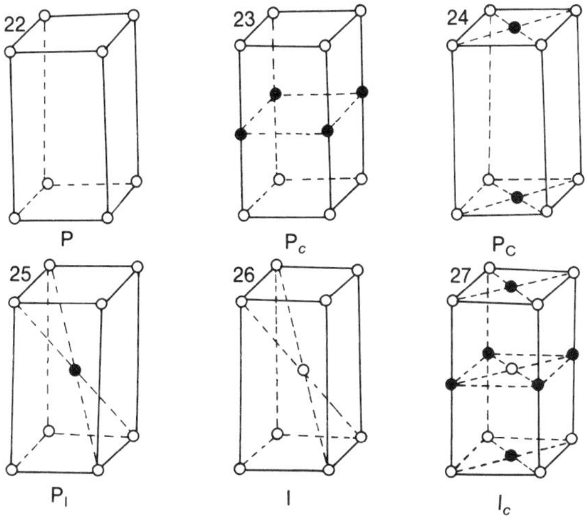

Hexagonal (rhombohedral) system

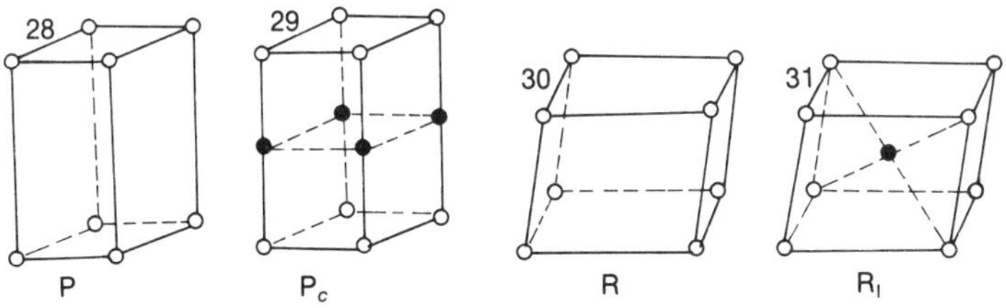

Cubic system

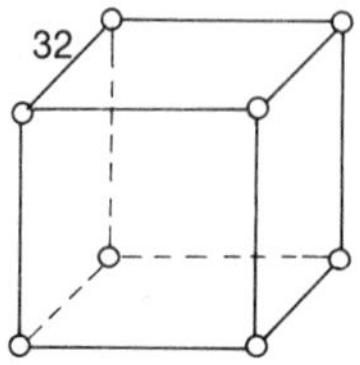

P

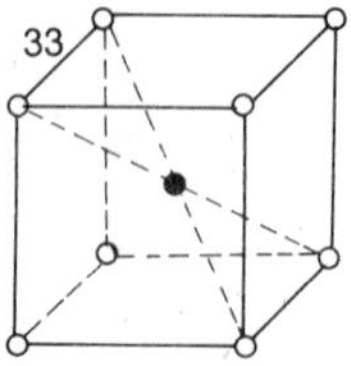

P_I

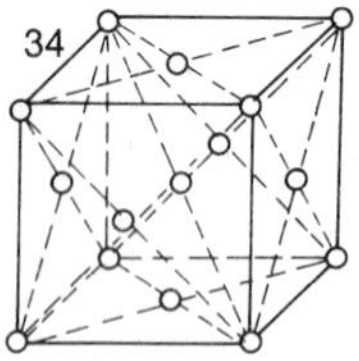

F

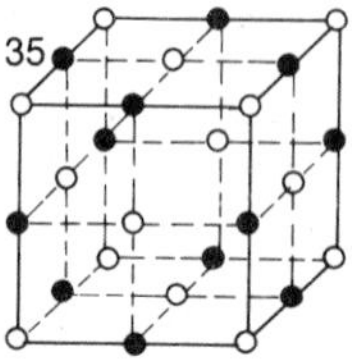

F_s

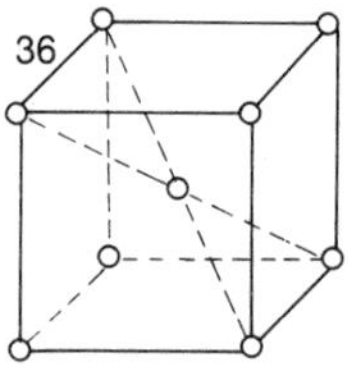

I

APPENDIX G

SUMMARY OF THE 1651 MAGNETIC SPACE GROUPS IN THE INTERNATIONAL NOTATION (adapted with modifications from Belov *et al* [88])

The total number of groups and lattices occurring in each crystal system is indicated at the bottom of the appropriate column by the number in parentheses. The coloured 'grey' groups are indicated by the addition of the symbol $\underline{1}$ for all groups except the cubic groups where the three-fold axis has been underlined to indicate the grey cubic groups. The Schoenflies notation for the ordinary point groups and space groups is also given in parentheses in the appropriate column immediately after the International notation.

Triclinic

Point Groups Ord.	Col.	Lattices Ord.	Col.	Space Groups Ord.	Col. (magnetic)
$1(C_1)$	$1, \underline{1}$	P	P, P_s	$P1(C_1^1)$	$P1, P\underline{1}, P_s1$
$\bar{1}(C_1)$	$\bar{1}, \bar{1}\underline{1}, \underline{\bar{1}}$			$P\bar{1}(C_1^1)$	$P\bar{1}, P\bar{1}\underline{1}, P\underline{\bar{1}}, P_s\bar{1}$
(2)	(5)	(1)	(2)	(2)	(7)

Monoclinic

Point Groups Ord.	Col.	Lattices Ord.	Col.	Space Groups Ord.	Col. (magnetic)
$2(C_2)$	$2, 2\underline{1}, \underline{2}$	P	P, P_a, P_b, P_C	$P2(C_2^1)$	$P2, P2\underline{1}, P\underline{2}, P_a2, P_b2, P_C2$
				$P2_1(C_2^2)$	$P2_1, P2_1\underline{1}, P\underline{2}_1, P_a2_1, P_b2_1, P_C2_1$
		C	$C, C_a(P_c)$	$C2(C_2^3)$	$C2, C2\underline{1}, C\underline{2}, C_c2, C_a2$
$m(C_s)$	$m, m\underline{1}, \underline{m}$		C_c	$Pm(C_s^1)$	$Pm, Pm\underline{1}, P\underline{m}, P_am, P_bm, P_Cm$
				$Pc(C_s^2)$	$Pc, Pc\underline{1}, P\underline{c}, P_ac, P_cc, P_bc, P_Cc, P_Ac$
				$Cm(C_s^3)$	$Cm, Cm\underline{1}, C\underline{m}, C_cm, C_am$
				$Cc(C_s^4)$	$Cc, Cc\underline{1}, C\underline{c}, C_cc, C_ac$
$2/m(C_{2h})$	$2/m, 2/m\underline{1}, \underline{2}/\underline{m}, 2/\underline{m}, \underline{2}/m$			$P2/m(C_{2h}^1)$	$P2/m, P2/m\underline{1}, P\underline{2}/m, P2/\underline{m}, P\underline{2}/\underline{m}, P_a2/m, P_b2/m, P_C2/m$
				$P2_1/m(C_{2h}^2)$	$P2_1/m, P2_1/m\underline{1}, P\underline{2}_1/m, P2_1/\underline{m}, P\underline{2}_1/\underline{m}, P_a2_1/m, P_b2_1/m, P_C2_1/m$
				$C2/m(C_{2h}^3)$	$C2/m, C2/m\underline{1}, C\underline{2}/m, C2/\underline{m}, C\underline{2}/\underline{m}, C_c2/m, C_a2/m$
				$P2/c(C_{2h}^4)$	$P2/c, P2/c\underline{1}, P\underline{2}/c, P2/\underline{c}, P\underline{2}/\underline{c}, P_a2/c, P_b2/c, P_c2/c, P_A2/c, P_C2/c$
				$P2_1/c(C_{2h}^5)$	$P2_1/c, P2_1/c\underline{1}, P\underline{2}_1/c, P2_1/\underline{c}, P\underline{2}_1/\underline{c}, P_a2_1/c, P_b2_1/c, P_c2_1/c, P_A2_1/c, P_C2_1/c$
				$C2/c(C_{2h}^6)$	$C2/c, C2/c\underline{1}, C\underline{2}/c, C2/\underline{c}, C\underline{2}/\underline{c}, C_c2/c, C_a2/c$
(3)	(11)	(2)	(7)	(13)	(91)

Orthorhombic

Point groups Ord.	Col.	Lattices Ord.	Col.	Space groups Ord.	Col. (magnetic)
$222\ (D_2)$	$222,\ 222\underline{1},\ 2\underline{22}$	P	$P,\ P_a$ or P_c	$P222\ (D_2^1)$	$P222,\ P222\underline{1},\ P\underline{222},\ P_a222,\ P_C222,\ P_I222$
			P_C or $P_A,\ P_I$	$P222_1(D_2^2)$	$P222_1,\ P222_1\underline{1},\ P\underline{222}_1,\ P2\underline{22}_1,\ P_a222_1,\ P_c222_1,\ P_A222_1,$ $P_C222_1,\ P_I222_1$
		C	C or A		
			C_c or A_a	$P2_12_12$	$P2_12_12,\ P2_12_12\underline{1},\ P\underline{2}_1\underline{2}_12,\ P2_1\underline{2}_1\underline{2},\ P_a2_12_12,\ P_c2_12_12,$
			C_a or A_c	(D_2^3)	$P_A2_12_12,\ P_C2_12_12,\ P_I2_12_12$
			C_A or A_C		
		I	$I,\ I_c$		
				$P2_12_12_1$	$P2_12_12_1,\ P2_12_12_1\underline{1},\ P\underline{2}_1\underline{2}_12_1,\ P_a2_12_12_1,$
		F	$F,\ F_s$	(D_2^4)	$P_C2_12_12_1,\ P_I2_12_12_1$
				$C222_1(D_2^5)$	$C222_1,\ C222_1\underline{1},\ C\underline{222}_1,\ C2\underline{22}_1,\ C_c222_1,\ C_a222_1,\ C_A222_1$
				$C222(D_2^6)$	$C222,\ C222\underline{1},\ C\underline{222},\ C2\underline{22},\ C_c222,\ C_a222,\ C_A222$
				$F222(D_2^7)$	$F222,\ F222\underline{1},\ F\underline{222},\ F_s222$
				$I222(D_2^8)$	$I222,\ I222\underline{1},\ I\underline{222},\ I_c222$
				$I2_12_12_1$ (D_2^9)	$I2_12_12_1,\ I2_12_12_1\underline{1},\ I\underline{2}_1\underline{2}_12_1,\ I_c2_12_12_1$
mm2	$mm2,\ mm2\underline{1}$			$Pmm2(C_{2v}^1)$	$Pmm2,\ Pmm2\underline{1},\ P\underline{m}m\underline{2},\ P\underline{mm}2$

(C_{2v})	$\underline{m}m\underline{2}$, $\underline{mm}2$		$P_c mm2$, $P_a mm2$, $P_C mm2$, $P_A mm2$, $P_I mm2$
		$Pmc2_1(C_{2v}^2)$	$Pmc2_1$, $Pmc2_1\underline{1}$, $P\underline{m}c\underline{2}_1$, $Pm\underline{c2}_1$, $P\underline{mc}2_1$, $P_a mc2_1$, $P_b mc2_1$, $P_c mc2_1$, $P_A mc2_1$, $P_B mc2_1$, $P_C mc2_1$, $P_I mc2_1$
		$Pcc2(C_{2v}^3)$	$Pcc2$, $Pcc2\underline{1}$, $P\underline{c}c\underline{2}$, $P\underline{cc}2$, $P_c cc2$, $P_a cc2$, $P_C cc2$, $P_A cc2$, $P_I cc2$
		$Pma2(C_{2v}^4)$	$Pma2$, $Pma2\underline{1}$, $P\underline{m}a\underline{2}$, $Pm\underline{a2}$, $P\underline{ma}2$, $P_a ma2$, $P_b ma2$, $P_c ma2$, $P_A ma2$, $P_B ma2$, $P_C ma2$, $P_I ma2$
		$Pca2_1(C_{2v}^5)$	$Pca2_1$, $Pca2_1\underline{1}$, $P\underline{c}a\underline{2}_1$, $Pc\underline{a2}_1$, $P\underline{ca}2_1$, $P_a ca2_1$, $P_b ca2_1$, $P_c ca2_1$, $P_A ca2_1$, $P_B ca2_1$, $P_C ca2_1$, $P_I ca2_1$
		$Pnc2(C_{2v}^6)$	$Pnc2$, $Pnc2\underline{1}$, $P\underline{n}c\underline{2}$, $Pn\underline{c2}$, $P\underline{nc}2$, $P_a nc2$, $P_b nc2$, $P_c nc2$, $P_A nc2$, $P_B nc2$, $P_C nc2$, $P_I nc2$
		$Pmn2_1(C_{2v}^7)$	$Pmn2_1$, $Pmn2_1\underline{1}$, $P\underline{m}n\underline{2}_1$, $Pm\underline{n2}_1$, $P\underline{mn}2_1$, $P_a mn2_1$, $P_b mn2_1$, $P_c mn2_1$, $P_A mn2_1$, $P_B mn2_1$, $P_C mn2_1$, $P_I mn2_1$
		$Pba2(C_{2v}^8)$	$Pba2$, $Pba2\underline{1}$, $P\underline{b}a\underline{2}$, $P\underline{ba}2$, $P_c ba2$, $P_a ba2$, $P_C ba2$, $P_A ba2$, $P_I ba2$
		$Pna2_1(C_{2v}^9)$	$Pna2_1$, $Pna2_1\underline{1}$, $P\underline{n}a\underline{2}_1$, $Pn\underline{a2}_1$, $P\underline{na}2_1$, $P_a na2_1$, $P_b na2_1$, $P_c na2_1$, $P_A na2_1$, $P_B na2_1$, $P_C na2_1$, $P_I na2_1$
		$Pnn2(C_{2v}^{10})$	$Pnn2$, $Pnn2\underline{1}$, $P\underline{n}n\underline{2}$, $P\underline{nn}2$, $P_a nn2$, $P_c nn2$, $P_A nn2$, $P_C nn2$, $P_I nn2$
		$Cmm2(C_{2v}^{11})$	$Cmm2$, $Cmm2\underline{1}$, $C\underline{m}m\underline{2}$, $C\underline{mm}2$, $C_c mm2$, $C_a mm2$, $C_A mm2$

Orthorhombic

Point groups Ord.	Col.	Lattices Ord.	Col.	Space groups Ord.	Col. (magnetic)
				$\mathrm{Cm}c2_1(\mathrm{C}_{2v}^{12})$	$\mathrm{Cm}c2_1$, $\mathrm{Cm}c2_1\underline{1}$, $\mathrm{C}\underline{\mathrm{m}}c\underline{2}_1$, $\mathrm{Cm}\underline{c2}_1$, $\mathrm{C}\underline{\mathrm{m}c}2_1$, $\mathrm{C}_c\mathrm{m}c2_1$, $\mathrm{C}_a\mathrm{m}c2_1$, $\mathrm{C}_\mathrm{A}\mathrm{m}c2_1$
				$\mathrm{C}cc2(\mathrm{C}_{2v}^{13})$	$\mathrm{C}cc2$, $\mathrm{C}cc2\underline{1}$, $\mathrm{C}\underline{c}c\underline{2}$, $\mathrm{C}\underline{cc}2$, $\mathrm{C}_c cc2$, $\mathrm{C}_a cc2$, $\mathrm{C}_\mathrm{A} cc2$
				$\mathrm{Amm2}(\mathrm{C}_{2v}^{14})$	$\mathrm{Amm2}$, $\mathrm{Amm2}\underline{1}$, $\mathrm{A}\underline{\mathrm{m}}\mathrm{m}\underline{2}$, $\mathrm{Am}\underline{\mathrm{m}2}$, $\mathrm{A}\underline{\mathrm{mm}}2$, $\mathrm{A}_a\mathrm{mm2}$, $\mathrm{A}_c\mathrm{mm2}$, $\mathrm{A}_\mathrm{C}\mathrm{mm2}$
				$\mathrm{A}b\mathrm{m2}(\mathrm{C}_{2v}^{15})$	$\mathrm{A}b\mathrm{m2}$, $\mathrm{A}b\mathrm{m2}\underline{1}$, $\mathrm{A}\underline{b}\mathrm{m}\underline{2}$, $\mathrm{A}b\underline{\mathrm{m2}}$, $\mathrm{A}\underline{b\mathrm{m}}2$, $\mathrm{A}_a b\mathrm{m2}$, $\mathrm{A}_c b\mathrm{m2}$, $\mathrm{A}_\mathrm{C} b\mathrm{m2}$
				$\mathrm{Am}a2(\mathrm{C}_{2v}^{16})$	$\mathrm{Am}a2$, $\mathrm{Am}a2\underline{1}$, $\mathrm{A}\underline{\mathrm{m}}a\underline{2}$, $\mathrm{Am}\underline{a2}$, $\mathrm{A}\underline{\mathrm{m}a}2$, $\mathrm{A}_a\mathrm{m}a2$, $\mathrm{A}_c\mathrm{m}a2$, $\mathrm{A}_\mathrm{C}\mathrm{m}a2$
				$\mathrm{A}ba2(\mathrm{C}_{2v}^{17})$	$\mathrm{A}ba2$, $\mathrm{A}ba2\underline{1}$, $\mathrm{A}\underline{b}a\underline{2}$, $\mathrm{A}b\underline{a2}$, $\mathrm{A}\underline{ba}2$, $\mathrm{A}_a ba2$, $\mathrm{A}_c ba2$, $\mathrm{A}_\mathrm{C} ba2$
				$\mathrm{Fmm2}(\mathrm{C}_{2v}^{18})$	$\mathrm{Fmm2}$, $\mathrm{Fmm2}\underline{1}$, $\mathrm{F}\underline{\mathrm{m}}\mathrm{m}\underline{2}$, $\mathrm{F}\underline{\mathrm{mm}}2$, $\mathrm{F}_\mathrm{s}\mathrm{mm2}$
				$\mathrm{Fdd2}(\mathrm{C}_{2v}^{19})$	$\mathrm{Fdd2}$, $\mathrm{Fdd2}\underline{1}$, $\mathrm{F}\underline{\mathrm{d}}\mathrm{d}\underline{2}$, $\mathrm{F}\underline{\mathrm{dd}}2$, $\mathrm{F}_\mathrm{s}\mathrm{dd2}$
				$\mathrm{Imm2}(\mathrm{C}_{2v}^{20})$	$\mathrm{Imm2}$, $\mathrm{Imm2}\underline{1}$, $\mathrm{I}\underline{\mathrm{m}}\mathrm{m}\underline{2}$, $\mathrm{I}\underline{\mathrm{mm}}2$, $\mathrm{I}_c\mathrm{mm2}$, $\mathrm{I}_a\mathrm{mm2}$
				$\mathrm{I}ba2(\mathrm{C}_{2v}^{21})$	$\mathrm{I}ba2$, $\mathrm{I}ba2\underline{1}$, $\mathrm{I}\underline{b}a\underline{2}$, $\mathrm{I}\underline{ba}2$, $\mathrm{I}_c ba2$, $\mathrm{I}_a ba2$
				$\mathrm{Im}a2(\mathrm{C}_{2v}^{22})$	$\mathrm{Im}a2$, $\mathrm{Im}a2\underline{1}$, $\mathrm{I}\underline{\mathrm{m}}a\underline{2}$, $\mathrm{Im}\underline{a2}$, $\mathrm{I}\underline{\mathrm{m}a}2$, $\mathrm{I}_c\mathrm{m}a2$, $\mathrm{I}_a\mathrm{m}a2$, $\mathrm{I}_b\mathrm{m}a2$
mmm (D_{2h})	mmm, $\mathrm{mmm}\underline{1}$, $\underline{\mathrm{mm}}\mathrm{m}$, $\mathrm{mm}\underline{\mathrm{m}}$, $\underline{\mathrm{mmm}}$			$\mathrm{Pmmm}(\mathrm{D}_{2h}^{1})$	Pmmm, $\mathrm{Pmmm}\underline{1}$, $\mathrm{P}\underline{\mathrm{m}}\mathrm{mm}$, $\mathrm{P}\underline{\mathrm{mm}}\mathrm{m}$, $\mathrm{P}\underline{\mathrm{mmm}}$, $\mathrm{P}_a\mathrm{mmm}$, $\mathrm{P}_\mathrm{C}\mathrm{mmm}$, $\mathrm{P}_\mathrm{I}\mathrm{mmm}$
				$\mathrm{P}nnn(\mathrm{D}_{2h}^{2})$	$\mathrm{P}nnn$, $\mathrm{P}nnn\underline{1}$, $\mathrm{P}\underline{n}nn$, $\mathrm{P}\underline{nn}n$, $\mathrm{P}\underline{nnn}$, $\mathrm{P}_a nnn$, $\mathrm{P}_\mathrm{C} nnn$, $\mathrm{P}_\mathrm{I} nnn$

$\mathrm{P}cc\mathrm{m}(\mathrm{D}_{2\mathrm{h}}^{3})$	$\mathrm{P}cc\mathrm{m}$, $\mathrm{P}cc\mathrm{m}\underline{1}$, $\mathrm{P}\underline{c}c\mathrm{m}$, $\mathrm{P}cc\underline{\mathrm{m}}$, $\mathrm{P}\underline{cc}\mathrm{m}$, $\mathrm{P}\underline{c}c\underline{\mathrm{m}}$, $\mathrm{P}\underline{cc\mathrm{m}}$, $\mathrm{P}_{a}cc\mathrm{m}$, $\mathrm{P}_{c}cc\mathrm{m}$, $\mathrm{P}_{\mathrm{A}}cc\mathrm{m}$, $\mathrm{P}_{\mathrm{C}}cc\mathrm{m}$, $\mathrm{P}_{\mathrm{I}}cc\mathrm{m}$
$\mathrm{P}ban(\mathrm{D}_{2\mathrm{h}}^{4})$	$\mathrm{P}ban$, $\mathrm{P}ban\underline{1}$, $\mathrm{P}\underline{b}an$, $\mathrm{P}ba\underline{n}$, $\mathrm{P}\underline{ba}n$, $\mathrm{P}\underline{b}a\underline{n}$, $\mathrm{P}\underline{ban}$, $\mathrm{P}_{a}ban$, $\mathrm{P}_{c}ban$, $\mathrm{P}_{\mathrm{A}}ban$, $\mathrm{P}_{\mathrm{C}}ban$, $\mathrm{P}_{\mathrm{I}}ban$
$\mathrm{Pmm}a(\mathrm{D}_{2\mathrm{h}}^{5})$	$\mathrm{Pmm}a$, $\mathrm{Pmm}a\underline{1}$, $\mathrm{P}\underline{\mathrm{m}}\mathrm{m}a$, $\mathrm{Pm}\underline{\mathrm{m}}a$, $\mathrm{Pmm}\underline{a}$, $\mathrm{P}\underline{\mathrm{mm}}a$, $\mathrm{Pm}\underline{\mathrm{m}a}$, $\mathrm{P}\underline{\mathrm{m}}\mathrm{m}\underline{a}$, $\mathrm{P}\underline{\mathrm{mm}a}$, $\mathrm{P}_{a}\mathrm{mm}a$, $\mathrm{P}_{b}\mathrm{mm}a$, $\mathrm{P}_{c}\mathrm{mm}a$, $\mathrm{P}_{\mathrm{A}}\mathrm{mm}a$, $\mathrm{P}_{\mathrm{B}}\mathrm{mm}a$, $\mathrm{P}_{\mathrm{C}}\mathrm{mm}a$, $\mathrm{P}_{\mathrm{I}}\mathrm{mm}a$
$\mathrm{P}nna(\mathrm{D}_{2\mathrm{h}}^{6})$	$\mathrm{P}nna$, $\mathrm{P}nna\underline{1}$, $\mathrm{P}\underline{n}na$, $\mathrm{P}n\underline{n}a$, $\mathrm{P}nn\underline{a}$, $\mathrm{P}\underline{nn}a$, $\mathrm{P}n\underline{na}$, $\mathrm{P}\underline{n}n\underline{a}$, $\mathrm{P}\underline{nna}$, $\mathrm{P}_{a}nna$, $\mathrm{P}_{b}nna$, $\mathrm{P}_{c}nna$, $\mathrm{P}_{\mathrm{A}}nna$, $\mathrm{P}_{\mathrm{B}}nna$, $\mathrm{P}_{\mathrm{C}}nna$, $\mathrm{P}_{\mathrm{I}}nna$
$\mathrm{Pm}na(\mathrm{D}_{2\mathrm{h}}^{7})$	$\mathrm{Pm}na$, $\mathrm{Pm}na\underline{1}$, $\mathrm{P}\underline{\mathrm{m}}na$, $\mathrm{Pm}\underline{n}a$, $\mathrm{Pm}n\underline{a}$, $\mathrm{P}\underline{\mathrm{m}n}a$, $\mathrm{Pm}\underline{na}$, $\mathrm{P}\underline{\mathrm{m}}n\underline{a}$, $\mathrm{P}\underline{\mathrm{m}na}$, $\mathrm{P}_{a}\mathrm{m}na$, $\mathrm{P}_{b}\mathrm{m}na$, $\mathrm{P}_{c}\mathrm{m}na$, $\mathrm{P}_{\mathrm{A}}\mathrm{m}na$, $\mathrm{P}_{\mathrm{B}}\mathrm{m}na$, $\mathrm{P}_{\mathrm{C}}\mathrm{m}na$, $\mathrm{P}_{\mathrm{I}}\mathrm{m}na$
$\mathrm{P}cca(\mathrm{D}_{2\mathrm{h}}^{8})$	$\mathrm{P}cca$, $\mathrm{P}cca\underline{1}$, $\mathrm{P}\underline{c}ca$, $\mathrm{P}c\underline{c}a$, $\mathrm{P}cc\underline{a}$, $\mathrm{P}\underline{cc}a$, $\mathrm{P}c\underline{ca}$, $\mathrm{P}\underline{c}c\underline{a}$, $\mathrm{P}\underline{cca}$, $\mathrm{P}_{a}cca$, $\mathrm{P}_{b}cca$, $\mathrm{P}_{c}cca$, $\mathrm{P}_{\mathrm{A}}cca$, $\mathrm{P}_{\mathrm{B}}cca$, $\mathrm{P}_{\mathrm{C}}cca$, $\mathrm{P}_{\mathrm{I}}cca$
$\mathrm{P}ba\mathrm{m}(\mathrm{D}_{2\mathrm{h}}^{9})$	$\mathrm{P}ba\mathrm{m}$, $\mathrm{P}ba\mathrm{m}\underline{1}$, $\mathrm{P}\underline{b}a\mathrm{m}$, $\mathrm{P}ba\underline{\mathrm{m}}$, $\mathrm{P}\underline{ba}\mathrm{m}$, $\mathrm{P}\underline{b}a\underline{\mathrm{m}}$, $\mathrm{P}\underline{ba\mathrm{m}}$, $\mathrm{P}_{a}ba\mathrm{m}$, $\mathrm{P}_{c}ba\mathrm{m}$, $\mathrm{P}_{\mathrm{A}}ba\mathrm{m}$, $\mathrm{P}_{\mathrm{C}}ba\mathrm{m}$, $\mathrm{P}_{\mathrm{I}}ba\mathrm{m}$
$\mathrm{P}ccn(\mathrm{D}_{2\mathrm{h}}^{10})$	$\mathrm{P}ccn$, $\mathrm{P}ccn\underline{1}$, $\mathrm{P}\underline{c}cn$, $\mathrm{P}cc\underline{n}$, $\mathrm{P}\underline{cc}n$, $\mathrm{P}\underline{c}n\underline{n}$, $\mathrm{P}\underline{ccn}$, $\mathrm{P}_{a}ccn$, $\mathrm{P}_{c}ccn$, $\mathrm{P}_{\mathrm{A}}ccn$, $\mathrm{P}_{\mathrm{C}}ccn$, $\mathrm{P}_{\mathrm{I}}ccn$
$\mathrm{P}bc\mathrm{m}(\mathrm{D}_{2\mathrm{h}}^{11})$	$\mathrm{P}bc\mathrm{m}$, $\mathrm{P}bc\mathrm{m}\underline{1}$, $\mathrm{P}\underline{b}c\mathrm{m}$, $\mathrm{P}b\underline{c}\mathrm{m}$, $\mathrm{P}bc\underline{\mathrm{m}}$, $\mathrm{P}\underline{bc}\mathrm{m}$, $\mathrm{P}b\underline{c\mathrm{m}}$, $\mathrm{P}\underline{b}c\underline{\mathrm{m}}$, $\mathrm{P}\underline{bc\mathrm{m}}$, $\mathrm{P}_{a}bc\mathrm{m}$, $\mathrm{P}_{b}bc\mathrm{m}$, $\mathrm{P}_{c}bc\mathrm{m}$, $\mathrm{P}_{\mathrm{A}}bc\mathrm{m}$, $\mathrm{P}_{\mathrm{B}}bc\mathrm{m}$, $\mathrm{P}_{\mathrm{C}}bc\mathrm{m}$, $\mathrm{P}_{\mathrm{I}}bc\mathrm{m}$

Orthorhombic

Point groups Ord.	Point groups Col.	Lattices Ord.	Lattices Col.	Space groups Ord.	Col. (magnetic)
				$Pnnm(D_{2h}^{12})$	$Pnnm$, $Pnnm\underline{1}$, $P\underline{n}nm$, $Pnn\underline{m}$, $P\underline{nn}m$, $Pn\underline{nm}$, $P\underline{nnm}$, $P_a nnm$, $P_c nnm$, $P_A nnm$, $P_C nnm$, $P_I nnm$
				$Pmmn(D_{2h}^{13})$	$Pmmn$, $Pmmn\underline{1}$, $P\underline{m}mn$, $Pmm\underline{n}$, $P\underline{mm}n$, $Pm\underline{mn}$, $P\underline{mmn}$, $P_a mmn$, $P_c mmn$, $P_A mmn$, $P_C mmn$, $P_I mmn$
				$Pbcn(D_{2h}^{14})$	$Pbcn$, $Pbcn\underline{1}$, $P\underline{b}cn$, $Pb\underline{c}n$, $Pbc\underline{n}$, $P\underline{bc}n$, $Pb\underline{cn}$, $P\underline{b}c\underline{n}$, $P\underline{bcn}$, $P_a bcn$, $P_b bcn$, $P_c bcn$, $P_A bcn$, $P_B bcn$, $P_C bcn$, $P_I bcn$
				$Pbca(D_{2h}^{15})$	$Pbca$, $Pbca\underline{1}$, $P\underline{b}ca$, $P\underline{bc}a$, $P\underline{bca}$, $P_a bca$, $P_C bca$, $P_I bca$
				$Pnma(D_{2h}^{16})$	$Pnma$, $Pnma\underline{1}$, $P\underline{n}ma$, $Pn\underline{m}a$, $Pnm\underline{a}$, $P\underline{nm}a$, $Pn\underline{ma}$, $P\underline{n}m\underline{a}$, $P\underline{nma}$, $P_a nma$, $P_b nma$, $P_c nma$, $P_A nma$, $P_B nma$, $P_C nma$, $P_I nma$
				$Cmcm(D_{2h}^{17})$	$Cmcm$, $Cmcm\underline{1}$, $C\underline{m}cm$, $Cm\underline{c}m$, $Cmc\underline{m}$, $C\underline{mc}m$, $Cm\underline{cm}$, $C\underline{m}c\underline{m}$, $C\underline{mcm}$, $C_a mcm$, $C_c mcm$, $C_A mcm$
				$Cmca(D_{2h}^{18})$	$Cmca$, $Cmca\underline{1}$, $C\underline{m}ca$, $Cm\underline{c}a$, $Cmc\underline{a}$, $C\underline{mc}a$, $Cm\underline{ca}$, $C\underline{m}c\underline{a}$, $C\underline{mca}$, $C_a mca$, $C_c mca$, $C_A mca$
				$Cmmm(D_{2h}^{19})$	$Cmmm$, $Cmmm\underline{1}$, $C\underline{m}mm$, $Cmm\underline{m}$, $C\underline{mm}m$, $Cm\underline{mm}$, $C\underline{mmm}$, $C_c mmm$, $C_a mmm$, $C_A mmm$
				$Cccm(D_{2h}^{20})$	$Cccm$, $Cccm\underline{1}$, $C\underline{c}cm$, $Ccc\underline{m}$, $C\underline{cc}m$, $Cc\underline{cm}$, $C\underline{ccm}$, $C_c ccm$, $C_a ccm$, $C_A ccm$

				$\mathrm{Cmm}a(\mathrm{D}_{2h}^{21})$	$\mathrm{Cmm}a$, $\mathrm{Cmm}\underline{a\mathrm{1}}$, $\mathrm{C}\underline{\mathrm{m}}\mathrm{m}a$, $\mathrm{Cmm}\underline{a}$, $\mathrm{C}\underline{\mathrm{mm}}a$, $\mathrm{Cm}\underline{\mathrm{m}a}$, $\mathrm{C}\underline{\mathrm{mm}a}$, $\mathrm{C}_c\mathrm{mm}a$, $\mathrm{C}_a\mathrm{mm}a$, $\mathrm{C_A mm}a$
				$\mathrm{C}cca(\mathrm{D}_{2h}^{22})$	$\mathrm{C}cca$, $\mathrm{C}cca\underline{\mathrm{1}}$, $\mathrm{C}\underline{c}ca$, $\mathrm{C}cc\underline{a}$, $\mathrm{C}\underline{cc}a$, $\mathrm{C}c\underline{ca}$, $\mathrm{C}\underline{cca}$, $\mathrm{C}_c cca$, $\mathrm{C}_a cca$, $\mathrm{C_A}cca$
				$\mathrm{Fmmm}(\mathrm{D}_{2h}^{23})$	Fmmm, $\mathrm{Fmmm\underline{1}}$, $\mathrm{F\underline{m}mm}$, $\mathrm{F\underline{mm}m}$, $\mathrm{F\underline{mmm}}$, $\mathrm{F_s mmm}$
				$\mathrm{Fddd}(\mathrm{D}_{2h}^{24})$	Fddd, $\mathrm{Fddd\underline{1}}$, $\mathrm{F\underline{d}dd}$, $\mathrm{F\underline{dd}d}$, $\mathrm{F\underline{ddd}}$, $\mathrm{F_s ddd}$
				$\mathrm{Immm}(\mathrm{D}_{2h}^{25})$	Immm, $\mathrm{Immm\underline{1}}$, $\mathrm{I\underline{m}m\underline{m}}$, $\mathrm{I\underline{mm}m}$, $\mathrm{I\underline{mmm}}$, $\mathrm{I}_c\mathrm{mmm}$
				$\mathrm{I}ba\mathrm{m}(\mathrm{D}_{2h}^{26})$	$\mathrm{I}ba\mathrm{m}$, $\mathrm{I}ba\mathrm{m\underline{1}}$, $\mathrm{I}\underline{b}a\mathrm{m}$, $\mathrm{I}ba\underline{\mathrm{m}}$, $\mathrm{I}\underline{ba}\mathrm{m}$, $\mathrm{I}b\underline{a\mathrm{m}}$, $\mathrm{I}\underline{ba\mathrm{m}}$, $\mathrm{I}_c ba\mathrm{m}$, $\mathrm{I}_a ba\mathrm{m}$
				$\mathrm{I}bca(\mathrm{D}_{2h}^{27})$	$\mathrm{I}bca$, $\mathrm{I}bca\underline{\mathrm{1}}$, $\mathrm{I}\underline{b}ca$, $\mathrm{I}\underline{bc}a$, $\mathrm{I}\underline{bca}$, $\mathrm{I}_c bca$
				$\mathrm{Imm}a(\mathrm{D}_{2h}^{28})$	$\mathrm{Imm}a$, $\mathrm{Imm}a\underline{\mathrm{1}}$, $\mathrm{I\underline{m}m}a$, $\mathrm{Imm}\underline{a}$, $\mathrm{I\underline{mm}}a$, $\mathrm{Im}\underline{\mathrm{m}a}$, $\mathrm{I}\underline{\mathrm{mm}a}$, $\mathrm{I}_c\mathrm{mm}a$, $\mathrm{I}_a\mathrm{mm}a$
(3)	(12)	(4)	(12)	(59)	(562)

Tetragonal

Point groups Ord.	Col.	Lattices Ord.	Col.	Space groups Ord.	Col. (magnetic)
$4(C_4)$	$4, 4\underline{1}, \underline{4}$	P	P, P_c, P_C, P_I	$P4(C_4^1)$	$P4, P4\underline{1}, P\underline{4}, P_c4, P_C4, P_I4$
		I	I, I_c	$P4_1(C_4^2)$	$P4_1, P4_1\underline{1}, P\underline{4}_1, P_c4_1, P_C4_1, P_I4_1$
				$P4_2(C_4^3)$	$P4_2, P4_2\underline{1}, P\underline{4}_2, P_c4_2, P_C4_2, P_I4_2$
				$P4_3(C_4^4)$	$P4_3, P4_3\underline{1}, P\underline{4}_3, P_c4_3, P_C4_3, P_I4_3$
				$I4(C_4^5)$	$I4, I4\underline{1}, I\underline{4}, I_c4$
				$I4_1(C_4^6)$	$I4_1, I4_1\underline{1}, I\underline{4}_1, I_c4_1$
$\bar{4}(S_4)$	$\bar{4}, \bar{4}\underline{1}\ \underline{\bar{4}}$			$P\bar{4}(S_4^1)$	$P\bar{4}, P\bar{4}\underline{1}, P\underline{\bar{4}}, P_c\bar{4}, P_C\bar{4}, P_I\bar{4}$
				$I\bar{4}(S_4^2)$	$I\bar{4}, I\bar{4}\underline{1}, I\underline{\bar{4}}, I_c\bar{4}$
$4/m(C_{4h})$	$4/m, 4/m\underline{1}$, $\underline{4}/m, 4/\underline{m}, \underline{4}/\underline{m}$			$P4/m(C_{4h}^1)$	$P4/m, P4/m\underline{1}, P\underline{4}/m, P4/\underline{m}, P\underline{4}/\underline{m}, P_c4/\underline{m}, P_C4/m$, P_I4/m
				$P4_2/m(C_{4h}^2)$	$P4_2/m, P4_2/m\underline{1}, P\underline{4}_2/m, P4_2/\underline{m}, P\underline{4}_2/\underline{m}, P_c4_2/m$, $P_C4_2/m, P_I4_2/m$
				$P4/n(C_{4h}^3)$	$P4/n, P4/n\underline{1}, P\underline{4}/n, P4/\underline{n}, P\underline{4}/\underline{n}, P_c4/n, P_C4/n, P_I4/n$
				$P4_2/n(C_{4h}^4)$	$P4_2/n, P4_2/n\underline{1}, P\underline{4}_2/n, P4_2/\underline{n}, P\underline{4}_2/\underline{n}, P_c4_2/n, P_C4_2/n$, P_I4_2/n
				$I4/m(C_{4h}^5)$	$I4/m, I4/m\underline{1}, I\underline{4}/m, I4/\underline{m}, I\underline{4}/\underline{m}, I_c4/m$
				$I4_1/a(C_{4h}^6)$	$I4_1/a, I4_1/a\underline{1}, I\underline{4}_1/a, I4_1/\underline{a}, I\underline{4}_1/\underline{a}, I_c4_1/a$

$422(D_4)$	$422, 422\underline{1}, \underline{4}2\underline{2}, 4\underline{22}$	$P422(D_4^1)$	$P422, P422\underline{1}, P\underline{4}2\underline{2}, P4\underline{22}, P\underline{42}2, P_c422, P_C422, P_I422$
		$P42_12(D_4^2)$	$P42_12, P42_12\underline{1}, P\underline{4}2_1\underline{2}, P4\underline{2}_1\underline{2}, P\underline{42}_12, P_c42_12, P_C42_12, P_I42_12$
		$P4_122(D_4^3)$	$P4_122, P4_122\underline{1}, P\underline{4}_12\underline{2}, P4_1\underline{22}, P\underline{4}_1\underline{2}2, P_c4_122, P_C4_122, P_I4_122$
		$P4_12_12(D_4^4)$	$P4_12_12, P4_12_12\underline{1}, P\underline{4}_12_1\underline{2}, P4_1\underline{2}_1\underline{2}, P\underline{4}_1\underline{2}_12, P_c4_12_12, P_C4_12_12, P_I4_12_12$
		$P4_222(D_4^5)$	$P4_222, P4_222\underline{1}, P\underline{4}_22\underline{2}, P4_2\underline{22}, P\underline{4}_2\underline{2}2, P_c4_222, P_C4_222, P_I4_222$
		$P4_22_12(D_4^6)$	$P4_22_12, P4_22_12\underline{1}, P\underline{4}_22_1\underline{2}, P4_2\underline{2}_1\underline{2}, P\underline{4}_2\underline{2}_12, P_c4_22_12, P_C4_22_12, P_I4_22_12$
		$P4_322(D_4^7)$	$P4_322, P4_322\underline{1}, P\underline{4}_32\underline{2}, P4_3\underline{22}, P\underline{4}_3\underline{2}2, P_c4_322, P_C4_322, P_I4_322$
		$P4_32_12(D_4^8)$	$P4_32_12, P4_32_12\underline{1}, P\underline{4}_32_1\underline{2}, P4_3\underline{2}_1\underline{2}, P\underline{4}_3\underline{2}_12, P_c4_32_12, P_C4_32_12, P_I4_32_12$
		$I422(D_4^9)$	$I422, I422\underline{1}, I\underline{4}2\underline{2}, I4\underline{22}, I\underline{42}2, I_c422$
		$I4_122(D_4^{10})$	$I4_122, I4_122\underline{1}, I\underline{4}_12\underline{2}, I4_1\underline{22}, I\underline{4}_1\underline{2}2, I_c4_122$
$4mm(C_{4v})$	$4mm, 4mm\underline{1}, \underline{4}m\underline{m}, 4\underline{mm}$	$P4mm(C_{4v}^1)$	$P4mm, P4mm\underline{1}, P\underline{4}m\underline{m}, P\underline{4}\underline{m}m, P4\underline{mm}, P_c4mm, P_C4mm, P_I4mm$
		$P4bm(C_{4v}^2)$	$P4bm, P4bm\underline{1}, P\underline{4}\underline{b}m, P\underline{4}b\underline{m}, P4\underline{bm}, P_c4bm, P_C4bm, P_I4bm$

Tetragonal

Point groups Ord.	Col.	Lattices Ord.	Col.	Space groups Ord.	Col. (magnetic)
				$P4_2cm(C_{4v}^3)$	$P4_2cm$, $P4_2cm\underline{1}$, $P\underline{4}_2\underline{c}m$, $P\underline{4}_2c\underline{m}$, $P4_2\underline{cm}$, P_c4_2cm, P_C4_2cm, P_I4_2cm
				$P4_2nm(C_{4v}^4)$	$P4_2nm$, $P4_2nm\underline{1}$, $P\underline{4}_2\underline{n}m$, $P\underline{4}_2n\underline{m}$, $P4_2\underline{nm}$, P_c4_2nm, P_C4_2nm, P_I4_2nm
				$P4cc(C_{4v}^5)$	$P4cc$, $P4cc\underline{1}$, $P\underline{4}\underline{c}c$, $P\underline{4}c\underline{c}$, $P4\underline{cc}$, P_c4cc, P_C4cc, P_I4cc
				$P4nc(C_{4v}^6)$	$P4nc$, $P4nc\underline{1}$, $P\underline{4}\underline{n}c$, $P\underline{4}n\underline{c}$, $P4\underline{nc}$, P_c4nc, P_C4nc, P_I4nc
				$P4_2mc(C_{4v}^7)$	$P4_2mc$, $P4_2mc\underline{1}$, $P\underline{4}_2\underline{m}c$, $P\underline{4}_2m\underline{c}$, $P4_2\underline{mc}$, P_c4_2mc, P_C4_2mc, P_I4_2mc
				$P4_2bc(C_{4v}^8)$	$P4_2bc$, $P4_2bc\underline{1}$, $P\underline{4}_2\underline{b}c$, $P\underline{4}_2b\underline{c}$, $P4_2\underline{bc}$, P_c4_2bc, P_C4_2bc, P_I4_2bc
				$I4mm(C_{4v}^9)$	$I4mm$, $I4mm\underline{1}$, $I\underline{4}\underline{m}m$, $I\underline{4}m\underline{m}$, $I4\underline{mm}$, I_c4mm
				$I4cm(C_{4v}^{10})$	$I4cm$, $I4cm\underline{1}$, $I\underline{4}\underline{c}m$, $I\underline{4}c\underline{m}$, $I4\underline{cm}$, I_c4cm
				$I4_1md(C_{4v}^{11})$	$I4_1md$, $I4_1md\underline{1}$, $I\underline{4}_1\underline{m}d$, $I\underline{4}_1m\underline{d}$, $I4_1\underline{md}$, I_c4_1md
				$I4_1cd(C_{4v}^{12})$	$I4_1cd$, $I4_1cd\underline{1}$, $I\underline{4}_1\underline{c}d$, $I\underline{4}_1c\underline{d}$, $I4_1\underline{cd}$, I_c4_1cd
$\bar{4}2m(D_{2d})$	$\bar{4}2m$, $\bar{4}2m\underline{1}$, $\underline{\bar{4}2}m$, $\underline{\bar{4}}2\underline{m}$, $\bar{4}\underline{2m}$			$P\bar{4}2m(D_{2d}^1)$	$P\bar{4}2m$, $P\bar{4}2m\underline{1}$, $P\underline{\bar{4}2}m$, $P\underline{\bar{4}}2\underline{m}$, $P\bar{4}\underline{2m}$, $P_c\bar{4}2m$, $P_C\bar{4}2m$, $P_I\bar{4}2m$
				$P\bar{4}2c(D_{2d}^2)$	$P\bar{4}2c$, $P\bar{4}2c\underline{1}$, $P\underline{\bar{4}2}c$, $P\underline{\bar{4}}2\underline{c}$, $P\bar{4}\underline{2c}$, $P_c\bar{4}2c$, $P_C\bar{4}2c$, $P_I\bar{4}2c$

		$P\bar{4}2_1m(D_{2d}^3)$	$P\bar{4}2_1m$, $P\bar{4}2_1m\underline{1}$, $P\underline{\bar{4}2_1}m$, $P\underline{\bar{4}}2_1\underline{m}$, $P\bar{4}\underline{2_1m}$, $P_c\bar{4}2_1m$, $P_C\bar{4}2_1m$, $P_I\bar{4}2_1m$
		$P\bar{4}2_1c(D_{2d}^4)$	$P\bar{4}2_1c$, $P\bar{4}2_1c\underline{1}$, $P\underline{\bar{4}2_1}c$, $P\underline{\bar{4}}2_1\underline{c}$, $P\bar{4}\underline{2_1c}$, $P_c\bar{4}2_1c$, $P_C\bar{4}2_1c$, $P_I\bar{4}2_1c$
		$P\bar{4}m2(D_{2d}^5)$	$P\bar{4}m2$, $P\bar{4}m2\underline{1}$, $P\underline{\bar{4}m}2$, $P\underline{\bar{4}}m\underline{2}$, $P\bar{4}\underline{m2}$, $P_c\bar{4}m2$, $P_C\bar{4}m2$, $P_I\bar{4}m2$
		$P\bar{4}c2(D_{2d}^6)$	$P\bar{4}c2$, $P\bar{4}c2\underline{1}$, $P\underline{\bar{4}c}2$, $P\underline{\bar{4}}c\underline{2}$, $P\bar{4}\underline{c2}$, $P_c\bar{4}c2$, $P_C\bar{4}c2$, $P_I\bar{4}c2$
		$P\bar{4}b2(D_{2d}^7)$	$P\bar{4}b2$, $P\bar{4}b2\underline{1}$, $P\underline{\bar{4}b}2$, $P\underline{\bar{4}}b\underline{2}$, $P\bar{4}\underline{b2}$, $P_c\bar{4}b2$, $P_C\bar{4}b2$, $P_I\bar{4}b2$
		$P\bar{4}n2(D_{2d}^8)$	$P\bar{4}n2$, $P\bar{4}n2\underline{1}$, $P\underline{\bar{4}n}2$, $P\underline{\bar{4}}n\underline{2}$, $P\bar{4}\underline{n2}$, $P_c\bar{4}n2$, $P_C\bar{4}n2$, $P_I\bar{4}n2$
		$I\bar{4}m2(D_{2d}^9)$	$I\bar{4}m2$, $I\bar{4}m2\underline{1}$, $I\underline{\bar{4}m}2$, $I\underline{\bar{4}}m\underline{2}$, $I\bar{4}\underline{m2}$, $I_c\bar{4}m2$
		$I\bar{4}c2(D_{2d}^{10})$	$I\bar{4}c2$, $I\bar{4}c2\underline{1}$, $I\underline{\bar{4}c}2$, $I\underline{\bar{4}}c\underline{2}$, $I\bar{4}\underline{c2}$, $I_c\bar{4}c2$
		$I\bar{4}2m(D_{2d}^{11})$	$I\bar{4}2m$, $I\bar{4}2m\underline{1}$, $I\underline{\bar{4}2}m$, $I\underline{\bar{4}}2\underline{m}$, $I\bar{4}\underline{2m}$, $I_c\bar{4}2m$
		$I\bar{4}2d(D_{2d}^{12})$	$I\bar{4}2d$, $I\bar{4}2d\underline{1}$, $I\underline{\bar{4}2}d$, $I\underline{\bar{4}}2\underline{d}$, $I\bar{4}\underline{2d}$, $I_c\bar{4}2d$
$4/mmm(D_{4h})$	$4/mmm$, $4/mmm\underline{1}$, $\underline{4}/mmm$, $4/m\underline{mm}$, $\underline{4}/\underline{mm}m$, $4/\underline{m}mm$, $4/\underline{mmm}$	$P4/mmm(D_{4h}^1)$	$P4/mmm$, $P4/mmm\underline{1}$, $P4/\underline{m}mm$, $P\underline{4}/m\underline{m}m$, $P\underline{4}/mm\underline{m}$, $P\underline{4}/\underline{mm}m$, $P4/m\underline{mm}$, $P\underline{4}/\underline{m}m\underline{m}$, $P4/\underline{mmm}$, P_c4/mmm, P_C4/mmm, P_I4/mmm
		$P4/mcc(D_{4h}^2)$	$P4/mcc$, $P4/mcc\underline{1}$, $P4/\underline{m}cc$, $P\underline{4}/m\underline{c}c$, $P\underline{4}/mc\underline{c}$, $P\underline{4}/\underline{mc}c$, $P4/m\underline{cc}$, $P\underline{4}/\underline{m}c\underline{c}$, $P4/\underline{mcc}$, P_c4/mcc, P_C4/mcc, P_I4/mcc

Tetragonal

Point groups Ord.	Col.	Lattices Ord.	Col.	Space groups Ord.	Col. (magnetic)
				$\mathrm{P}4/nb\mathrm{m}(\mathrm{D}_{4\mathrm{h}}^{3})$	$\mathrm{P}4/nb\mathrm{m}$, $\mathrm{P}4/nb\mathrm{m}\underline{1}$, $\mathrm{P}4/\underline{n}b\mathrm{m}$, $\mathrm{P}\underline{4}/n\underline{b}\mathrm{m}$, $\mathrm{P}\underline{4}/nb\underline{\mathrm{m}}$, $\mathrm{P}\underline{4}/\underline{nb}\mathrm{m}$, $\mathrm{P}4/n\underline{b\mathrm{m}}$, $\mathrm{P}\underline{4}/\underline{n}b\underline{\mathrm{m}}$, $\mathrm{P}4/\underline{nb\mathrm{m}}$, $\mathrm{P}_c4/nb\mathrm{m}$, $\mathrm{P}_\mathrm{C}4/nb\mathrm{m}$, $\mathrm{P}_\mathrm{I}4/nb\mathrm{m}$
				$\mathrm{P}4/nnc(\mathrm{D}_{4\mathrm{h}}^{4})$	$\mathrm{P}4/nnc$, $\mathrm{P}4/nnc\underline{1}$, $\mathrm{P}4/\underline{n}nc$, $\mathrm{P}\underline{4}/n\underline{n}c$, $\mathrm{P}\underline{4}/nn\underline{c}$, $\mathrm{P}\underline{4}/\underline{nn}c$, $\mathrm{P}4/n\underline{nc}$, $\mathrm{P}\underline{4}/\underline{n}n\underline{c}$, $\mathrm{P}4/\underline{nnc}$, P_c4/nnc, $\mathrm{P}_\mathrm{C}4/nnc$, $\mathrm{P}_\mathrm{I}4/nnc$
				$\mathrm{P}4/\mathrm{m}b\mathrm{m}(\mathrm{D}_{4\mathrm{h}}^{5})$	$\mathrm{P}4/\mathrm{m}\underline{b}\mathrm{m}$, $\mathrm{P}4/\mathrm{m}b\mathrm{m}\underline{1}$, $\mathrm{P}4/\underline{\mathrm{m}}b\mathrm{m}$, $\mathrm{P}\underline{4}/\mathrm{m}\underline{b}\mathrm{m}$, $\mathrm{P}\underline{4}/\mathrm{m}b\underline{\mathrm{m}}$, $\mathrm{P}\underline{4}/\underline{\mathrm{m}b}\mathrm{m}$, $\mathrm{P}4/\mathrm{m}\underline{b\mathrm{m}}$, $\mathrm{P}\underline{4}/\underline{\mathrm{m}}b\underline{\mathrm{m}}$, $\mathrm{P}4/\underline{\mathrm{m}b\mathrm{m}}$, $\mathrm{P}_c4/\mathrm{m}b\mathrm{m}$, $\mathrm{P}_\mathrm{C}4/\mathrm{m}b\mathrm{m}$, $\mathrm{P}_\mathrm{I}4/\mathrm{m}b\mathrm{m}$
				$\mathrm{P}4/\mathrm{m}nc(\mathrm{D}_{4\mathrm{h}}^{6})$	$\mathrm{P}4/\mathrm{m}nc$, $\mathrm{P}4/\mathrm{m}nc\underline{1}$, $\mathrm{P}4/\underline{\mathrm{m}}nc$, $\mathrm{P}\underline{4}/\mathrm{m}\underline{n}c$, $\mathrm{P}\underline{4}/\mathrm{m}n\underline{c}$, $\mathrm{P}\underline{4}/\underline{\mathrm{m}n}c$, $\mathrm{P}4/\mathrm{m}\underline{nc}$, $\mathrm{P}\underline{4}/\underline{\mathrm{m}}n\underline{c}$, $\mathrm{P}4/\underline{\mathrm{m}nc}$, $\mathrm{P}_c4/\mathrm{m}nc$, $\mathrm{P}_\mathrm{C}4/\mathrm{m}nc$, $\mathrm{P}_\mathrm{I}4/\mathrm{m}nc$
				$\mathrm{P}4/n\mathrm{mm}(\mathrm{D}_{4\mathrm{h}}^{7})$	$\mathrm{P}4/n\mathrm{mm}$, $\mathrm{P}4/n\mathrm{mm}\underline{1}$, $\mathrm{P}4/\underline{n}\mathrm{mm}$, $\mathrm{P}\underline{4}/n\underline{\mathrm{m}}\mathrm{m}$, $\mathrm{P}\underline{4}/n\mathrm{m}\underline{\mathrm{m}}$, $\mathrm{P}\underline{4}/\underline{n\mathrm{m}}\mathrm{m}$, $\mathrm{P}4/n\underline{\mathrm{mm}}$, $\mathrm{P}\underline{4}/\underline{n}\mathrm{m}\underline{\mathrm{m}}$, $\mathrm{P}4/\underline{n\mathrm{mm}}$, $\mathrm{P}_c4/n\mathrm{mm}$, $\mathrm{P}_\mathrm{C}4/n\mathrm{mm}$, $\mathrm{P}_\mathrm{I}4/n\mathrm{mm}$
				$\mathrm{P}4/ncc(\mathrm{D}_{4\mathrm{h}}^{8})$	$\mathrm{P}4/ncc$, $\mathrm{P}4/ncc\underline{1}$, $\mathrm{P}4/\underline{n}cc$, $\mathrm{P}\underline{4}/n\underline{c}c$, $\mathrm{P}\underline{4}/nc\underline{c}$, $\mathrm{P}\underline{4}/\underline{nc}c$, $\mathrm{P}4/n\underline{cc}$, $\mathrm{P}\underline{4}/\underline{n}c\underline{c}$, $\mathrm{P}4/\underline{ncc}$, P_c4/ncc, $\mathrm{P}_\mathrm{C}4/ncc$, $\mathrm{P}_\mathrm{I}4/ncc$

$P4_2/mmc(D_{4h}^{9})$	$P4_2/mmc$, $P4_2/mmc\underline{1}$, $P4_2/\underline{m}mc$, $P\underline{4}_2/m\underline{m}c$, $P\underline{4}_2/mm\underline{c}$, $P\underline{4}_2/\underline{mm}c$, $P4_2/m\underline{mc}$, $P\underline{4}_2/\underline{m}m\underline{c}$, $P4_2/\underline{mmc}$, P_c4_2/mmc, P_C4_2/mmc, P_I4_2/mmc
$P4_2/mcm(D_{4h}^{10})$	$P4_2/mcm$, $P4_2/mcm\underline{1}$, $P4_2/\underline{m}cm$, $P\underline{4}_2/m\underline{c}m$, $P\underline{4}_2/mc\underline{m}$, $P\underline{4}_2/\underline{mc}m$, $P4_2/m\underline{cm}$, $P\underline{4}_2/\underline{m}c\underline{m}$, $P4_2/\underline{mcm}$, P_c4_2/mcm, P_C4_2/mcm, P_I4_2/mcm
$P4_2/nbc(D_{4h}^{11})$	$P4_2/nbc$, $P4_2/nbc\underline{1}$, $P4_2/\underline{n}bc$, $P\underline{4}_2/n\underline{b}c$, $P\underline{4}_2/nb\underline{c}$, $P\underline{4}_2/\underline{nb}c$, $P4_2/n\underline{bc}$, $P\underline{4}_2/\underline{n}b\underline{c}$, $P4_2/\underline{nbc}$, P_c4_2/nbc, P_C4_2/nbc, P_I4_2/nbc
$P4_2/nnm(D_{4h}^{12})$	$P4_2/nnm$, $P4_2/nnm\underline{1}$, $P4_2/\underline{n}nm$, $P\underline{4}_2/n\underline{n}m$, $P\underline{4}_2/nn\underline{m}$, $P\underline{4}_2/\underline{nn}m$, $P4_2/n\underline{nm}$, $P\underline{4}_2/\underline{n}n\underline{m}$, $P4_2/\underline{nnm}$, P_c4_2/nnm, P_C4_2/nnm, P_I4_2/nnm
$P4_2/mbc(D_{4h}^{13})$	$P4_2/mbc$, $P4_2/mbc\underline{1}$, $P4_2/\underline{m}bc$, $P\underline{4}_2/m\underline{b}c$, $P\underline{4}_2/mb\underline{c}$, $P\underline{4}_2/\underline{mb}c$, $P4_2/m\underline{bc}$, $P\underline{4}_2/\underline{m}b\underline{c}$, $P4_2/\underline{mbc}$, P_c4_2/mbc, P_C4_2/mbc, P_I4_2/mbc
$P4_2/mnm(D_{4h}^{14})$	$P4_2/mnm$, $P4_2/mnm\underline{1}$, $P4_2/\underline{m}nm$, $P\underline{4}_2/m\underline{n}m$, $P\underline{4}_2/mn\underline{m}$, $P\underline{4}_2/\underline{mn}m$, $P4_2/m\underline{nm}$, $P\underline{4}_2/\underline{m}n\underline{m}$, $P4_2/\underline{mnm}$, P_c4_2/mnm, P_C4_2/mnm, P_I4_2/mnm
$P4_2/nmc(D_{4h}^{15})$	$P4_2/nmc$, $P4_2/nmc\underline{1}$, $P4_2/\underline{n}mc$, $P\underline{4}_2/n\underline{m}c$, $P\underline{4}_2/nm\underline{c}$, $P\underline{4}_2/\underline{nm}c$, $P4_2/n\underline{mc}$, $P\underline{4}_2/\underline{n}m\underline{c}$, $P4_2/\underline{nmc}$, P_c4_2/nmc, P_C4_2/nmc, P_I4_2/nmc

Tetragonal

Point groups Ord.	Col.	Lattices Ord.	Col.	Space groups Ord.	Col. (magnetic)
				$P4_2/ncm(D_{4h}^{16})$	$P4_2/ncm$, $P4_2/ncm\underline{1}$, $P4_2/\underline{n}cm$, $P\underline{4}_2/n\underline{c}m$, $P\underline{4}_2/nc\underline{m}$, $P\underline{4}_2/\underline{n}cm$, $P4_2/n\underline{cm}$, $P\underline{4}_2/\underline{n}c\underline{m}$, $P4_2/\underline{ncm}$, P_c4_2/ncm, P_C4_2/ncm, P_I4_2/ncm
				$I4/mm(D_{4h}^{17})$	$I4/mmm$, $I4/mmm\underline{1}$, $I4/\underline{m}mm$, $I\underline{4}/m\underline{m}m$, $I\underline{4}/mm\underline{m}$, $I\underline{4}/\underline{m}mm$, $I4/m\underline{mm}$, $I\underline{4}/\underline{m}m\underline{m}$, $I4/\underline{mmm}$, I_c4/mmm
				$I4/mcm(D_{4h}^{18})$	$I4/mcm$, $I4/mcm\underline{1}$, $I4/\underline{m}cm$, $I\underline{4}/m\underline{c}m$, $I\underline{4}/mc\underline{m}$, $I\underline{4}/\underline{m}cm$, $I4/m\underline{cm}$, $I\underline{4}/\underline{m}c\underline{m}$, $I4/\underline{mcm}$, I_c4/mcm
				$I4_1/amd(D_{4h}^{19})$	$I4_1/amd$, $I4_1/amd\underline{1}$, $I4_1/\underline{a}md$, $I\underline{4}_1/a\underline{m}d$, $I\underline{4}_1/am\underline{d}$, $I\underline{4}_1/\underline{a}md$, $I4_1/a\underline{md}$, $I\underline{4}_1/\underline{a}m\underline{d}$, $I4_1/\underline{amd}$, I_c4_1/amd
				$I4_1/acd(D_{4h}^{20})$	$I4_1/acd$, $I4_1/acd\underline{1}$, $I4_1/\underline{a}cd$, $I\underline{4}_1/a\underline{c}d$, $I\underline{4}_1/ac\underline{d}$, $I\underline{4}_1/\underline{a}cd$, $I4_1/a\underline{cd}$, $I\underline{4}_1/\underline{a}c\underline{d}$, $I4_1/\underline{acd}$, I_c4_1/acd
(7)	(31)	(2)	(6)	(68)	(570)

Hexagonal

Point groups Ord.	Col.	Lattices Ord.	Col.	Space groups Ord.	Col. (magnetic)
$3(C_3)$	$3, 3\underline{1}$	P	P, P_c	$P3(C_3^1)$	$P3, P3\underline{1}, P_c3$
		R	R, R_I	$P3_1(C_3^2)$	$P3_1, P3_1\underline{1}, P_c3_1$
				$P3_2(C_3^3)$	$P3_2, P3_2\underline{1}, P_c3_2$
				$R3(C_3^4)$	$R3, R3\underline{1}, R_I3$
$\bar{3}(S_6)$	$\bar{3}, \bar{3}\underline{1}, \underline{\bar{3}}$			$P\bar{3}(S_6^1)$	$P\bar{3}, P\bar{3}\underline{1}, P\underline{\bar{3}}, P_c\bar{3}$
				$R\bar{3}(S_6^2)$	$R\bar{3}, R\bar{3}\underline{1}, R\underline{\bar{3}}, R_I\bar{3}$
$312(D_3)$	$312, 3\underline{1}2, 31\underline{2}$			$P312(D_3^1)$	$P312, P3\underline{1}2, P31\underline{2}, P_c312$
				$P321(D_3^2)$	$P321, P32\underline{1}, P3\underline{2}1, P_c321$
				$P3_112(D_3^3)$	$P3_112, P3_1\underline{1}2, P3_11\underline{2}, P_c3_112$
				$P3_121(D_3^4)$	$P3_121, P3_12\underline{1}, P3_1\underline{2}1, P_c3_121$
				$P3_212(D_3^5)$	$P3_212, P3_2\underline{1}2, P3_21\underline{2}, P_c3_212$
				$P3_221(D_3^6)$	$P3_221, P3_22\underline{1}, P3_2\underline{2}1, P_c3_221$
				$R32(D_3^7)$	$R32, R32\underline{1}, R3\underline{2}, R_I32$
$3m1(C_{3v})$	$3m1, 3m\underline{1}, 3\underline{m}1$			$P3m1(C_{3v}^1)$	$P3m1, P3m\underline{1}, P3\underline{m}1, P_c3m1$
				$P31m(C_{3v}^2)$	$P31m, P3\underline{1}m, P31\underline{m}, P_c31m$
				$P3c1(C_{3v}^3)$	$P3c1, P3c\underline{1}, P3\underline{c}1, P_c3c1$
				$P31c(C_{3v}^4)$	$P31c, P3\underline{1}c, P31\underline{c}, P_c31c$
				$R3m(C_{3v}^5)$	$R3m, R3m\underline{1}, R3\underline{m}, R_I3m$
				$R3c(C_{3v}^6)$	$R3c, R3c\underline{1}, R3\underline{c}, R_I3c$

Hexagonal

Point groups Ord.	Col.	Lattices Ord.	Lattices Col.	Space groups Ord.	Col. (magnetic)
$\mathrm{\bar{3}1m(D_{3d})}$	$\mathrm{\bar{3}1m}$, $\mathrm{\bar{3}\underline{1}m}$,			$\mathrm{P\bar{3}1m(D_{3d}^{1})}$	$\mathrm{P\bar{3}1m}$, $\mathrm{P\bar{3}\underline{1}m}$, $\mathrm{P\underline{\bar{3}}1m}$, $\mathrm{P\underline{\bar{3}}1\underline{m}}$, $\mathrm{P\bar{3}1\underline{m}}$, $\mathrm{P_{\mathit{c}}\bar{3}1m}$
	$\mathrm{\bar{3}1\underline{m}}$, $\mathrm{\underline{\bar{3}}1m}$,			$\mathrm{P\bar{3}1}c\mathrm{(D_{3d}^{2})}$	$\mathrm{P\bar{3}1}c$, $\mathrm{P\bar{3}\underline{1}}c$, $\mathrm{P\underline{\bar{3}}1}c$, $\mathrm{P\underline{\bar{3}}1}\underline{c}$, $\mathrm{P\bar{3}1}\underline{c}$, $\mathrm{P_{\mathit{c}}\bar{3}1}c$
	$\mathrm{\underline{\bar{3}}1\underline{m}}$			$\mathrm{P\bar{3}m1(D_{3d}^{3})}$	$\mathrm{P\bar{3}m1}$, $\mathrm{P\bar{3}m\underline{1}}$, $\mathrm{P\underline{\bar{3}}m1}$, $\mathrm{P\underline{\bar{3}m}1}$, $\mathrm{P\bar{3}\underline{m}1}$, $\mathrm{P_{\mathit{c}}\bar{3}m1}$
				$\mathrm{P\bar{3}}c\mathrm{1(D_{3d}^{4})}$	$\mathrm{P\bar{3}}c\mathrm{1}$, $\mathrm{P\bar{3}}c\mathrm{\underline{1}}$, $\mathrm{P\underline{\bar{3}}}c\mathrm{1}$, $\mathrm{P}\underline{\bar{3}c}\mathrm{1}$, $\mathrm{P\bar{3}}\underline{c}\mathrm{1}$, $\mathrm{P_{\mathit{c}}\bar{3}}c\mathrm{1}$
				$\mathrm{R\bar{3}m(D_{3d}^{5})}$	$\mathrm{R\bar{3}m}$, $\mathrm{R\bar{3}m\underline{1}}$, $\mathrm{R\underline{\bar{3}}m}$, $\mathrm{R\underline{\bar{3}m}}$, $\mathrm{R\bar{3}\underline{m}}$, $\mathrm{R_{I}\bar{3}m}$
				$\mathrm{R\bar{3}}c\mathrm{(D_{3d}^{6})}$	$\mathrm{R\bar{3}}c$, $\mathrm{R\bar{3}}c\mathrm{\underline{1}}$, $\mathrm{R\underline{\bar{3}}}c$, $\mathrm{R}\underline{\bar{3}c}$, $\mathrm{R\bar{3}}\underline{c}$, $\mathrm{R_{I}\bar{3}}c$
$\mathrm{6(C_{6})}$	6, $\mathrm{6\underline{1}}$, $\underline{6}$			$\mathrm{P6(C_{6}^{1})}$	P6, $\mathrm{P6\underline{1}}$, $\mathrm{P\underline{6}}$, $\mathrm{P_{\mathit{c}}6}$
				$\mathrm{P6_{1}(C_{6}^{2})}$	$\mathrm{P6_{1}}$, $\mathrm{P6_{1}\underline{1}}$, $\mathrm{P\underline{6}_{1}}$, $\mathrm{P_{\mathit{c}}6_{1}}$
				$\mathrm{P6_{2}(C_{6}^{3})}$	$\mathrm{P6_{2}}$, $\mathrm{P6_{2}\underline{1}}$, $\mathrm{P\underline{6}_{2}}$, $\mathrm{P_{\mathit{c}}6_{2}}$
				$\mathrm{P6_{3}(C_{6}^{4})}$	$\mathrm{P6_{3}}$, $\mathrm{P6_{3}\underline{1}}$, $\mathrm{P\underline{6}_{3}}$, $\mathrm{P_{\mathit{c}}6_{3}}$
				$\mathrm{P6_{4}(C_{6}^{5})}$	$\mathrm{P6_{4}}$, $\mathrm{P6_{4}\underline{1}}$, $\mathrm{P\underline{6}_{4}}$, $\mathrm{P_{\mathit{c}}6_{4}}$
				$\mathrm{P6_{5}(C_{6}^{6})}$	$\mathrm{P6_{5}}$, $\mathrm{P6_{5}\underline{1}}$, $\mathrm{P\underline{6}_{5}}$, $\mathrm{P_{\mathit{c}}6_{5}}$
$\mathrm{\bar{6}(C_{3h})}$	$\bar{6}$, $\mathrm{\bar{6}\underline{1}}$, $\underline{\bar{6}}$			$\mathrm{P\bar{6}(C_{3h}^{1})}$	$\mathrm{P\bar{6}}$, $\mathrm{P\bar{6}\underline{1}}$, $\mathrm{P\underline{\bar{6}}}$, $\mathrm{P_{\mathit{c}}\bar{6}}$
$\mathrm{6/m(C_{6h})}$	6/m, $\mathrm{6/m\underline{1}}$,			$\mathrm{P6/m(C_{6h}^{1})}$	P6/m, $\mathrm{P6/m\underline{1}}$, $\mathrm{P\underline{6}/m}$, $\mathrm{P6/\underline{m}}$, $\mathrm{P\underline{6}/\underline{m}}$, $\mathrm{P_{\mathit{c}}6/m}$
	$\mathrm{\underline{6}/\underline{m}}$, $\mathrm{\underline{6}/m}$, $\mathrm{6/\underline{m}}$			$\mathrm{P6_{3}/m(C_{6h}^{2})}$	$\mathrm{P6_{3}/m}$, $\mathrm{P6_{3}/m\underline{1}}$, $\mathrm{P\underline{6}_{3}/m}$, $\mathrm{P6_{3}/\underline{m}}$, $\mathrm{P\underline{6}_{3}/\underline{m}}$, $\mathrm{P_{\mathit{c}}6_{3}/m}$
$\mathrm{622(D_{6})}$	622, $\mathrm{622\underline{1}}$,			$\mathrm{P622(D_{6}^{1})}$	P622, $\mathrm{P622\underline{1}}$, $\mathrm{P\underline{62}2}$, $\mathrm{P\underline{6}2\underline{2}}$, $\mathrm{P6\underline{22}}$, $\mathrm{P_{\mathit{c}}622}$
	$\mathrm{\underline{6}22}$, $\mathrm{6\underline{2}2}$			$\mathrm{P6_{1}22(D_{6}^{2})}$	$\mathrm{P6_{1}22}$, $\mathrm{P6_{1}22\underline{1}}$, $\mathrm{P\underline{6}_{1}\underline{2}2}$, $\mathrm{P\underline{6}_{1}2\underline{2}}$, $\mathrm{P6_{1}\underline{22}}$, $\mathrm{P_{\mathit{c}}6_{1}22}$
				$\mathrm{P6_{2}22(D_{6}^{3})}$	$\mathrm{P6_{2}22}$, $\mathrm{P6_{2}22\underline{1}}$, $\mathrm{P\underline{6}_{2}\underline{2}2}$, $\mathrm{P\underline{6}_{2}2\underline{2}}$, $\mathrm{P6_{2}\underline{22}}$, $\mathrm{P_{\mathit{c}}6_{2}22}$

				$P6_322(D_6^4)$	$P6_322$, $P6_322\underline{1}$, $P\underline{6}_3\underline{2}2$, $P\underline{6}_32\underline{2}$, $P6_3\underline{22}$, P_c6_322
				$P6_422(D_6^5)$	$P6_422$, $P6_422\underline{1}$, $P\underline{6}_4\underline{2}2$, $P\underline{6}_42\underline{2}$, $P6_4\underline{22}$, P_c6_422
				$P6_522(D_6^6)$	$P6_522$, $P6_522\underline{1}$, $P\underline{6}_5\underline{2}2$, $P\underline{6}_52\underline{2}$, $P6_5\underline{22}$, P_c6_522
$6mm(C_{6v})$	$6mm$, $6mm\underline{1}$, $\underline{6m}m$, $6\underline{mm}$			$P6mm(C_{6v}^1)$	$P6mm$, $P6mm\underline{1}$, $P\underline{6m}m$, $P\underline{6}m\underline{m}$, $P6\underline{mm}$, P_c6mm
				$P6cc(C_{6v}^2)$	$P6cc$, $P6cc\underline{1}$, $P\underline{6c}c$, $P\underline{6}c\underline{c}$, $P6\underline{cc}$, P_c6cc
				$P6_3cm(C_{6v}^3)$	$P6_3cm$, $P6_3cm\underline{1}$, $P\underline{6}_3\underline{c}m$, $P\underline{6}_3c\underline{m}$, $P6_3\underline{cm}$, P_c6_3cm
				$P6_3mc(C_{6v}^4)$	$P6_3mc$, $P6_3mc\underline{1}$, $P\underline{6}_3\underline{m}c$, $P\underline{6}_3m\underline{c}$, $P6_3\underline{mc}$, P_c6_3mc
$\bar{6}m2(D_{3h})$	$\bar{6}m2$, $\bar{6}m2\underline{1}$, $\underline{\bar{6}}m\underline{2}$, $\underline{\bar{6}m}2$, $\bar{6}\underline{m2}$			$P\bar{6}m2(D_{3h}^1)$	$P\bar{6}m2$, $P\bar{6}m2\underline{1}$, $P\underline{\bar{6}m}2$, $P\underline{\bar{6}}m\underline{2}$, $P\bar{6}\underline{m2}$, $P_c\bar{6}m2$
				$P\bar{6}c2(D_{3h}^2)$	$P\bar{6}c2$, $P\bar{6}c2\underline{1}$, $P\underline{\bar{6}c}2$, $P\underline{\bar{6}}c\underline{2}$, $P\bar{6}\underline{c2}$, $P_c\bar{6}c2$
				$P\bar{6}2m(D_{3h}^3)$	$P\bar{6}2m$, $P\bar{6}2m\underline{1}$, $P\underline{\bar{6}2}m$, $P\underline{\bar{6}}2\underline{m}$, $P\bar{6}\underline{2m}$, $P_c\bar{6}2m$
				$P\bar{6}2c(D_{3h}^4)$	$P\bar{6}2c$, $P\bar{6}2c\underline{1}$, $P\underline{\bar{6}2}c$, $P\underline{\bar{6}}2\underline{c}$, $P\bar{6}\underline{2c}$, $P_c\bar{6}2c$
$6/mmm(D_{6h})$	$6/mmm$, $6/mmm\underline{1}$, $\underline{6}/\underline{mm}m$, $6/m\underline{mm}$, $\underline{6}/m\underline{m}m$, $\underline{6}/m\underline{m}m$, $6/\underline{m}mm$			$P6/mmm(D_{6h}^1)$	$P6/mmm$, $P6/mmm\underline{1}$, $P6/\underline{m}mm$, $P\underline{6}/m\underline{m}m$, $P\underline{6}/mm\underline{m}$, $P\underline{6}/\underline{mm}m$, $P\underline{6}/\underline{m}m\underline{m}$, $P6/m\underline{mm}$, $P6/\underline{mmm}$, P_c6/mmm
				$P6/mcc(D_{6h}^2)$	$P6/mcc$, $P6/mcc\underline{1}$, $P6/\underline{m}cc$, $P\underline{6}/m\underline{c}c$, $P\underline{6}/mc\underline{c}$, $P\underline{6}/\underline{mc}c$, $P\underline{6}/\underline{m}c\underline{c}$, $P6/m\underline{cc}$, $P6/\underline{mcc}$, P_c6/mcc
				$P6_3/mcm(D_{6h}^3)$	$P6_3/mcm$, $P6_3/mcm\underline{1}$, $P6_3/\underline{m}cm$, $P\underline{6}_3/m\underline{c}m$, $P\underline{6}_3/mc\underline{m}$, $P\underline{6}_3/\underline{mc}m$, $P\underline{6}_3/\underline{m}c\underline{m}$, $P6_3/m\underline{cm}$, $P6_3/\underline{mcm}$, P_c6_3/mcm
	$6/\underline{mmm}$			$P6_3/mmc(D_{6h}^4)$	$P6_3/mmc$, $P6_3/mmc\underline{1}$, $P6_3/\underline{m}mc$, $P\underline{6}_3/m\underline{m}c$, $P\underline{6}_3/mm\underline{c}$, $P\underline{6}_3/\underline{mm}c$, $P\underline{6}_3/\underline{m}m\underline{c}$, $P6_3/m\underline{mc}$, $P6_3/\underline{mmc}$, P_c6_3/mmc
(12)	(47)	(2)	(4)	(52)	(272)

Cubic

Point groups Ord.	Col.	Lattices Ord.	Col.	Space groups Ord.	Col. (magnetic)
23(T)	23	P	P, P_I	$P23(T^1)$	$P23$, $P2\underline{3}$, P_I23
	$2\underline{3}$	F	F, F_s	$F23(T^2)$	$F23$, $F2\underline{3}$, F_s23
		I	I	$I23(T^3)$	$I23$, $I2\underline{3}$
				$P2_13(T^4)$	$P2_13$, $P2_1\underline{3}$, P_I2_13
				$I2_13(T^5)$	$I2_13$, $I2_1\underline{3}$
$m3(T_h)$	m3, $m\underline{3}$, $\underline{m}3$			$Pm3(T_h^1)$	$Pm3$, $Pm\underline{3}$, $P\underline{m}3$, P_Im3
				$Pn3(T_h^2)$	$Pn3$, $Pn\underline{3}$, $P\underline{n}3$, P_In3
				$Fm3(T_h^3)$	$Fm3$, $Fm\underline{3}$, $F\underline{m}3$, F_sm3
				$Fd3(T_h^4)$	$Fd3$, $Fd\underline{3}$, $F\underline{d}3$, F_sd3
				$Im3(T_h^5)$	$Im3$, $Im\underline{3}$, $I\underline{m}3$
				$Pa3(T_h^6)$	$Pa3$, $Pa\underline{3}$, $P\underline{a}3$, P_Ia3
				$Ia3(T_h^7)$	$Ia3$, $Ia\underline{3}$, $I\underline{a}3$
432(O)	432, $4\underline{3}2$, $\underline{4}32$			$P432(O^1)$	$P432$, $P4\underline{3}2$, $P\underline{4}3\underline{2}$, P_I432
				$P4_232(O^2)$	$P4_232$, $P4_2\underline{3}2$, $P\underline{4}_23\underline{2}$, P_I4_232
				$F432(O^3)$	$F432$, $F4\underline{3}2$, $F\underline{4}3\underline{2}$, F_s432
				$F4_132(O^4)$	$F4_132$, $F4_1\underline{3}2$, $F\underline{4}_13\underline{2}$, F_s4_132
				$I432(O^5)$	$I432$, $I4\underline{3}2$, $I\underline{4}3\underline{2}$
				$P4_332(O^6)$	$P4_332$, $P4_3\underline{3}2$, $P\underline{4}_33\underline{2}$, P_I4_332

				$P4_132(O^7)$	$P4_132$, $P4_1\underline{3}2$, $P\underline{4}_13\underline{2}$, P_I4_132
				$I4_132(O^8)$	$I4_132$, $I4_1\underline{3}2$, $I\underline{4}_13\underline{2}$
$\bar{4}3m(T_d)$	$\bar{4}3m$, $\bar{4}\underline{3}m$, $\underline{\bar{4}}3\underline{m}$			$P\bar{4}3m(T_d^1)$	$P\bar{4}3m$, $P\bar{4}\underline{3}m$, $P\underline{\bar{4}}3\underline{m}$, $P_I\bar{4}3m$
				$F\bar{4}3m(T_d^2)$	$F\bar{4}3m$, $F\bar{4}\underline{3}m$, $F\underline{\bar{4}}3\underline{m}$, $F_s\bar{4}3m$
				$I\bar{4}3m(T_d^3)$	$I\bar{4}3m$, $I\bar{4}\underline{3}m$, $I\underline{\bar{4}}3\underline{m}$
				$P\bar{4}3n(T_d^4)$	$P\bar{4}3n$, $P\bar{4}\underline{3}n$, $P\underline{\bar{4}}3\underline{n}$, $P_I\bar{4}3n$
				$F\bar{4}3c(T_d^5)$	$F\bar{4}3c$, $F\bar{4}\underline{3}c$, $F\underline{\bar{4}}3\underline{c}$, $F_s\bar{4}3c$
				$I\bar{4}3d(T_d^6)$	$I\bar{4}3d$, $I\bar{4}\underline{3}d$, $I\underline{\bar{4}}3\underline{d}$
$m3m(O_h)$	$m3m$, $m\underline{3}m$, $m3\underline{m}$, $\underline{m}3m$, $\underline{m}3\underline{m}$			$Pm3m(O_h^1)$	$Pm3m$, $Pm\underline{3}m$, $P\underline{m}3m$, $Pm3\underline{m}$, $P\underline{m}3\underline{m}$, P_Im3m
				$Pn3n(O_h^2)$	$Pn3n$, $Pn\underline{3}n$, $P\underline{n}3n$, $Pn3\underline{n}$, $P\underline{n}3\underline{n}$, P_In3n
				$Pm3n(O_h^3)$	$Pm3n$, $Pm\underline{3}n$, $P\underline{m}3n$, $Pm3\underline{n}$, $P\underline{m}3\underline{n}$, P_Im3n
				$Pn3m(O_h^4)$	$Pn3m$, $Pn\underline{3}m$, $P\underline{n}3m$, $Pn3\underline{m}$, $P\underline{n}3\underline{m}$, P_In3m
				$Fm3m(O_h^5)$	$Fm3m$, $Fm\underline{3}m$, $F\underline{m}3m$, $Fm3\underline{m}$, $F\underline{m}3\underline{m}$, F_sm3m
				$Fm3c(O_h^6)$	$Fm3c$, $Fm\underline{3}c$, $F\underline{m}3c$, $Fm3\underline{c}$, $F\underline{m}3\underline{c}$, F_sm3c
				$Fd3m(O_h^7)$	$Fd3m$, $Fd\underline{3}m$, $F\underline{d}3m$, $Fd3\underline{m}$, $F\underline{d}3\underline{m}$, F_sd3m
				$Fd3c(O_h^8)$	$Fd3c$, $Fd\underline{3}c$, $F\underline{d}3c$, $Fd3\underline{c}$, $F\underline{d}3\underline{c}$, F_sd3c
				$Im3m(O_h^9)$	$Im3m$, $Im\underline{3}m$, $I\underline{m}3m$, $Im3\underline{m}$, $I\underline{m}3\underline{m}$
				$Ia3d(O_h^{10})$	$Ia3d$, $Ia\underline{3}d$, $I\underline{a}3d$, $Ia3\underline{d}$, $I\underline{a}3\underline{d}$
(5)	(16)	(3)	(5)	(36)	(149)
Total					
(32)	(122)	(14)	(36)	(230)	(1651)

HINTS AND SOLUTIONS TO THE EXERCISES

Chapter 1

1.1 (a) The three H and CN in CH_3CN are at the corners of a tetrahedron. The C is at the centre of the tetrahedron. This belongs to the point group C_{3v} with symmetry elements E, C_3, $3\sigma_v$. The $3\sigma_v$ contain C_3.

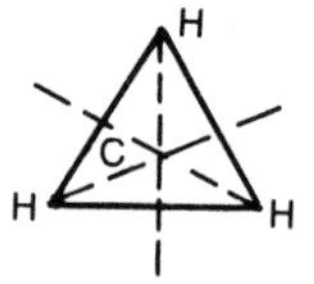

(b)

This belongs to the group C_{2v}. Operations present: E, C_2, σ_v, $\sigma_{v'}$. σ_v and $\sigma_{v'}$ contain C_2.

(c)

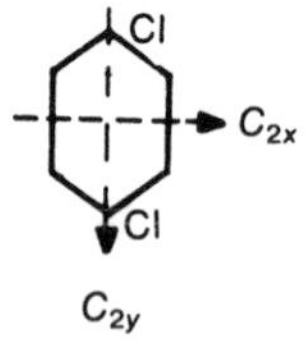

This belongs to the group D_{2h}; E, C_{2x}, C_{2y}, C_{2z}, I, σ_{xy}, σ_{xz}, σ_{yz}.

(d)

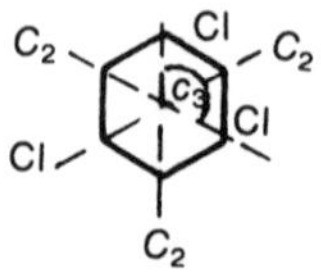

This belongs to the group D_{3h}; E, C_3, C_3^2, $3C_2$, σ_h in the plane of the paper, $3\sigma_v$ perpendicular to the plane of the paper and containing C_2, $2S_3$.

(e)

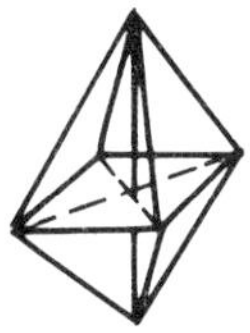

UF_6: octahedral arrangement with U at the centre and F at the corners. Contains E, $8C_3$, $6C_2$, $6C_4$, $3C_2(=3C_4^2)$, I, $8S_6$, $6S_4$, $6\sigma_d$; group = O_h.

1.2 (a) This set forms a group. Verify the group properties by forming the multiplication table:

	1	−1	i	−i
1	1	−1	i	−i
−1	−1	1	−i	i
i	i	−i	−1	1
−i	−i	i	1	−1

Note that the group is Abelian since $1 \times (-1) = -1 \times 1$; $i \times 1 = 1 \times i$; etc.

(b) Easy to see that this set does not form a group under arithmetic addition since $1 + 1 = 2$ which is not an element of the group etc.

(c) This set forms a group. Verify by forming the multiplication table:

	E	A	B	C	D	F
E	E	A	B	C	D	F
A	A	B	E	F	C	D
B	B	E	A	D	F	C
C	C	D	F	E	A	B
D	D	F	C	B	E	A
F	F	C	D	A	B	E

The identity element is

$$E = \begin{pmatrix} 1 & 0 \\ 0 & 1 \end{pmatrix}.$$

The group is not Abelian.

(d) Show, for example, $\phi_5\phi_3 = \phi_2$ as follows:

$$(\phi_5\phi_3)(x) = \phi_5\phi_3(x) = \phi_5 \frac{x}{x-1} = \phi_5(X) \quad \text{where } X = \frac{x}{x-1}$$

given

$$\phi_5(X) = \frac{1}{1-X} = \frac{1}{1-x/(x-1)} = 1 - x = \phi_3(x).$$

Therefore $\phi_5\phi_3 = \phi_2$ etc. By finding the multiplication table notice that the group is isomorphic to C_{3v}.

1.3 Easy to see that there are six possible permutations which could be labelled as

$$E = \begin{pmatrix} a & b & c \\ a & b & c \end{pmatrix} \quad A = \begin{pmatrix} a & b & c \\ b & c & a \end{pmatrix} \quad B = \begin{pmatrix} a & b & c \\ c & a & b \end{pmatrix}, \text{ etc.}$$

Denote the possible states as $\phi_E = (a \quad b \quad c)$, $\phi_A = (b \quad c \quad a)$ and so on. Then show that $A\phi_A = \phi_B$ etc and work out the multiplication table.

1.4 (a) The symmetry operations in the alternative convention are as follows:

E: the identity operation

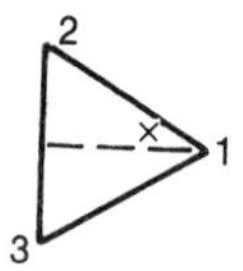

C_3: anticlockwise rotation of 120° about the z axis

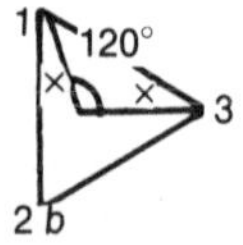

C_3^{-1}: clockwise rotation of 120° about the z axis

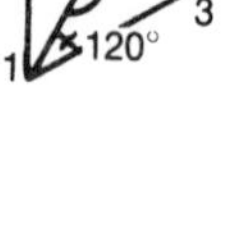

σ_{v_1}, σ_{v_2}, σ_{v_3}: reflection operations in the planes through the three vertices 1, 2 and 3 respectively

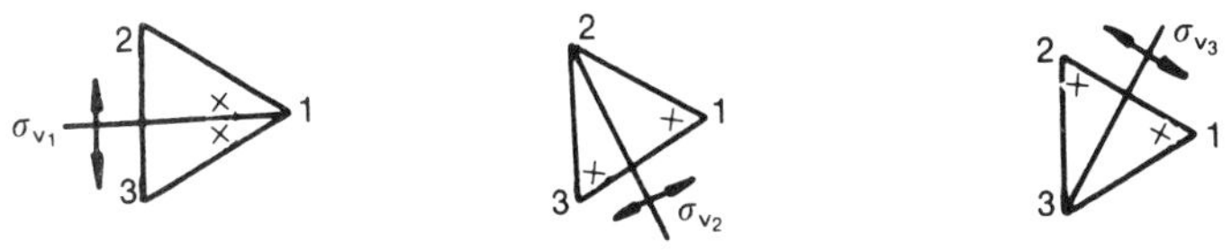

(b) E: the identity operation

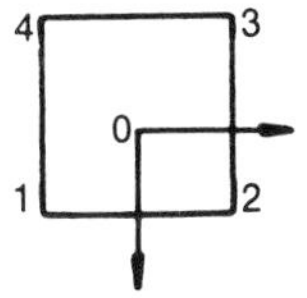

C_{4z}^{+}: anticlockwise rotation by $+90°$ about the z axis through 0

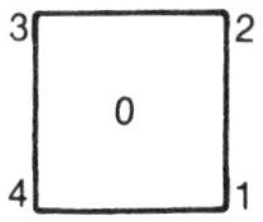

Similarly draw and describe the following operations present: C_{2z}, C_{4z}^{-}, σ_{v_x}, σ_{v_y}, σ_{13}, σ_{24}. Note: Here σ_{13} and σ_{24} denote reflections in diagonal planes joining points 1, 3 and 2, 4 respectively.

1.5 Obtain the following table:

C_{4v}	E	C_{4z}^{+}	C_{2z}	C_{4z}^{-}	σ_{v_x}	σ_{v_y}	σ_{13}	σ_{24}
E	E	C_{4z}^{+}	C_{2z}	C_{4z}^{-}	σ_{v_x}	σ_{v_y}	σ_{13}	σ_{24}
C_{4z}^{+}	C_{4z}^{+}	C_{2z}	C_{4z}^{-}	E	σ_{13}	σ_{24}	σ_{v_y}	σ_{v_x}
C_{2z}	C_{2z}	C_{4z}^{-}	E	C_{4z}^{+}	σ_{v_y}	σ_{v_x}	σ_{24}	σ_{13}
C_{4z}^{-}	C_{4z}^{-}	E	C_{4z}^{+}	C_{2x}	σ_{24}	σ_{13}	σ_{v_x}	σ_{v_y}
σ_{v_x}	σ_{v_x}	σ_{24}	σ_{v_y}	σ_{13}	E	C_{2z}	C_{4z}^{-}	C_{4z}^{+}
σ_{v_y}	σ_{v_y}	σ_{13}	σ_{v_x}	σ_{24}	C_{2z}	E	C_{4z}^{+}	C_{4z}^{-}
σ_{13}	σ_{13}	σ_{v_x}	σ_{24}	σ_{v_y}	C_{4z}^{+}	C_{4z}^{-}	E	C_{2z}
σ_{24}	σ_{24}	σ_{v_y}	σ_{13}	σ_{v_x}	C_{4z}^{-}	C_{4z}^{+}	C_{2z}	E

1.6 Hint: One subgroup of order 4 is E, C_{4z}^{+}, C_{2z}, C_{4z}^{-}; one subgroup of order 2 is E, C_{2z}, etc. You must verify that these subgroups satisfy all the group properties (i)–(iv) of §1.3.4.

1.7 Obtain the following five classes using the table given in exercise 1.5: $C_1 = E$; $C_2 = C_{2z}$; $C_3 = \{C_{4z}^{+}, C_{4z}^{-}\}$, $C_4 = \{\sigma_{v_x}, \sigma_{v_y}\}$; $C_5 = \{\sigma_{13}, \sigma_{24}\}$.

1.8 Obtain the coefficients as follows: $c_{12,2} = 1$; $c_{23,3} = 1$; $c_{33,2} = 2$; $c_{45,2} = 2$; $c_{55,1} = 2$.

Chapter 2

2.1 One possible set of matrices:

$$E = \begin{pmatrix} 1 & 0 \\ 0 & 1 \end{pmatrix} \quad C_{4z}^{+} = \begin{pmatrix} 0 & -1 \\ 1 & 0 \end{pmatrix} \quad C_{4z}^{-} = \begin{pmatrix} 0 & 1 \\ -1 & 0 \end{pmatrix} \quad C_{2z} = \begin{pmatrix} -1 & 0 \\ 0 & -1 \end{pmatrix}$$

$$\sigma_{v_x} = \begin{pmatrix} -1 & 0 \\ 0 & 1 \end{pmatrix} \quad \sigma_{v_y} = \begin{pmatrix} 1 & 0 \\ 0 & -1 \end{pmatrix} \quad \sigma_{13} = \begin{pmatrix} 0 & -1 \\ -1 & 0 \end{pmatrix} \quad \sigma_{24} = \begin{pmatrix} 0 & 1 \\ 1 & 0 \end{pmatrix}.$$

Show that these matrices multiply in the same way as that given in exercise 1.5. Similarly find a set of 3×3 matrices for the group.

2.2 Observe that $C_{2z} = (C_{4z}^{+})^2$; $C_{4z}^{-} = (C_{4z}^{+})^3$; $\sigma_{13} = C_{4z}^{+}\sigma_{v_x}$; etc. Use this method to obtain all the eight operations of the group. You may use the set of 2×2 matrices given in exercise 2.1.

2.3 Rearrange the group multiplication table so that the identity element E always appears along the diagonal, then proceed as in §2.5. For example, for C_{4z}^{+} the matrix would be as follows:

$$\begin{pmatrix} 0 & 1 & 0 & 0 & 0 & 0 & 0 & 0 \\ 0 & 0 & 1 & 0 & 0 & 0 & 0 & 0 \\ 0 & 0 & 0 & 1 & 0 & 0 & 0 & 0 \\ 1 & 0 & 0 & 0 & 0 & 0 & 0 & 0 \\ 0 & 0 & 0 & 0 & 0 & 0 & 0 & 1 \\ 0 & 0 & 0 & 0 & 0 & 0 & 1 & 0 \\ 0 & 0 & 0 & 0 & 1 & 0 & 0 & 0 \\ 0 & 0 & 0 & 0 & 0 & 1 & 0 & 0 \end{pmatrix}.$$

2.4 Γ^* is also a representation of G. Show this by finding the complex conjugates of the matrices of E, A, B, C, D, F and obtaining the multiplication table. Γ^{-1} is not a representation of G. Show this by finding the inverse of the matrices E, A, B, C, D, F and obtaining the multiplication table.

2.5 Find the adjoint of $\mathbf{U}$, that is $\mathbf{U}^{\dagger} = (\tilde{\mathbf{U}})^*$ where $(\tilde{\mathbf{U}})^*$ denotes the transpose complex conjugate of the elements of $\mathbf{U}$ and show that

$$\mathbf{U}\mathbf{U}^{\dagger} = \begin{pmatrix} 1 & 0 & 0 \\ 0 & 1 & 0 \\ 0 & 0 & 1 \end{pmatrix}.$$

2.6 Determine how $\hat{x}$, $\hat{y}$, $\hat{z}$ transform with respect to the operations of the group. For example (see figure) for the operation C_3 we can write

$$C_3(\hat{x}) = \hat{x}' = -\tfrac{1}{2}\hat{x} + \sqrt{\tfrac{3}{4}}\hat{y}$$

$$C_3(\hat{y}) = \hat{y}' = -\sqrt{\tfrac{3}{4}}\hat{x} - \tfrac{1}{2}\hat{y}$$

$$C_3(\hat{z}) = \hat{z}' = \hat{z}.$$

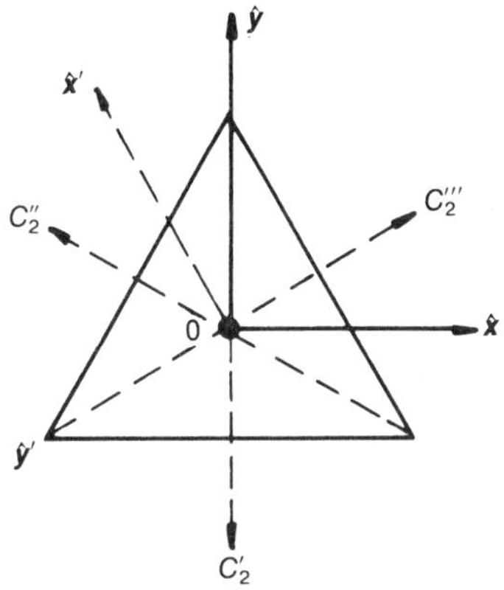

Therefore the matrix representing this operation would be

$$\mathbf{D}(C_3) = \begin{pmatrix} -\frac{1}{2} & -\sqrt{\frac{3}{4}} & 0 \\ \sqrt{\frac{3}{4}} & -\frac{1}{2} & 0 \\ 0 & 0 & 1 \end{pmatrix}.$$

Similarly, find the matrices for the other operations.

The 2×2 and 1×1 representations are as follows:

E	C_3	C_3^2	C_2'	C_2''	C_2'''
$\begin{pmatrix} 1 & 0 \\ 0 & 1 \end{pmatrix}$	$\begin{pmatrix} -\frac{1}{2} & -\sqrt{\frac{3}{4}} \\ \sqrt{\frac{3}{4}} & -\frac{1}{2} \end{pmatrix}$	$\begin{pmatrix} -\frac{1}{2} & \sqrt{\frac{3}{4}} \\ -\sqrt{\frac{3}{4}} & -\frac{1}{2} \end{pmatrix}$	$\begin{pmatrix} -1 & 0 \\ 0 & 1 \end{pmatrix}$	$\begin{pmatrix} \frac{1}{2} & -\sqrt{\frac{3}{4}} \\ -\sqrt{\frac{3}{4}} & -\frac{1}{2} \end{pmatrix}$	$\begin{pmatrix} \frac{1}{2} & \sqrt{\frac{3}{4}} \\ \sqrt{\frac{3}{4}} & -\frac{1}{2} \end{pmatrix}$
1	1	1	−1	−1	−1

Chapter 3

3.1 Construction of character table for C_{3v}:

The method is based on the following equations:

$$\sum_i n_i^2 = g \qquad (1)$$

that is, the sum of the squares of IR is equal to the order of group g

$$\sum_i r_i |\chi_i|^2 = g \tag{2}$$

that is, the sum of squares of characters in any IR equals g. Here r is the order of class, that is number of elements in the class.

$$C_i C_j = \sum_k c_{ij,k} C_k \tag{3a}$$

that is, class multiplication rule

$$r_i r_j \chi_i \chi_j = n \sum_k c_{ij,k} r_k \chi_k \tag{3b}$$

that is, character multiplication rule.

Let us apply these equations to find the character table for the C_{3v} group. Here $g = 6$, so that using (1) we see there can be only three IRS of dimensions $n_1 = 1$, $n_2 = 1$, $n_3 = 2$, such that

$$\sum n_i^2 = n_1^2 + n_2^2 + n_3^2 = 1^2 + 1^2 + 2^2 = 6.$$

C_{3v} has three classes defined as follows:

$$C_1 = E \qquad C_2 = C_3, C_3^2 \qquad C_3 = \sigma_1, \sigma_2, \sigma_3$$

that is, $r_1 = 1$, $r_2 = 2$, $r_3 = 3$. Applying (3a) gives $C_1C_1 = C_1$, $C_2C_2 = C_2$, $C_1C_3 = C_3$, $C_2C_1 = C_2$, $C_2C_2 = 2C_1 + 1C_2$, $C_2C_3 = 2C_3$, $C_3C_1 = C_3$, $C_3C_2 = 2C_3$, $C_3C_3 = 3C_1 + 3C_2$.

The coefficients $c_{ij,k}$ are defined as follows. Since $C_1 = E$ (identity) all $C_iC_j = C_jC_i$ involving C_1 are left out since they are simply identity. Now $c_{22,1} = 2$, $c_{22,2} = 1$, $c_{23,3} = 2$, $c_{32,3} = 2$, $c_{33,1} = 3$, $c_{33,2} = 3$. Using (3b) for $i = j = 2$ we get

$$r_2 r_2 \chi_2 \chi_2 = n(c_{22,1} r_1 \chi_1 + c_{22,2} r_2 \chi_2 + c_{22,3} r_3 \chi_3).$$

Now $r_2 = 2$, $r_1 = 1$, $c_{22,1} = 2$, $\chi_1 = n$ (note: $\chi_1 = \chi(E)$ always, which gives the dimension of representation n), $c_{22,2} = 1$, $c_{22,3} = 0$ so that $4\chi_2^2 = n(2n + 2\chi_2)$ or $\chi_2 = n$ or $-n/2$.

For $i = j = 3$ we get

$$r_3^2 \chi_3^2 = n(c_{33,1} r_1 \chi_1 + c_{33,2} r_2 \chi_2 + c_{33,3} r_3 \chi_3) \quad \text{or} \quad 9\chi_3^2 = n(3n + 6\chi_2) \tag{4}$$

For $i = 2$, $j = 3$ we get

$$r_2 r_3 \chi_2 \chi_3 = n(c_{23,1} r_1 \chi_1 + c_{23,2} r_2 \chi_2 + c_{23,3} r_3 \chi_3) \quad \text{or} \quad \chi_2 = n, \chi_3 = 0.$$

Taking first the one-dimensional representation $n = 1$, we have $\chi_2 = 1$ or $-\frac{1}{2}$. When $\chi_2 = 1$, we get $9\chi_3^2 = 9$ or $\chi_3 = \pm 1$; when $\chi_2 = -\frac{1}{2}$, $9\chi_3^2 = 0$ or $\chi_3 = 0$. We note that for one-dimensional representations the characters can never be zero. Hence the only solutions when $n = 1$ are $\chi_1 = 1, \chi_2 = 1, \chi_3 = 1$; and $\chi_1 = 1, \chi_2 = 1, \chi_3 = -1$.

When $n = 2$, $\chi_1 = 2$, $\chi_2 = 2$ or -1, and χ_3 is found from (4), that is $9\chi_3^2 = 2(6 + 12)$ or $\chi_3^2 = 4$ or $\chi_3 = \pm 2$. The other solution for $\chi_2 = -1$ is $\chi_3 = 0$. Now, if we take for $n = 2$, $\chi_1 = 2$, $\chi_2 = 2$ and $\chi_3 = \pm 2$, we get using $\sum_i r_i |\chi_i|^2 = 4 + 8 + 12 = 24 \neq g$ and hence this combination is not allowed. The only other acceptable combination is $\chi_1 = 2$, $\chi_2 = -1$ and $\chi_3 = 0$. The character table is therefore as shown below for the C_{3v} group:

C_{3v}	C_1	C_2	C_3
Γ_1	1	1	1
Γ_2	1	1	-1
Γ_3	2	-1	0

For C_{4v} see [5]. There are five classes (see exercise 1.7) and therefore we expect five representations as follows:

C_{4v}	$C_1 = E$	$C_2 = C_{4z}^+, C_{4z}^-$	$C_3 = C_{2z}$	$C_4 = \sigma_{v_x}, \sigma_{v_y}$	$C_5 = \sigma_{13}, \sigma_{24}$
Γ_1	1	1	1	1	1
Γ_2	1	1	1	-1	-1
Γ_3	1	-1	1	1	-1
Γ_4	1	-1	1	-1	1
Γ_5	2	0	-2	0	0

3.2 Verify that

$$\sum_{\text{all classes}} g(C)\chi_i(C)\chi_j^*(C) = 0. \tag{1}$$

Here $g(C)$ is the number of elements in the class C and $\chi_i(C)\chi_j^*(C)$ are the characters in the class C for Γ_i and Γ_j respectively. Take Γ_4 and Γ_5 as a concrete example to verify (1).

3.3 See properties (1–3) of irreducible representations (§3.3).

3.4 $\Gamma = a_1\Gamma_1 + a_2\Gamma_2 + a_3\Gamma_3 + a_4\Gamma_4 + a_5\Gamma_5$, where Γ is the reducible representation and Γ_i ($i = 1, 2, 3, 4, 5$) is the IR:

$$a_i = \text{coefficients} = \frac{1}{h}\sum_R \chi(R)\chi^*(R)g_R.$$

The symbols are explained as in equation (3.13); h, the order of the group, equals 24. Obtain $a_1 = 1$, $a_2 = 0$, $a_3 = 1$, $a_4 = 0$, $a_5 = 2$. Thus $\Gamma = \Gamma_1 + \Gamma_3 + 2\Gamma_5$.

3.5 See example given under property 3.

3.6 Consider the classes $2C_3 = \{C_3, C_3^2\}$ and $3\sigma_v = \{\sigma_{v_x}, \sigma_{v_y}, \sigma_{v_z}\}$. Then show, for instance, that $[C_3, \sigma_{v_x}] = [C_3, \sigma_{v_y}] = [C_3, \sigma_{v_z}] = 0$, where the commutator $[A, B] = AB - BA$. For example

$$\begin{aligned}[C_3, \sigma_{v_x}](xyz) &= C_3\sigma_{v_x}(xyz) - \sigma_{v_x}C_3(xyz)\\ &= C_3(xzy) - \sigma_{v_x}(yzx)\\ &= (zyx) - (zyx)\\ &= 0 \text{ etc.}\end{aligned}$$

3.7 Since D_4 has five classes as follows: $C_1 = E$, $C_2 = \{C_{4z}^1, C_{4z}^3\}$, $C_3 = C_{4z}^2$, $C_4 = \{C'_{2x}, C'_{2y}\}$, $C_5 = \{C''_{2d_1}, C''_{2d_2}\}$, the group D_{4h} which is the product of the group D_4 with the group I consisting of two classes ($C_1 = E$, $C_2 = I$), would have 10 classes as follows: C_1 to C_5 as defined above, plus $C_6 = C_1 \times 1$, $C_7 = C_2 \times \text{I}$, $C_8 = C_3 \times \text{I}$, $C_9 = C_4 \times \text{I}$, $C_{10} = C_5 \times \text{I}$. The group D_{4h} would therefore have 10 IRS which can be readily obtained by multiplying the table for D_4 with the character table for the group I, which is simply:

I	E	I
Γ_1	1	1
Γ_2	1	-1

We should thus get the table for D_{4h} as given below:

	D_{4h}	C_1	C_2	C_3	C_4	C_5	C_6	C_7	C_8	C_9	C_{10}	
(I)	Γ_1	1	1	1	1	1	1	1	1	1	1	(II)
	Γ_2	1	1	1	-1	-1	1	1	1	-1	-1	
	Γ_3	1	-1	1	1	-1	1	-1	1	1	-1	
	Γ_4	1	-1	1	-1	1	1	-1	1	-1	1	
	Γ_5	2	0	-2	0	0	2	0	-2	0	0	
(III)	Γ_6	1	1	1	1	1	-1	-1	-1	-1	-1	(IV)
	Γ_7	1	1	1	-1	-1	-1	-1	-1	1	1	
	Γ_8	1	-1	1	1	-1	-1	1	-1	-1	1	
	Γ_9	1	-1	1	-1	1	-1	1	-1	1	-1	
	Γ_{10}	2	0	-2	0	0	-2	0	2	0	0	

Observe carefully how the character table for D_{4h} has been constructed from D_4 using the table for I.

The characters in the block labelled (I) are characters of $D_4 \times \chi_1(\Gamma_1^E)$, i.e. $D_4 \times 1 = D_4$ (repeated).
The characters in the block labelled (II) are characters of $D_4 \times \chi_1(\Gamma_1^I)$, i.e. $D_4 \times 1 = D_4$ (repeated).
The characters in the block labelled (III) are characters of $D_4 \times \chi_1(\Gamma_2^E)$, i.e. $D_4 \times 1 = D_4$ (repeated).
The characters in the block labelled (IV) are characters of $D_4 \times \chi_1(\Gamma_2^I)$, i.e. $D_4 \times -1 = -D_4$ (all characters of D_4 multiplied by -1).

Note that this method is quite general and widely used to obtain the character table of those larger groups which are the product of two smaller groups and whose character tables are known.

Chapter 4

4.1

	Compound	Lattice type	Fractional coordinates
(a)	NaCl	FCC	Four building units/cell; Na: $(\frac{1}{2}, 0, 0)$, $(0, \frac{1}{2}, 0)$, $(0, 0, \frac{1}{2})$, $(\frac{1}{2}, \frac{1}{2}, \frac{1}{2})$; Cl: $(0, 0, 0)$, $(\frac{1}{2}, \frac{1}{2}, 0)$, $(\frac{1}{2}, 0, \frac{1}{2})$, $(0, \frac{1}{2}, \frac{1}{2})$
(b)	CsCl	SC	One building unit/cell; $(0, 0, 0)$ and $(\frac{1}{2}, \frac{1}{2}, \frac{1}{2})$
(c)	ZnS	FCC	Four building units per cell; Zn: $(\frac{1}{4}, \frac{1}{4}, \frac{1}{4})$, $(\frac{1}{4}, \frac{3}{4}, \frac{3}{4})$, $(\frac{3}{4}, \frac{1}{4}, \frac{3}{4})$, $(\frac{3}{4}, \frac{3}{4}, \frac{1}{4})$; S: $(0, 0, 0)$, $(0, \frac{1}{2}, \frac{1}{2})$, $(\frac{1}{2}, 0, \frac{1}{2})$, $(\frac{1}{2}, \frac{1}{2}, 0)$
(d)	Cu	FCC	Four Cu atoms/cell; $(0, 0, 0)$, $(\frac{1}{2}, \frac{1}{2}, 0)$, $(0, \frac{1}{2}, \frac{1}{2})$, $(\frac{1}{2}, 0, \frac{1}{2})$

4.2 (a) $\langle 100 \rangle$; (b) $\langle 111 \rangle$; (c) $\langle 110 \rangle$.

4.3

$$\langle 111 \rangle \equiv (1,1,1),(\bar{1},1,1),(1,\bar{1},1),(1,1,\bar{1}),(\bar{1},\bar{1},1),(\bar{1},1,\bar{1}),(1,\bar{1},\bar{1}),(\bar{1},\bar{1},\bar{1}).$$

$$\langle 110 \rangle \equiv (1,1,0),(1,0,1),(0,1,1),(\bar{1},\bar{1},0),(\bar{1},0,\bar{1}),(0,\bar{1},\bar{1}),(\bar{1},1,0),(\bar{1},0,1),$$
$$(0,\bar{1},1),(1,\bar{1},0),(1,0,\bar{1}),(0,1,\bar{1}).$$

4.4 Figures: see text; nearest neighbours: 6, 8 and 12 for SC, BCC and FCC respectively.

4.5

$$f = \frac{\text{vol. of spheres} \times \text{no. of spheres/cell}}{\text{vol. of cell}}.$$

Example for simple cubic: Atoms touch along $\langle 100 \rangle$ directions. Therefore, $a = 2R$; volume of cell is $a^3 = 8R^3$; volume of each sphere is $\frac{4}{3}\pi R^3$. For simple cubic arrangement, number of spheres/cell = 1. Hence $f = \frac{4}{3}\pi R^3 \times 1/(8R^3) = \frac{1}{6}\pi = 52.4\%$.

Similar calculations for (b) and (c).

4.6 See figure 4.5 of text.

4.7 (a) A Bravais lattice satisfies the important condition or constraint given in §4.2 and hence there are only 14 Bravais lattices. A mathematical space lattice has no constraints and therefore there is an infinite number of mathematical space lattices.

(b) See [36, 39].

(c) Crystal class refers to the 32 distinct symmetry point groups into which all crystals found in nature can be classified. Crystal systems refer to the 7 types of crystal architecture which are determined by the lattice constants a, b, c and the interaxial angles α, β, γ (see table 4.2).

(d) A Wigner–Seitz cell is a symmetrical primitive cell in direct lattice space. A Brillouin zone is a symmetrical primitive cell in reciprocal lattice space.

4.8 (a) Note that a is the length of primitive translation vector and not side of cube edge. Therefore, obtain

$$\boldsymbol{t}_1 = \frac{a}{\sqrt{3}}(\hat{\boldsymbol{i}} + \hat{\boldsymbol{j}} - \hat{\boldsymbol{k}}) \qquad \boldsymbol{t}_2 = \frac{a}{\sqrt{3}}(\hat{\boldsymbol{i}} - \hat{\boldsymbol{j}} + \hat{\boldsymbol{k}}) \qquad \boldsymbol{t}_3 = \frac{a}{\sqrt{3}}(-\hat{\boldsymbol{i}} + \hat{\boldsymbol{j}} + \boldsymbol{k}).$$

(b) Obtain

$$\boldsymbol{t}_1 = a\sqrt{2}(\hat{\boldsymbol{i}} + \hat{\boldsymbol{j}}) \qquad \boldsymbol{t}_2 = a\sqrt{2}(\hat{\boldsymbol{i}} + \hat{\boldsymbol{k}}) \qquad \boldsymbol{t}_3 = a\sqrt{2}(\hat{\boldsymbol{j}} + \hat{\boldsymbol{k}}).$$

4.9 Use equation (4.5): $\boldsymbol{a}_i \cdot \boldsymbol{a}_j^* = \delta_{ij}$ gives, for example,

$$\boldsymbol{a}_1 \cdot \boldsymbol{a}_1^* = 1. \qquad (1)$$

The volume of unit cell in direct lattice space is

$$V = \boldsymbol{a}_1 \cdot (\boldsymbol{a}_2 \times \boldsymbol{a}_3) \quad \text{or} \quad 1 = \boldsymbol{a}_1 \cdot (\boldsymbol{a}_2 \times \boldsymbol{a}_3)/V. \qquad (2)$$

Comparing (1) and (2) gives $\boldsymbol{a}_1^* = (\boldsymbol{a}_2 \times \boldsymbol{a}_3)/V$ etc.

4.10 Use equation (4.5) to obtain the reciprocal lattice vectors. For example, for BCC lattice: $\boldsymbol{a}_1 = a(\hat{\boldsymbol{i}} + \hat{\boldsymbol{j}} - \hat{\boldsymbol{k}})$, $\boldsymbol{a}_2 = a(\hat{\boldsymbol{i}} - \hat{\boldsymbol{j}} + \hat{\boldsymbol{k}})$, $\boldsymbol{a}_3 = a(-\hat{\boldsymbol{i}} + \hat{\boldsymbol{j}} + \hat{\boldsymbol{k}})$. Then using $\boldsymbol{a}_i \cdot \boldsymbol{a}_i^* = \delta_{ij}$, where $\boldsymbol{a}_j^*$ are the reciprocal lattice vectors, obtain $\boldsymbol{a}_1^* = \frac{1}{2}a(\hat{\boldsymbol{i}} + \hat{\boldsymbol{j}})$, $\boldsymbol{a}_2^* = \frac{1}{2}a(\hat{\boldsymbol{i}} + \hat{\boldsymbol{k}})$, $\boldsymbol{a}_3^* = \frac{1}{2}a(\hat{\boldsymbol{j}} + \hat{\boldsymbol{k}})$, which defines an FCC lattice. Similarly start with the direct lattice vectors for an FCC lattice and obtain the reciprocal lattice vectors which would define a BCC lattice.

4.11 See figure 4.9 and [1].

4.12 (a)

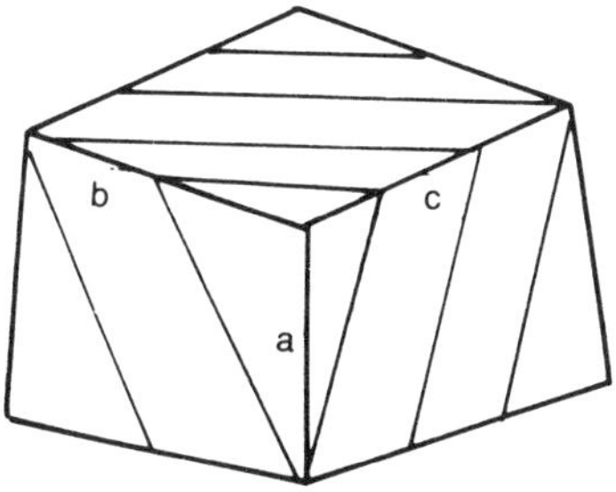

Similarly, draw figures for (b), (c) and (d).

4.13 Several methods possible:

Method 1: Face normal unit vector is

$$\boldsymbol{F} = d\left(\frac{h}{a}\hat{\boldsymbol{i}} + \frac{k}{a}\hat{\boldsymbol{j}} + \frac{l}{a}\hat{\boldsymbol{k}}\right).$$

Hence length is

$$\boldsymbol{F}\cdot\boldsymbol{F} = 1 = d^2\left(\frac{h^2}{a^2} + \frac{k^2}{a^2} + \frac{l^2}{a^2}\right)$$

or

$$\frac{1}{d^2} = \frac{h^2 + k^2 + l^2}{a^2}.$$

Method 2: Consider in terms of reciprocal lattice vectors (RLV):

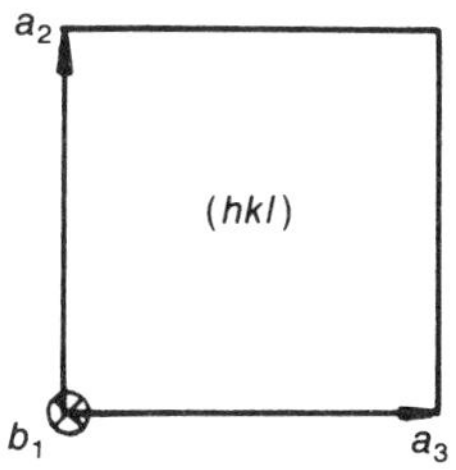

The RLV b_1 (see figure) is perpendicular to set of planes (hkl) of direct lattice. Therefore, if d is the distance between the planes (hkl) in the direct lattice, then by definition magnitude of $\boldsymbol{b}_1$ is

$$\boldsymbol{b}_1\cdot\boldsymbol{b}_1 = \frac{1}{d}\cdot\frac{1}{d} \quad \text{i.e.} \quad b_1^2 = \frac{1}{d^2}. \qquad (1)$$

For a simple cube if $\boldsymbol{a}_1 = a(\hat{\boldsymbol{i}}h + \hat{\boldsymbol{j}}k + \hat{\boldsymbol{k}}l)$ its RLV is

$$\boldsymbol{b}_1 = \frac{1}{a}(\hat{\boldsymbol{i}}h + \hat{\boldsymbol{j}}k + \hat{\boldsymbol{k}}l) \quad \text{and} \quad b_1^2 = \frac{1}{a^2}(h^2 + k^2 + l^2) = \frac{1}{d^2}$$

using (1). Hence

$$\frac{1}{d^2} = \frac{h^2 + k^2 + l^2}{a^2}.$$

Method 3: Use direction cosines $\cos\alpha = x/r$, $\cos\beta = y/r$, $\cos\gamma = z/r$; here $x = h/a$, $y = k/a$, $z = l/a$ and $r = 1/d$.

Use $\cos^2\alpha + \cos^2\beta + \cos^2\gamma = 1$ to give

$$\frac{1}{d^2} = \frac{h^2 + k^2 + l^2}{a^2}.$$

4.14 (a) Obtain length $= a\sqrt{3}$.

(b) To determine angle between vectors, use $\boldsymbol{A}\cdot\boldsymbol{B} = AB\cos\theta$. Obtain angle $= 109.5°$ or $70°$.

4.15 Use the rule given by equation (4.9) to get: (a) $\{E|(100)\}$; (b) $\{C_{2y}|(xyz)\}$. Here E is the identity element.

4.16 Obviously they do not. From exercise 4.15(a), for example,

$$\{C_{2x}|(\tfrac{1}{2}\ \tfrac{1}{2}\ 0)\}\{C_{2x}|(\tfrac{1}{2}\ \tfrac{1}{2}\ 0)\} = \{E|(1\ 0\ 0)\},$$

which is not a given member of the group.

4.17

$$\begin{aligned}\{C_{2x}|(xyz)\}^{-1}\{\mathrm{I}|0\}\{C_{2x}|(xyz)\} &= \{C_{2x}^{-1}| -(x\bar{y}\bar{z})\}\{IC_{2x}|(\bar{x}\bar{y}\bar{z})\} \\ &= \{I|(\bar{x}yz) + (\bar{x}yz)\} \\ &= \{I|(2\bar{x} + 2y + 2z)\} \\ &= \{I|(\boldsymbol{t}_1 + \boldsymbol{t}_2 + \boldsymbol{t}_3)\} \\ &= \{I|\boldsymbol{T}\}.\end{aligned}$$

4.18 Inverse is given by equation (4.10):

$$\begin{aligned}\{C_{2x}|\tfrac{1}{2}(xyz)\}^{-1} &= \{C_{2x}^{-1}| -C_{2x}^{-1}\ \tfrac{1}{2}(xyz)\} \\ &= \{C_{2x}| -C_{2x}\ \tfrac{1}{2}(xyz)\} \\ &= \{C_{2x}| -\tfrac{1}{2}(x\bar{y}\bar{z})\} \\ &= \{C_{2x}|\tfrac{1}{2}(\bar{x}yz)\}.\end{aligned}$$

Verify

$$\begin{aligned}\{C_{2x}|\tfrac{1}{2}(xyz)\}\{C_{2x}|\tfrac{1}{2}(xyz)\}^{-1} &= \{C_{2x}|\tfrac{1}{2}(xyz)\}\{C_{2x}|\tfrac{1}{2}(\bar{x}yz)\}\\ &= \{E|C_{2x}\,\tfrac{1}{2}(\bar{x}yz)+\tfrac{1}{2}(xyz)\}\\ &= \{E|\tfrac{1}{2}(\bar{x}\bar{y}\bar{z})+\tfrac{1}{2}(xyz)\}\\ &= \{E|0\}.\end{aligned}$$

Chapter 5

5.1 Consider the operation of $R_z(\phi)$ on Cartesian coordinates (x, y) in the plane of the circle and obtain

$$R_z(\phi) = \begin{pmatrix} \cos\phi & \sin\phi \\ -\sin\phi & \cos\phi \end{pmatrix}.$$

5.2 Use the matrix given in exercise 5.1 to verify that the law of composition is $R(\phi)R(\theta) = R(\phi+\theta)$.

(a) The identity element is $R(0)$.

(b) The inverse element is $R_z(2\pi-\phi)$.

5.3 From exercise 5.1 it follows readily that for O_3

$$R_z(\phi) = \begin{pmatrix} \cos\phi & \sin\phi & 0 \\ -\sin\phi & \cos\phi & 0 \\ 0 & 0 & 1 \end{pmatrix}.$$

5.4 Spinor group is a group including spin operations also. It therefore has double the number of elements of the corresponding related point group. Note that for even integral spin $\chi(\alpha) = \sin(J+\tfrac{1}{2})\alpha/\sin\alpha/2$ and $\chi(E) = (2J+1)$. For the case of half-integral spin a new element R corresponding to a rotation of 2π has to be introduced such that $\chi(R) = -(2J+1)$. Thus E is rotation through 4π and $\chi(E) = (2J+1)$ whilst R is rotation through 2π and $\chi(R) = -(2J+1)$. The spinor group $\bar{\mathrm{O}}$ (read as O bar) is then obtained by forming a direct product of the ordinary parent group O with $(R \times E)$. (See §5.3 and [2]).

5.5 Obtain the multiplication table for ${}^{d}C_2$ as given below:

${}^{d}C_2$	E	C_{2z}	$\bar{E}$	$\bar{C}_{2z}$
E	E	C_{2z}	$\bar{E}$	$\bar{C}_{2z}$
C_{2z}	C_{2z}	$\bar{E}$	$\bar{C}_{2z}$	E
$\bar{E}$	$\bar{E}$	$\bar{C}_{2z}$	E	C_{2z}
$\bar{C}_{2z}$	$\bar{C}_{2z}$	E	C_{2z}	$\bar{E}$

5.6 Obtain the first quarter of the table as follows:

${}^{d}C_{4v}$	E	C_{4z}^{+}	C_{2z}	C_{4z}^{-}	σ_{v_x}	σ_{v_y}	σ_{13}	σ_{24}
E	E	C_{4z}^{+}	C_{2z}	C_{4z}^{-}	σ_{v_x}	σ_{v_y}	σ_{13}	σ_{24}
C_{4z}^{+}	C_{4z}^{+}	C_{2z}	$\bar{C}_{4z}^{-}$	E	σ_{13}	σ_{24}	σ_{v_y}	$\bar{\sigma}_{v_x}$
C_{2z}	C_{2z}	$\bar{C}_{4z}^{-}$	$\bar{E}$	C_{4z}^{+}	σ_{v_y}	$\bar{\sigma}_{v_x}$	σ_{24}	$\bar{\sigma}_{13}$
C_{4z}^{-}	C_{4z}^{-}	E	C_{4z}^{+}	$\bar{C}_{2z}$	$\bar{\sigma}_{24}$	σ_{13}	σ_{v_x}	σ_{v_y}
σ_{v_x}	σ_{v_x}	$\bar{\sigma}_{24}$	$\bar{\sigma}_{v_y}$	σ_{13}	$\bar{E}$	C_{2z}	$\bar{C}_{4z}^{-}$	C_{4z}^{+}
σ_{v_y}	σ_{v_y}	σ_{13}	σ_{v_x}	σ_{24}	$\bar{C}_{2z}$	$\bar{E}$	$\bar{C}_{4z}^{+}$	$\bar{C}_{4z}^{-}$
σ_{13}	σ_{13}	σ_{v_x}	$\bar{\sigma}_{24}$	σ_{v_y}	$\bar{C}_{4z}^{+}$	$\bar{C}_{4z}^{-}$	$\bar{E}$	C_{2z}
σ_{24}	σ_{24}	σ_{v_y}	σ_{13}	$\bar{\sigma}_{v_x}$	C_{4z}^{-}	$\bar{C}_{4z}^{+}$	$\bar{C}_{2z}$	$\bar{E}$

For complete table apply rule as follows:

(1) For the product of one unbarred and one barred operator, barred and unbarred operators are to be interchanged in the products.
(2) The product of two barred operators is identical to that of two unbarred operators.

Example: $C_{4z}^{+} \cdot C_{2z} = \bar{C}_{4z}^{-}$ (from table)

$\bar{C}_{4z}^{+} \cdot C_{2z} = C_{4z}^{-}$

$\bar{C}_{4z}^{+} \cdot \bar{C}_{2z} = \bar{C}_{4z}^{-}$.

Chapter 6

6.1 See equation (6.2) of text; all the symbols have been explained.

6.2 (a) Weak crystal field case: here spin–orbit coupling dominates. This is prevalent in the 4f rare earth ionic compounds. The 4f ions experience a weak crystal field since the 4f shell is 'shielded' from the direct crystalline field by the 5s and 5p shells.

(b) Medium crystal field case: here the interelectronic repulsion term is greatest; this often leads to a reduction or even to an elimination of the orbital magnetic moment. This is prevalent in 3d ions, most common being the hydrated ions of the iron group.

(c) Strong crystal field case: the ions lie on the surface and hence experience a stronger crystalline field than the shielded 4f rare earths (case (a)). $\boldsymbol{l} \cdot \boldsymbol{s}$ coupling is broken, the number of states is greatly reduced and the spins usually get paired. The total spin is therefore usually less than the weak field case. This is prevalent in complexes of the 4d and 5d groups and complex cyanides of the 3d group.

6.3 We can use the table for the symmetry group D_4 since it would give the same result as with D_{4h} and is much easier to work with.

For ^3G ion, $L = 4$; the representation to be reduced (use (6.3)) is as follows:

D_4	E	$2C_4$	C_2	$2C_2$	$2C_2$
$\Gamma_{^3G}$	9	1	1	1	1

Obtain $\Gamma_{^3G} = 2\Gamma_1 + \Gamma_2 + \Gamma_3 + \Gamma_4 + 2\Gamma_5$. Check: since $L = 4$, we expect $(2L + 1) = 9$ levels. Writing down the degeneracy of the Γ_i levels we find $\Gamma_{^3G} = 2 \times 1 + 1 + 1 + 1 + 2 \times 2 = 9$.

6.4 Ti^{3+}: $3d^1$ configuration. First find the ground state term using Hund's rule to be ^2D. The splittings can then be readily obtained as follows:

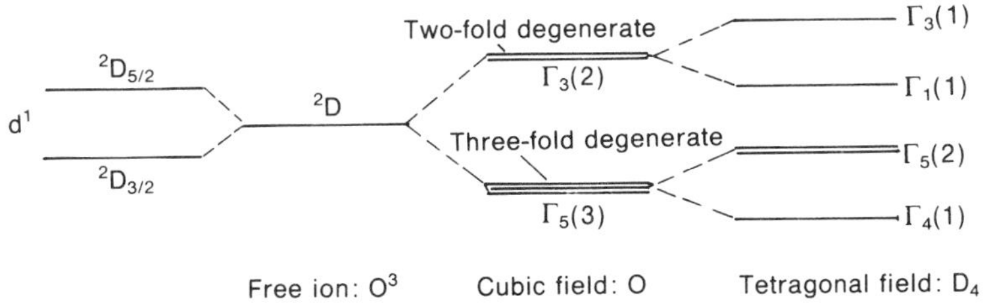

Note: For ^2D term we have $L = 2$, $S = \frac{1}{2}$ and therefore $J = 5/2$. Since this is half an odd integer one should strictly be using double groups. In the above treatment we have neglected spins; the two spin levels ($^2D_{5/2}$, $^2D_{3/2}$) are shown on the extreme left.

6.5 Use Hund's rule: (1) maximum multiplicity $(2S + 1)$ followed by (2) maximum M_L value without violating the Pauli principle.

Case	Ground state
d^1	^2D
d^2	^3F
d^3	^4F
d^4	^5D
d^5	^6S
d^6	^5D
d^7	^4F
d^8	^3F
d^9	^2D

Note: $d^n = d^{10-n}$. Thus $d^6 = d^4 = {}^5D$ state, etc.

Example: Case of d^4: $l_1 = 2$, $l_2 = 2$, $l_3 = 2$, $l_4 = 2$. Therefore $L = 8$, $s_1 = \frac{1}{2}$, $s_2 = \frac{1}{2}$, $s_3 = \frac{1}{2}$, $s_4 = \frac{1}{2}$; then $S = 2$. Max. multiplicity $= (2S + 1) = 5$. M_L can take values from $+8$ to -8 in steps of 1, so arrange the electrons in 'boxes' in accordance with the Pauli principle. Note that an arrangement like 2^+, 2^+, 2^+, 2^+ violates the Pauli principle and thus $M_L = 8$ arrangement is not possible. The only possible M_L arrangement for the four electrons giving maximum multiplicity is 2^+, 1^+, 0^+, -1^+, for which $L = 2$, hence D state. Therefore ground state $= {}^5D$.

6.6 The figure below shows schematically the splittings of a ${}^4F$ state by cubic crystal fields of various strengths. The numbers in the parentheses gives the degeneracies of the levels:

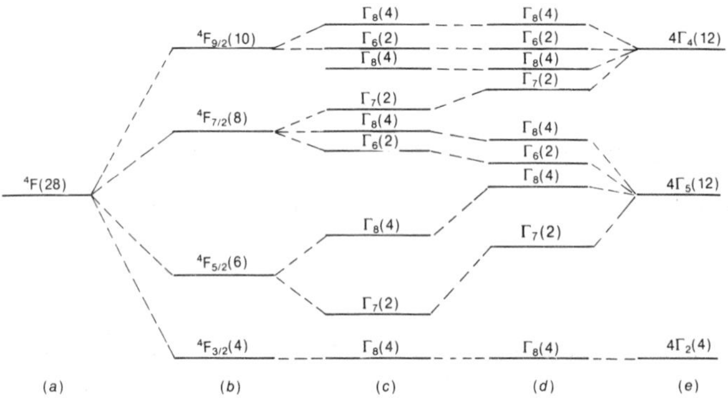

(a) free ion; (b) no cubic field but spin–orbit coupling present; (c) weak cubic field 0; (d) strong cubic field 0; (e) no spin–orbit coupling but crystal field present.

6.7 The Brillouin zone of a simple square lattice with side a is also a simple square but with side $2\pi/a$. The special points and lines of symmetry are as indicated in the figure below:

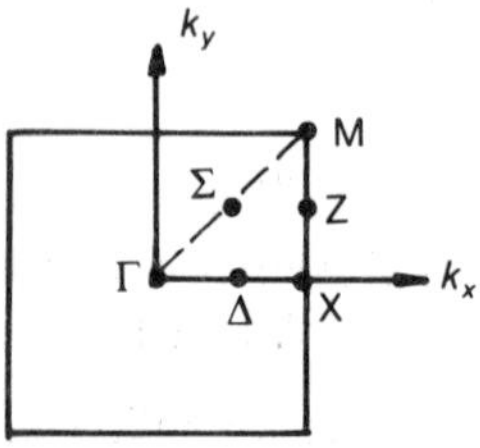

The $\boldsymbol{k}$ vectors and their little groups for the points Γ, Σ, M, Δ, X, Z are summarized in the table below:

Symmetry point or line	$\boldsymbol{k}$	Little group
Γ	(0, 0)	C_{4v}
Σ	$\frac{2\pi}{a}(\alpha, \alpha)$	C_{1h}
M	$\frac{2\pi}{a}(\frac{1}{2}, \frac{1}{2})$	C_{4v}
Δ	$\frac{2\pi}{a}(\alpha, 0)$	C_{1h}
X	$\frac{2\pi}{a}(\frac{1}{2}, 0)$	C_{2v}
Z	$\frac{2\pi}{a}(\alpha, \frac{1}{2})$	C_{1h}

6.8 Obtain the free electron energy bands along ΓX and ΓM directions as follows:

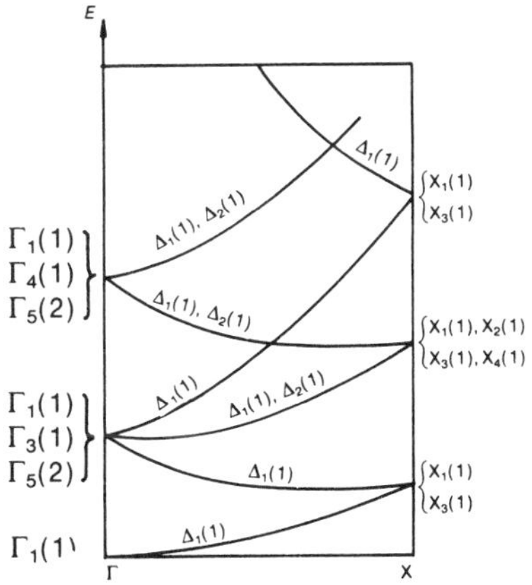

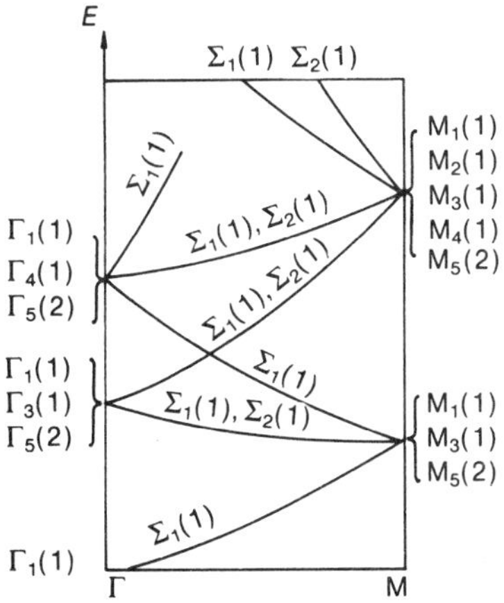

6.9 See [47] and [8] for solution.

6.10 Use table 6.6(a) and Koster *et al* [10] in writing the character tables for the little groups at Γ, Δ, X, Λ and R.

Example: For Γ and Δ we have the following tables:

Γ: O_h	E	$3C_4^2 = 3C_2$	$6C_4$	$6C_2$	$8C_3$	I	$3IC_4^2 = 3\sigma_h$	$6IC_4 = 6S_4$	$6IC_2 = 6\sigma_d$	$8IC_3 = 8S_6$
Γ_1	1	1	1	1	1	1	1	1	1	1
Γ_2	1	1	−1	−1	1	1	1	−1	−1	1
Γ_3	2	2	0	0	−1	2	2	0	0	−1
Γ_4	3	−1	1	−1	0	3	−1	1	−1	0
Γ_5	3	−1	−1	1	0	3	−1	−1	1	0
Γ_6	1	1	1	1	1	−1	−1	−1	−1	−1
Γ_7	1	1	−1	−1	1	−1	−1	1	1	−1
Γ_8	2	2	0	0	−1	−2	−2	0	0	1
Γ_9	3	−1	1	−1	0	−3	1	−1	1	0
Γ_{10}	3	−1	−1	1	0	−3	1	1	−1	0

Δ: C_{4v}	E	$C_4^2 = C_2$	$2C_4$	$2\sigma_v = 2IC_4^2$	$2\sigma_d = 2IC_2$
Δ_1	1	1	1	1	1
Δ_2	1	1	−1	1	−1
Δ_3	1	1	−1	−1	1
Δ_4	1	1	1	−1	−1
Δ_5	2	−2	0	0	0

In a similar way proceed to write the tables for the points X, Λ and R. Obtain the compatibility relations as shown in text (§6.3, table 6.8). For example, the compatibility relations between Γ and Δ would be as follows:

Γ: O_h	Γ_1	Γ_2	Γ_3	Γ_4	Γ_5	Γ_6	Γ_7	Γ_8	Γ_9	Γ_{10}
Δ: C_{4v}	Δ_1	Δ_2	$\Delta_1\Delta_2$	$\Delta_4\Delta_5$	$\Delta_3\Delta_5$	Δ_4	Δ_3	$\Delta_3\Delta_4$	$\Delta_1\Delta_5$	$\Delta_2\Delta_5$

Sketch of the free-electron bands along the Δ axis and Λ axis:

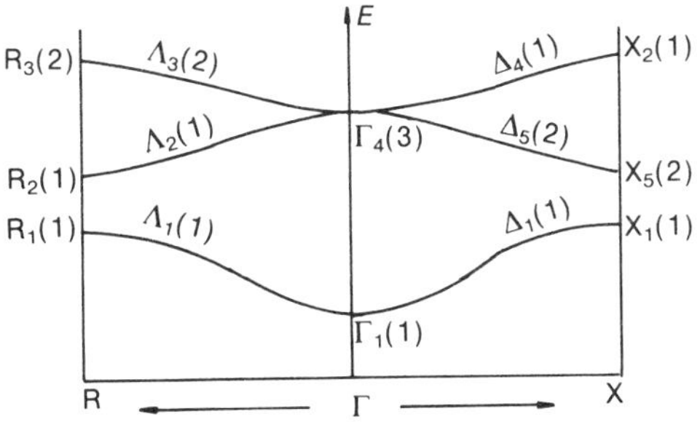

6.11 Use the table given by Koster *et al* [10] to obtain the compatibility relations as follows:

O_h	Γ_1^+	Γ_2^+	Γ_3^+	Γ_4^+	Γ_5^+	Γ_1^-	Γ_2^-	Γ_3^-	Γ_4^-	Γ_5^-	Γ_6^+	Γ_7^+	Γ_8^+	Γ_6^-	Γ_7^-	Γ_8^-
O	Γ_1	Γ_2	Γ_3	Γ_4	Γ_5	Γ_1	Γ_2	Γ_3	Γ_4	Γ_5	Γ_6	Γ_7	Γ_8	Γ_6	Γ_7	Γ_8

6.12 Solution given in [47]; see also [8].

6.13 For infrared spectra there must be changes in the electric dipole moments. For Raman spectra there must be changes in the polarizability.

In group theoretical terms: (i) for infrared spectra, the operator corresponding to infrared transitions must transform like the basis functions x, y, z, or R_x, R_y, R_z, in the group. These are shown in the character tables (see [10]); (ii) for Raman spectra, the polarizability operator must transform as x^2, y^2, z^2, or xy, yz, zx. These are also shown in the character tables of [10].

6.14 The point-group symmetry of NH_3 molecule is C_{3v}. Using the tables of Koster *et al* [10] it is easy to see the transformations of functions as follows:

$$\Gamma_1: x^2 + y^2, z, z^2 \qquad \Gamma_3: (x, y)(zx, zy).$$

Thus Γ_1 and Γ_3 are active for both infrared and Raman transitions.

6.15 Use tables of representation given by Koster *et al* [10]. From the tables observe the assignment of functions as follows:

$$\Gamma_1: x^2 + y^2 + z^2$$

$$\Gamma_3: (2z^2 - x^2 - y^2), (x^2 - y^2)$$

$$\Gamma_5: (x, y, z)(xy, yz, zx).$$

Hence Γ_5 mode is infrared active whilst $\Gamma_1, \Gamma_3, \Gamma_5$ modes are Raman active.

6.16 The point-group symmetry for H_2O molecule is C_{2v}. Look at its character table in [10]. By observing the transformation properties of the basis functions it is obvious that all the modes can be excited from the ground state by Raman active transitions.

6.17 For $\langle 000 \rangle$, $n = 1$; for $\langle 100 \rangle$, $n = 6$; for $\langle 110 \rangle$, $n = 12$. Thus total number of waves is 19.

6.18 Use Jones' symbol: $\{C_{4z}|0\} P(xyz) = P'(y\bar{x}z)$.

Hence $\boldsymbol{K} = (100)$ becomes $(0\bar{1}0)$ and for diamond structure $\boldsymbol{r}$ becomes $P'(y, [\bar{x} - a/4], [z - a/4])$. Therefore $(\boldsymbol{K} \cdot \boldsymbol{r}) = a/4$ so that

$$E = \exp\left(\frac{2\pi \mathrm{i}}{a} \frac{a}{4}\right) = \exp\left(\frac{\mathrm{i}\pi}{2}\right) = \mathrm{i}.$$

6.19 The figure below shows the free-electron energy bands for Ge along the ΓX direction:

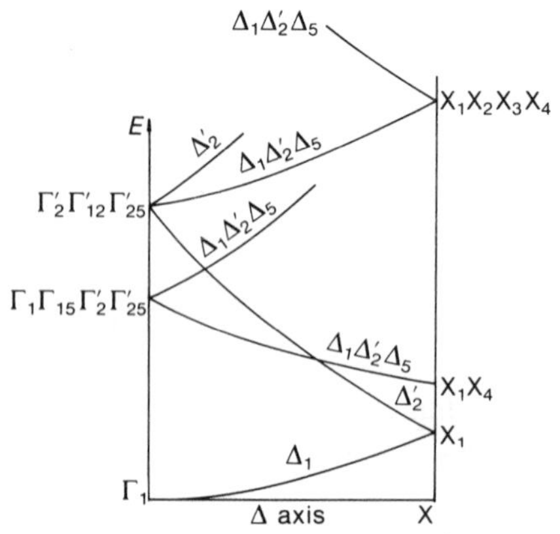

Note: We have used the notation of Bouckaert *et al* [47].

6.20 Calculate the determinants as follows. For Γ'_{25}:

$$\Gamma'_{25} = \begin{vmatrix} 3\left(\frac{2\pi}{a}\right)^2 - V_8^s - E & \sqrt{2}(V_3^s - V_{11}^s) \\ \sqrt{2}(V_3^s - V_{11}^s) & 4\left(\frac{2\pi}{a}\right)^2 - E \end{vmatrix} = 0.$$

Convert all quantities to atomic units, substitute and solve to get

$$E_1 = 1.64 \text{ Ryd}, E_2 = 0.75 \text{ Ryd}.$$

For Γ'_2: $E_1 = 1.69$ Ryd, $E_2 = 0.79$ Ryd. (1 Ryd = 13.6 eV.)

6.21 The orthogonal matrices are obtained by forming symmetric combinations of plane waves. They are:

(a)
$$\begin{pmatrix} 1 & -1 & -1 & -1 & 1 & -1 & -1 & -1 \\ 1 & -1 & -1 & -1 & -1 & 1 & 1 & 1 \\ 1 & 1 & 1 & -1 & 1 & 1 & 1 & -1 \\ 1 & 1 & -1 & 1 & 1 & 1 & -1 & 1 \\ 1 & -1 & 1 & 1 & 1 & -1 & 1 & 1 \\ 1 & 1 & 1 & -1 & -1 & -1 & -1 & 1 \\ 1 & 1 & -1 & 1 & -1 & -1 & 1 & -1 \\ 1 & -1 & 1 & 1 & -1 & 1 & -1 & -1 \end{pmatrix}$$

(b)
$$\begin{pmatrix} \frac{1}{\sqrt{6}} & \frac{1}{\sqrt{6}} & \frac{1}{\sqrt{6}} & -\frac{1}{\sqrt{6}} & -\frac{1}{\sqrt{6}} & -\frac{1}{\sqrt{6}} \\ 0 & \frac{1}{2} & -\frac{1}{2} & 0 & -\frac{1}{2} & \frac{1}{2} \\ -\frac{1}{\sqrt{3}} & \frac{1}{2\sqrt{3}} & \frac{1}{2\sqrt{3}} & \frac{1}{\sqrt{3}} & -\frac{1}{2\sqrt{3}} & -\frac{1}{2\sqrt{3}} \\ 0 & 0 & \frac{1}{\sqrt{2}} & 0 & 0 & \frac{1}{\sqrt{2}} \\ 0 & \frac{1}{\sqrt{2}} & 0 & 0 & \frac{1}{\sqrt{2}} & 0 \\ \frac{1}{\sqrt{2}} & 0 & 0 & \frac{1}{\sqrt{2}} & 0 & 0 \end{pmatrix}.$$

REFERENCES

This book is intended to provide a workmanlike knowledge of group theory and its applications to solid state physics. Students are therefore strongly advised to consult the following references for fundamental theorems and proofs and for further reading on the topics discussed here.

Part I

Prescribed texts

[1] Lax M 1974 *Symmetry Principles in Solid State and Molecular Physics* (New York: Wiley)

[2] Tinkham M 1964 *Group Theory and Quantum Mechanics* (New York: McGraw-Hill)

[3] Wigner E P 1959 *Group Theory and its Applications to the Quantum Mechanics of Atomic Spectra* (New York: Academic Press)

Recommended reading

[4] Cotton F A 1965 *Chemical Applications of Group Theory* (New York: Interscience)

[5] Cracknell A P 1967 *Applied Group Theory* (Oxford: Pergamon)

[6] Elliott J P and Dawber P G 1986 *Symmetry in Physics* vols I and II (London: Macmillan); see also Cornwell J F 1986 *Group Theory in Physics* vols I and II (New York: Academic Press)

[7] Heine V 1960 *Group Theory in Quantum Mechanics* (Oxford: Pergamon)

[8] Jones H 1960 *The Theory of Brillouin Zones in Crystals* (Amsterdam: North-Holland)

[9] Koster G F 1957 *Space Groups and their Representations, Solid State Physics* vol. 5 reprint (New York: Academic Press)

[10] Koster G F, Dimmock J O, Wheeler R G and Statz H 1963 *Properties of the Thirty-two Point-Groups* (Cambridge, MA: MIT Press)

[11] Knox R S and Gold A 1964 *Symmetry in the Solid State* (New York: Benjamin)

Peripheral reading

[12] Bhagavantam S 1966 *Crystal Symmetry and Physical Properties* (New York: Academic Press)

[13] Birman J L 1984 *Theory of Crystal Space Groups and Lattice Dynamics* (Berlin: Springer)

[14] Falicov L M 1966 *Group Theory and its Physical Applications* (Chicago: Chicago University Press)
[15] Hamermesh M 1962 *Group Theory* (Reading, MA: Addison-Wesley)
[16] Joshi A W 1978 *Elements of Group Theory for Physicists* (New Delhi: Wiley Eastern)
[17] Mariot L 1962 *Group Theory and Solid State Physics* (Englewood Cliffs, NJ: Prentice-Hall)
[18] Slater J C 1965 *Quantum Theory of Molecules and Solids* vol. II *Symmetry and Energy Bands in Crystals* (New York: McGraw-Hill)

Mathematical books

[19] Bradley C J and Cracknell A P 1972 *The Mathematical Theory of Symmetry in Solids* (Oxford: Clarendon)
[20] Burnside W 1955 *Theory of Groups of Finite Order* (New York: Dover)
[21] Heading J 1958 *Matrix Theory for Physicists* (London: Longmans)
[22] Hohn F E 1964 *Elementary Matrix Algebra* (London: Macmillan)
[23] Littlewood D E 1950 *The Theory of Group Characters and Matrix Representations of Groups* (Oxford: Oxford University Press)
[24] Lomont J S 1959 *Applications of Finite Groups* (New York: Academic Press)
[25] Murnagham F D 1938 *The Theory of Group Representations* (Baltimore, MD: Johns Hopkins University Press)

Atomic spectra and crystal field theory

[26] Ballhausen C J 1962 *Introduction to Ligand Field Theory* (New York: McGraw-Hill)
[27] Dunn T M, McClure D S and Pearson R G 1965 *Crystal Field Theory* (New York: Harper and Row)
[28] Griffith J S 1961 *The Theory of Transition Metal Ions* (Cambridge: Cambridge University Press)
[29] Herzfeld C M and Meijer P H E 1961 *Group Theory and Crystal Field Theory, Solid State Physics* vol. 12 (New York: Academic Press)
[30] Jorgensen C R 1962 *Absorption Spectra and Chemical Bonding in Complexes* (Reading, MA: Addison-Wesley)
[31] Slater J C 1960 *Quantum Theory and Atomic Structure* (New York: McGraw-Hill)
[32] McClure D S 1959 *Electronic Spectra of Molecules and Ions in Crystals, Solid State Physics* vol. 9 (New York: Academic Press)

Solid state physics and crystallography

[33] Buerger M J 1956 *Elementary Crystallography* (New York: Wiley)
[34] Kittel C 1976 *Introduction to Solid State Physics* 5th edn (New York: Wiley)
[35] Nye J F 1957 *Physical Properties of Crystals* (Oxford: Clarendon)
[36] Phillips F C 1963 *An Introduction to Crystallography* (London: Longmans)
[37] Steadman R 1982 *Crystallography* (New York: Van Nostrand Reinhold)
[38] Verma A R and Srivastava O N 1982 *Crystallography for Solid State Physics* (New Delhi: Wiley Eastern)

[39] Wert C A and Thomson R M 1964 *Physics of Solids* (New York: McGraw-Hill)
[40] Ziman J M 1960 *Electrons and Phonons* (Oxford: Oxford University Press)

Magnetic symmetry

[41] Birss R R 1964 *Symmetry and Magnetism* (Amsterdam: North-Holland)
[42] Bradley C J and Davies B L 1968 *Rev. Mod. Phys.* **40** 359
[43] Cracknell A P 1969 *Rep. Prog. Phys.* **32** 633
[44] Cracknell A P 1974 *Magnetism in Crystalline Materials* (Oxford: Pergamon)
[45] Dimmock J O and Wheeler R G 1956 *Mathematics of Physics and Chemistry* vol. II, ed. G M Murphy and H Marganeau (New York: Van Nostrand)

Useful review articles and papers

[46] Bethe H A 1929 Splittings of terms in crystals *Ann. Phys.* **3** 133; English translation by Consultants Bureau, 227 West 17th Street, New York, NY 10011, USA
[47] Bouckaert L P, Smoluchowski R and Wigner E 1936 *Phys. Rev.* **50** 58
[48] Cohen M L and Bergstresser T K 1966 *Phys. Rev.* **141** 789
[49] Cracknell A P 1974 *Adv. Phys.* **23** 673
[50] Eckart C 1930 *Rev. Mod. Phys.* **2** 344
[51] Herman F 1954 *Phys. Rev.* **93** 1214
[52] Herring C 1940 *Phys. Rev.* **57** 1169
[53] Luehrmann A W 1968 *Adv. Phys.* **17** 1
[54] Lukes T, Morgan D J and Joshua S J 1971 *J. Phys. C: Solid State Phys.* **4** 2623
[55] Mitra S S 1962 *Solid State Phys.* **13** 1 (New York: Academic Press)
[56] Joshua S J and Morgan D 1973 *Aust. J. Phys.* **26** 501

Note

(1) Important specific references pertaining to illustrative examples are indicated at the beginning of each application (see Chapter 6).
(2) Additional references for the topic *Symmetry and the Magnetic State* are given below with Part II references.

Part II

[1] Bhagavantam S and Venkatarayudu T 1969 *Theory of Groups and its Application to Physical Problems* (New York: Academic Press)
[2] Nye J F 1957 *Physical Properties of Crystals* (Oxford: Clarendon)
[3] Hamermesh M 1962 *Group Theory and its Application to Physical Problems* (Reading, MA: Addison-Wesley)
[4] Knox R S and Gold A 1964 *Symmetry in the Solid State* (New York: Benjamin)
[5] Tinkham M 1964 *Group Theory and Quantum Mechanics* (New York: McGraw-Hill)
[6] Jones H 1960 *The Theory of Brillouin Zones and Electronic States in Crystals* (Amsterdam: North-Holland)

[7] Cracknell A P 1968 *Applied Group Theory* (Oxford: Pergamon)

[8] Lax M 1974 *Symmetry Principles in Solid State and Molecular Physics* (New York: Wiley)

[9] Cracknell A P and Wong K C 1973 *The Fermi Surface: Its Concept, Determination and Use in the Physics of Metals* (Oxford: Clarendon)

[10] Bradley C J and Cracknell A P 1970 *The Mathematical Theory of Symmetry in Solids: Representation Theory for Point Groups and Space Groups* (Oxford: Clarendon)

[11] Shubnikov A V and Belov N V 1964 *Coloured Symmetry* (Oxford: Pergamon)

[12] Dimmock J O and Wheeler R G 1962 *Phys. Rev.* **127** 391; see also *The Mathematics of Physics and Chemistry* vol. 2, ed. G M Murphy and H Margenau (New York: Van Nostrand) p 275

[13] Dimmock J O and Wheeler R G 1962 *J. Phys. Chem. Solids* **23** 729

[14] Birss R R 1964 *Symmetry and Magnetism* (Amsterdam: North-Holland); 1963, *Rep. Prog. Phys.* **26** 307

[15] Loudon R 1968 *Adv. Phys.* **17** 243

[16] Cracknell A P 1969 *Rep. Prog. Phys.* **32** 633

[17] Cracknell A P 1974 *Adv. Phys.* **23** 673

[18] Cracknell A P and Joshua S J 1969 *Proc. Camb. Philos. Soc.* **66** 493

[19] Cracknell A P and Joshua S J 1969 *Phys. Status Solidi* **36** 737

[20] Cracknell A P and Joshua S J 1970 *Proc. Camb. Philos. Soc.* **67** 647

[21] Joshua S J and Cracknell A P 1969 *J. Phys. C: Solid State Phys.* **2** 24

[22] Joshua S J 1970 *Phys. Status Solidi* **38** 643

[23] Joshua S J 1970 *Phys. Status Solidi* **41** 309

[24] Morgan F D and Joshua S J 1973 *Phys. Status Solidi* (b) **58** 803

[25] Joshua S J and Morgan F D 1973 *Phys. Status Solidi* (b) **59** 269

[26] Deonarine S and Joshua S J 1973 *Phys. Status Solidi* (a) **20** 595

[27] Joshua S J 1974 *Acta Crystallogr.* **30A** 353

[28] Deonarine S and Joshua S J 1976 *Phys. Status Solidi* (b) **74** 659

[29] Bertaut E F 1968 *Acta Crystallogr.* **24A** 217

[30] Bradley and Davies B L 1968 *Rev. Mod. Phys.* **40** 359

[31] Opechowski W and Guccione R 1965 *Magnetism* vol. 2A, ed. G T Rado and H Suhl (New York: Academic Press) p 105

[32] Buerger M J 1956 *Elementary Crystallography* (New York: Wiley)

[33] Phillips F C 1963 *An Introduction to Crystallography* (London: Longmans)

[34] International Union of Crystallography 1952 *International Tables for X-ray Crystallography* (Birmingham: Kynock)

[35] Koster G F, Dimmock J O, Wheeler R G and Statz H 1963 *Properties of the Thirty-Two Point Groups* (Cambridge, MA: MIT Press)

[36] Kovalev O V 1965 *Irreducible Representations of the Space Groups* (New York: Gordon and Breach)

[37] Tavger B A and Zaitsev V M 1956 *Sov. Phys.–JETP* **3** 430

[38] Wigner E P 1959 *Group Theory and its Applications to the Quantum Mechanics of Atomic Spectra* (New York: Academic Press)

[39] Shubnikov A V 1951 *Symmetry and Antisymmetry of Finite Figures* (Moscow: USSR Academy of Sciences)

[40] Cracknell A P 1974 *Magnetism in Crystalline Materials* (Oxford: Pergamon)

[41] Bouckaert L P, Smoluchowski R and Wigner E 1936 *Phys. Rev.* **50** 58

[42] Koster G F 1957 *Solid State Phys.* **5** 329 (New York: Academic Press)
[43] Herring C 1937 *Phys. Rev.* **52** 361
[44] Heesch H 1930 *Z. Kristallogr.* **73** 725
[45] de Jongh L J and Miedema A R 1974 *Experiments on Simple Magnetic Model Systems* (London: Taylor and Francis)
[46] Price P B, Shirk E K, Osborne W Z and Pinsky L S 1975 *Phys. Rev.Lett.* **36** 487
[47] Haefner K, Stout J W and Barrett C S 1966 *J. Appl. Phys.* **37** 449
[48] Cracknell A P 1967 *Prog. Theor. Phys. Jpn* **38** 1252
[49] Faddeyev D K 1964 *Tables of the Principal Unitary Representations of Fedorov Groups* (Oxford: Pergamon)
[50] Nagamiya T, Yosida K and Kubo R 1955 *Adv. Phys.* **4** 1
[51] Roth W L 1958 *Phys. Rev.* **110** 1333
[52] Roth W L 1958 *Phys. Rev.* **111** 772
[53] Shull C G, Strauser W A and Wollan E O 1951 *Phys. Rev.* **83** 333
[54] Rooksby H P 1948 *Acta Crystallogr.* **1** 26
[55] Uchida E, Fukuoka N, Kondoh H, Takeda T, Nakazumi Y and Nagamiya T 1967 *J. Phys. Soc. Jpn* **23** 1197
[56] Yamada T 1966 *J. Phys. Soc. Jpn* **31** 664
[57] Yamada T, Saito S and Shimomura Y 1966 *J. Phys. Soc. Jpn* **21** 672
[58] Slack G A 1960 *J. Appl. Phys.* **31** 1571
[59] Cracknell A P 1966 *Prog. Theor. Phys. Jpn* **35** 196
[60] Cracknell and Wong K C 1967 *Aust. J. Phys.* **20** 173
[61] Kittel C 1963 *Quantum Theory of Solids* (New York: Wiley)
[62] Elliott R J and Loudon R 1960 *J. Phys. Chem. Solids* **15** 146
[63] Lax M and Hopfield J J 1961 *Phys. Rev.* **124** 115
[64] Zak J 1962 *J. Math. Phys.* **3** 1278
[65] Mitra S S 1962 *Solid. State Phys.* **13** 1 (New York: Academic Press)
[66] Birman J L 1963 *Phys. Rev.* **131** 1489
[67] Loudon R 1964 *Adv. Phys.* **13** 423; 1965 *Phys. Rev.* **137** A1784
[68] Lax M 1965 *Phys. Rev.* **138** A793
[69] Burnstein E, Johnson F A and Loudon R 1965 *Phys. Rev.* **139** A1239
[70] Birman J L, Lax M and Loudon R 1966 *Phys. Rev.* **145** 620
[71] Bradley C J 1966 *J. Math. Phys.* **7** 1145
[72] Bethe H A 1929 *Ann. Phys.* **3** 133
[73] Herzfeld C M and Meijer P H E 1961 *Solid State Phys.* **12** 1 (New York: Academic Press)
[74] Cotton F A 1963 *Chemical Applications of Group Theory* (New York: Wiley)
[75] Dunn T M, McClure D S and Pearson R G 1965 *Some Aspects of Crystal Field Theory* (New York: Harper and Row)
[76] Cracknell A P 1968 *Adv. Phys.* **17** 367
[77] Kanomori J 1963 *Magnetism* ed. G T Rado and H Suhl (New York: Academic Press)
[78] Landau L D and Lifshitz E M 1958 *Statistical Physics* (Oxford: Pergamon)
[79] Lyubarskii G Y 1960 *The Application of Group Theory in Physics* (Oxford: Pergamon)
[80] Anderson P W and Blount E I 1965 *Phys. Rev. Lett.* **14** 217
[81] Haas C 1965 *Phys. Rev.* **140** A863
[82] Barker A S and Loudon R 1967 *Phys. Rev.* **158** 433

[83] Birman J L 1962 *Phys. Rev.* **127** 1093
[84] Jahn H A and Teller E 1937 *Proc. R. Soc.* **A161** 220
[85] Kittel C and Galt J K 1956 *Solid State Phys.* **3** 439 (New York: Academic Press)
[86] Dillon J F 1963 *Magnetism* ed. G T Rado and H Suhl, vol. 3 (New York: Academic Press) p 415
[87] Craik D J and Tebble R S 1961 *Rep. Prog. Phys.* **24** 116
[88] Belov N V, Neronova N N and Smirnova T S 1957 *Soc. Phys.–Crystallogr.* **2** 311

INDEX